2013 中国环境统计年报

ANNUAL STATISTIC REPORT ON ENVIRONMENT IN CHINA

中华人民共和国环境保护部 编

MINISTRY OF ENVIRONMENTAL PROTECTION
OF THE PEOPLE'S REPUBLIC OF CHINA

中国环境出版社·北京

图书在版编目（CIP）数据

2013 年中国环境统计年报/中华人民共和国环境保护部编.
—北京：中国环境出版社，2014.12
ISBN 978-7-5111-2157-8

Ⅰ. ①2… Ⅱ. ①中… Ⅲ. ①环境统计学—统计资料—
中国—2013—年报 Ⅳ. ①X508.2-66

中国版本图书馆 CIP 数据核字（2014）第 292056 号

出 版 人	王新程
责任编辑	殷玉婷
责任校对	尹 芳
封面设计	彭 杉

出版发行　**中国环境出版社**
　　　　　（100062　北京市东城区广渠门内大街 16 号）
　　　　　网　　址：http://www.cesp.com.cn
　　　　　电子邮箱：bjgl@cesp.com.cn
　　　　　联系电话：010-67112765（编辑管理部）
　　　　　　　　　　010-67187041（学术著作图书出版中心）
　　　　　发行热线：010-67125803，010-67113405（传真）

印　　刷	北京中科印刷有限公司
经　　销	各地新华书店
版　　次	2014 年 12 月第 1 版
印　　次	2014 年 12 月第 1 次印刷
开　　本	889×1194　1/16
印　　张	28
字　　数	500 千字
定　　价	196.00 元

《中国环境统计年报·2013》编委会

主　任	翟　青					
副主任	刘炳江	陈　斌	于　飞	傅德黔		
编　委						
	马　磊	王军霞	王　鑫	王冬朴	毛玉如	安海蓉	吕　卓
	韦洪莲	孙振世	李红兵	刘通浩	杜会杰	陈　默	胡　明
	张太真	张玉军	张　欣	郭从容	范志国	周　囝	封　雪
	赵　莹	赵银慧	高　颖	钟妮华	裴晓菲	唐桂刚	董广霞
	董文福	景立新	谢光轩	樊　彤			
主　编	陈　斌					
副主编	毛玉如	景立新	董文福	唐桂刚		
编　辑						
	乔淑芳	刘新平	李金香	于　艳	王东莲	毛树声	刘佳泓
	冯军会	王　佳	苏海燕	班惠昭	高小武	盛若虹	莫晓莲
	李　勤	董　杰	陈　鑫	高　峰	李肖肖	刘世刚	马成伟
	刘继莉	谭海涛	李肇全	赵新宇	韩　铮	罗海林	魏　峻
	王秀臣	施　磊	汪小英	俞　鹏	蔡笑涯	周文娟	韩福敏
	季　晃	苏雷霖	胡清华	陈兴涛	刘颢刚	杨　斌	贾　曼
	王琳琳	李玉华	陈　静	许鹏飞	申　剑	杨　华	黄　霞
	施敏芳	刘　芳	罗　鑫	张　敏	何海敬	林志凌	兰波涛
	何金华	黄睿智	黄　春	杨朝晖	胡国林	刘泓辉	詹　忠
	王　飞	尹　萍	赵　磊	王思扬	王　炜	黎惠卉	郑　兵
	王　健	黄成超	巫鹏飞	拉　平	尼玛卓玛	廖慧彬	
	郭　琦	马　祥	姜泳波	杨　波	毛国辉	余淑娟	魏永邦
	历凌辉	党惠雯	马　柯	锁玉琴	王海林	刘端阳	李万刚
	罗洪明	韦志疆	胡新利				

主持单位　环境保护部污染物排放总量控制司

主编单位　中　国　环　境　监　测　总　站

资料提供　各省、自治区、直辖市环境保护厅（局）

批　　准　中　华　人　民　共　和　国　环　境　保　护　部

目录

综述　/1

1　统计调查企业基本情况　3～10

1.1　工业企业调查基本情况　/5
1.2　农业源调查基本情况　/7
1.3　集中式污染治理设施调查基本情况　/7

2　废水　11～70

2.1　废水及主要污染物排放情况　/13
2.2　各地区废水及主要污染物排放情况　/15
2.3　工业行业废水及主要污染物排放情况　/20
2.4　重点流域废水及主要污染物排放与治理情况　/28
2.5　沿海地区废水及主要污染物排放情况　/64

3　废气　71～98

3.1　废气及废气中主要污染物排放情况　/73
3.2　各地区废气中主要污染物排放情况　/74
3.3　工业行业废气中主要污染物排放情况　/76
3.4　大气污染防治重点区域（三区十群）废气污染物排放及处理情况　/89
3.5　机动车主要污染物排放情况　/94

4　工业固体废物　99～110

4.1　一般工业固体废物产生及处理情况　/101
4.2　工业危险废物产生和处理情况　/105

5　环境污染治理投资　111～118

5.1　总体情况　/113
5.2　城市环境基础设施建设　/115
5.3　老工业污染源治理投资　/116

5.4　建设项目"三同时"环保投资　/117

6　环境管理　　　　　　　　　　　　　　119～124

6.1　环保机构建设　/121

6.2　环境法制及环境信访情况　/121

6.3　环境监测　/122

6.4　自然生态保护　/122

6.5　环境影响评价及环保验收　/123

6.6　污染源控制及管理　/124

6.7　污染源自动监控及排污收费　/124

6.8　环境宣教　/124

7　全国辐射环境水平　　　　　　　　　125～129

7.1　环境电离辐射　/127

7.2　运行核电厂周围环境电离辐射　/128

7.3　民用研究堆周围环境电离辐射　/128

7.4　核燃料循环设施和废物处置设施周围环境电离辐射　/128

7.5　铀矿冶周围环境电离辐射　/129

7.6　电磁辐射　/129

简要说明　/130

8　各地区环境统计　　　　　　　　　　　133～184

各地区主要污染物排放情况　/135

各地区工业废水排放及处理情况　/141

各地区工业废气排放及处理情况　/147

各地区一般工业固体废物产生及处置利用情况　/151

各地区工业危险废物产生及处置利用情况　/152

各地区汇总工业企业概况　/153

各地区工业污染防治投资情况　/156

各地区农业污染排放情况　/160

各地区城镇生活污染排放及处理情况　/165

各地区机动车污染排放情况　/168

各地区城镇污水处理情况　/169

各地区生活垃圾处理情况　/173

各地区危险（医疗）废物集中处置情况　/177

各地区环境污染治理投资情况　/182

各地区城市环境基础设施建设投资情况　/183

9　重点城市环境统计　185～286

重点城市工业废水排放及处理情况　/187

重点城市工业废气排放及处理情况　/202

重点城市工业固体废物产生及处置利用情况　/208

重点城市工业危险废物产生及处置利用情况　/211

重点城市汇总工业企业概况　/214

重点城市工业污染防治投资情况　/223

重点城市农业污染排放情况　/232

重点城市城镇生活污染排放及处理情况　/235

重点城市机动车污染排放情况　/244

重点城市污水处理情况　/247

重点城市生活垃圾处理情况　/259

重点城市危险（医疗）废物集中处置情况　/271

10　各工业行业环境统计　277～299

按行业分重点调查工业废水排放及处理情况　/279

按行业分重点调查工业废气排放及处理情况　/284

按行业分重点调查一般工业固体废物产生及处置利用情况　/286

按行业分重点调查危险废物产生及处置利用情况　/287

按行业分重点调查工业汇总情况　/288

各地区独立火电厂污染排放及处理情况　/290

各地区水泥行业污染排放及处理情况　/293

各地区钢铁行业污染排放及处理情况　/295

各地区制浆及造纸行业污染排放及处理情况　/298

11 流域及入海陆源废水排放统计

十大水系接纳工业废水及处理情况 /303

十大水系工业废水污染防治投资情况 /318

十大水系农业污染排放情况 /321

十大水系城镇生活污染排放及处理情况 /324

重点流域接纳工业废水及处理情况 /330

重点流域工业废水污染防治投资情况 /340

重点流域农业污染排放情况 /342

重点流域城镇生活污染排放及处理情况 /344

湖泊水库接纳工业废水及处理情况 /348

湖泊水库工业污染防治投资情况 /353

湖泊水库农业污染排放情况 /354

湖泊水库城镇生活污染排放及处理情况 /355

三峡地区接纳工业废水及处理情况 /357

三峡地区工业污染防治投资情况 /359

三峡地区农业污染排放情况 /360

三峡地区城镇生活污染排放及处理情况 /361

入海陆源接纳工业废水及处理情况 /362

入海陆源工业废水污染防治投资情况 /367

入海陆源农业污染排放情况 /368

入海陆源城镇生活污染排放及处理情况 /369

12 大气污染防治重点区域废气排放统计

大气防治重点区域工业废气排放及处理情况 /373

大气防治重点区域工业废气污染防治投资情况 /377

大气防治重点区域城镇生活废气排放情况 /379

大气防治重点区域机动车废气排放情况 /380

大气防治重点区域生活垃圾处理厂废气排放情况 /381

大气防治重点区域危险（医疗）废物处置厂废气排放情况 /382

13 环境管理统计   383～408

各地区环保机构情况　/385

各地区环境信访与环境法制情况　/391

各地区环境保护能力建设投资情况　/393

各地区环境污染源控制与管理情况　/395

各地区环境监测情况　/396

各地区污染源自动监控情况　/399

各地区排污费征收情况　/400

各地区自然生态保护与建设情况　/401

各地区环境影响评价　/404

各地区建设项目竣工环境保护验收情况　/405

各地区突发环境事件　/407

各地区环境宣教情况　/408

14 附表   409～416

全国行政区划　/411

国民经济与社会发展总量指标摘要　/412

国民经济与社会发展结构指标摘要　/414

自然资源状况　/415

15 主要统计指标解释   417～435

综述

2013年是全面贯彻落实党的十八大精神的开局之年，在党中央、国务院的正确领导下，各地区、各部门围绕大气、水体和土壤污染治理三项重点工作，加大环境保护工作力度，着力解决突出环境问题，主要污染物总量减排工作扎实推进，大气污染防治取得新进展，生态文明建设不断深化。与2012年相比，化学需氧量（COD）排放量下降2.93%、氨氮排放量下降3.12%、二氧化硫（SO_2）排放量下降3.48%、氮氧化物（NO_x）排放量下降4.72%。但是环境形势依然严峻，环境风险不断凸显，污染治理任务仍然艰巨。

全国废水排放总量695.4亿吨。其中，工业废水排放量209.8亿吨、城镇生活污水排放量485.1亿吨。废水中化学需氧量排放量2 352.7万吨，其中，工业源化学需氧量排放量为319.5万吨、农业源化学需氧量排放量为1 125.8万吨、城镇生活化学需氧量排放量为889.8万吨。废水中氨氮排放量245.7万吨。其中，工业源氨氮排放量为24.6万吨、农业源氨氮排放量为77.9万吨、城镇生活氨氮排放量为141.4万吨。

全国废气中二氧化硫排放量2 043.9万吨。其中，工业二氧化硫排放量为1 835.2万吨、城镇生活二氧化硫排放量为208.5万吨。全国废气中氮氧化物排放量2 227.4万吨。其中，工业氮氧化物排放量为1 545.6万吨、城镇生活氮氧化物排放量为40.7万吨、机动车氮氧化物排放量为640.6万吨。全国废气中烟（粉）尘排放量1 278.1万吨。其中，工业烟（粉）尘排放量为1 094.6万吨、城镇生活烟尘排放量为123.9万吨、机动车烟（粉）尘排放量为59.4万吨。

全国一般工业固体废物产生量32.8亿吨，综合利用量20.6亿吨，贮存量4.3亿吨，处置量8.3亿吨，倾倒丢弃量129.3万吨，全国一般工业固体废物综合利用率为62.2%。全国工业危险废物产生量3 156.9万吨，综合利用量1 700.1万吨，贮存量810.8万吨，处置量701.2万吨，全国工业危险废物综合利用处置率为74.8%。

全国共调查统计工业企业147 657家，其中火电行业3 102家，共排放二氧化硫782.7万吨，氮氧化物964.6万吨，烟（粉）尘218.8万吨。其中独立火电厂1 853家，拥有4 825台机组，共有脱硫设施3 547套，脱硝设施1 076套，除尘设施5 140套，排放二氧化硫634.1万吨，氮氧化物861.8万吨，烟（粉）尘183.9万吨。自备电厂1 249家，有2 690台机组，排放二氧化硫148.6万吨，氮氧化物102.8万吨，烟（粉）尘34.9万吨。

调查统计水泥制造企业3 679家，其中有熟料生产的水泥企业1 899家。共有脱硝设施538套，除尘设施46 592套。排放氮氧化物196.9万吨，烟（粉）尘64.9万吨。

调查统计黑色金属冶炼和压延加工4 202家，其中有烧结机或球团设备的钢铁企业740家，共拥有烧结机1 258台，球团设备598套。共排放二氧化硫199.3万吨，氮氧化物55.5

万吨，烟（粉）尘 61.9 万吨。

调查统计造纸和纸制品企业 4 856 家，有制浆和（或）造纸生产工艺的企业 3 866 家，共拥有废水治理设施 5 122 套。排放废水 28.5 亿吨，化学需氧量 53.3 万吨，氨氮 1.8 万吨。

调查统计规模化畜禽养殖场 138 730 家，规模化畜禽养殖小区 9 420 家，排放化学需氧量 312.1 万吨，氨氮 31.3 万吨，总氮 140.9 万吨，总磷 23.5 万吨。

调查统计城镇污水处理厂 5 364 座，设计处理能力达到 1.7 亿吨/日，全年共处理污水 456.1 亿吨；生活垃圾处理厂（场）2 135 座，全年共处理生活垃圾 2.06 亿吨，其中采用填埋方式处置的共 1.79 亿吨，采用堆肥方式处置的共 0.04 亿吨，采用焚烧方式处置的共 0.23 亿吨；危险废物集中处理（置）厂（场）767 座，医疗废物集中处理（置）厂（场）243 座，全年共综合利用危险废物 457.8 万吨，处置危险废物 282.3 万吨。

1

统计调查企业基本情况

ANNUAL STATISTIC REPORT ON ENVIRONMENT IN CHINA
2013

1.1 工业企业调查基本情况

2013 年，全国重点调查了 147 657 家工业企业，其中，有废水及废水污染物排放的企业有 90 884 家，有废气及废气污染物排放的企业有 109 512 家，有一般工业固体废物产生的企业有 102 634 家，有工业危险废物产生的企业有 23 871 家。对其他工业企业的污染排放量按比率进行估算。

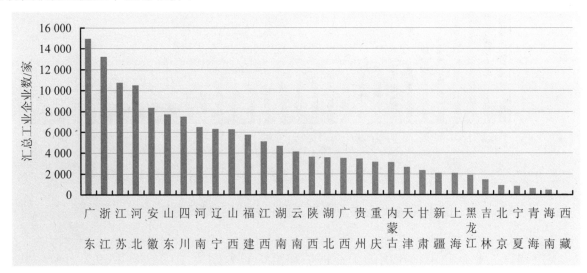

图 1-1　各地区汇总重点调查工业企业数量分布情况

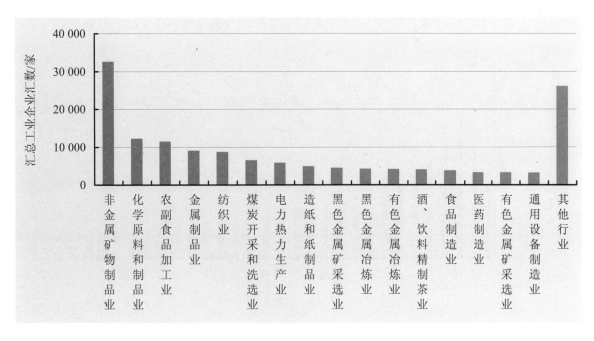

图 1-2　各行业汇总重点调查工业企业数量分布情况

重点调查工业企业中，共有 80 298 套废水治理设施，形成了 25 642 万吨/日的废水处理能力，投入运行费用 628.7 亿元。共处理 492.5 亿吨工业废水，去除化学需氧量 1 804.9 万吨，氨氮 109.6 万吨，石油类 24.4 万吨，挥发酚 6.9 万吨，氰化物 0.6 万吨。

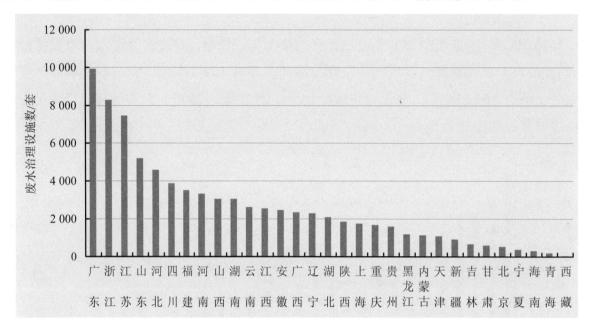

图 1-3　各地区重点调查工业企业废水治理设施数量

重点调查工业企业中，在用的工业锅炉和窑炉数分别为 10.2 万台和 9.4 万台，共安装 234 316 套废气治理设施（其中，脱硫设施 21 677 套，脱硝设施 2 152 套，除尘设施 180 154 套），形成了 143.5 亿米^3/小时的废气处理能力，投入运行费用 1 497.8 亿元。共去除二氧化硫 4 382.0 万吨，氮氧化物 348.6 万吨，烟（粉）尘 72 560.7 万吨。

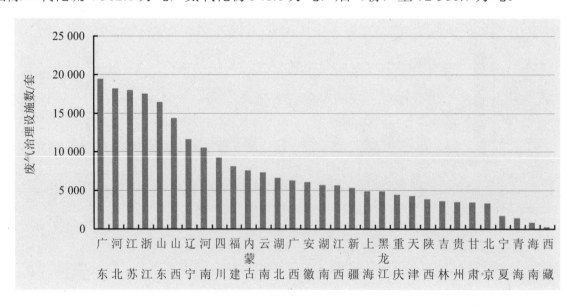

图 1-4　各地区重点调查工业企业废气治理设施数量

1.2　农业源调查基本情况

2013 年，重点调查了 138 730 家规模化畜禽养殖场，9 420 家规模化畜禽养殖小区，对种植业、水产养殖业和其他养殖专业户按产排污强度等进行了核算。

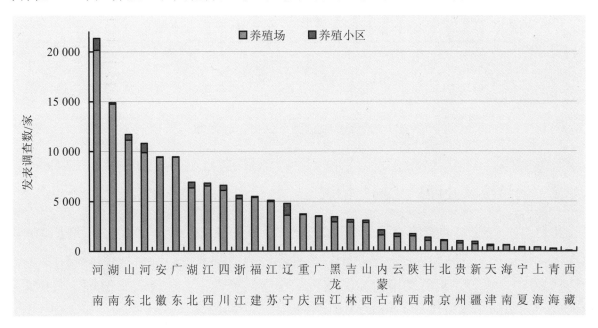

图 1-5　各地区农业源重点调查规模化养殖场和规模化养殖小区数量分布情况

1.3　集中式污染治理设施调查基本情况

1.3.1　城镇生活污水集中处理厂情况

2013 年，全国共调查统计 5 364 座城市污水处理厂，比 2012 年增加 736 座；设计处理能力为 16 574 万吨/日，比 2012 年新增 1 260 万吨/日；运行费用为 393.6 亿元，比 2012 年增加 45.4 亿元。全年共处理废水 456.1 亿吨，比 2012 年增加 39.9 亿吨，其中，处理生活污水 397.7 亿吨，占总处理水量的 87.2%。再生水生产量 24.4 亿吨，再生水利用量 15.4 亿吨。共去除化学需氧量 1 109.7 万吨，氨氮 100.3 万吨，油类 4.7 万吨，总氮 81.3 万吨，总磷 11.4 万吨，挥发酚 970 吨，氰化物 1 267 吨。污水处理厂的污泥产生量为 2 635.8 万吨，污泥处置量为 2 635.8 万吨。

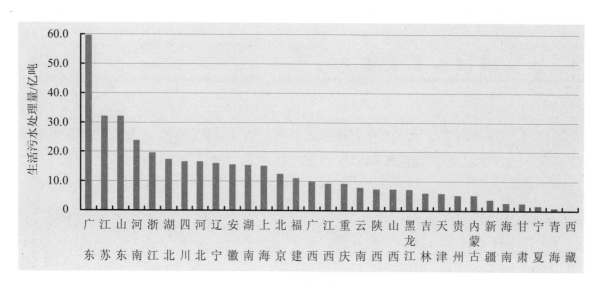

图 1-6　各地区城镇生活污水处理量

1.3.2　生活垃圾处理厂（场）情况

2013 年，全国共调查统计了生活垃圾处理厂（场）2 135 座，比 2012 年增加 10 座；填埋设计容量达到 334 519 万米³；堆肥设计处理能力达到 23 925 吨/日；焚烧设计处理能力达到 80 017 吨/日；运行费用为 86.6 亿元。全年共处理生活垃圾 2.06 亿吨，其中采用填埋方式处置的生活垃圾共 1.79 亿吨，采用堆肥方式处置的共 0.03 亿吨，采用焚烧方式处置的共 0.24 亿吨。

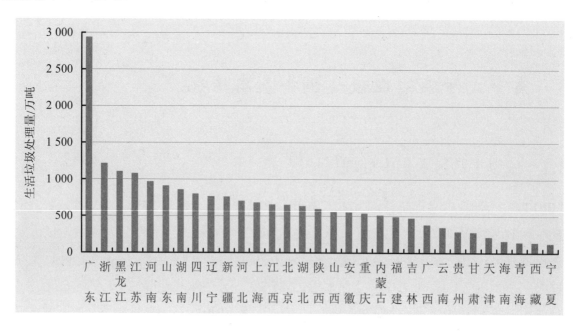

图 1-7　各地区生活垃圾处理量

8

1.3.3 危险废物（医疗废物）集中处理（置）厂（场）情况

2013 年，全国共调查统计危险废物集中处理（置）厂（场）767 座，比 2012 年增加 45 座；医疗废物集中处理（置）厂（场）243 座，比 2012 年增加 7 座；危险废物设计处置能力达到 85 886 吨/日；运行费用为 58.7 亿元，比 2012 年增加 4.8 亿元。全年共综合利用危险废物 457.8 万吨。全年共处置危险废物 282.3 万吨，其中工业危险废物 197.8 万吨，医疗废物 59.7 万吨。采用填埋方式处置的危险废物共 88.2 万吨，采用焚烧方式处置的危险废物共 149.0 万吨。

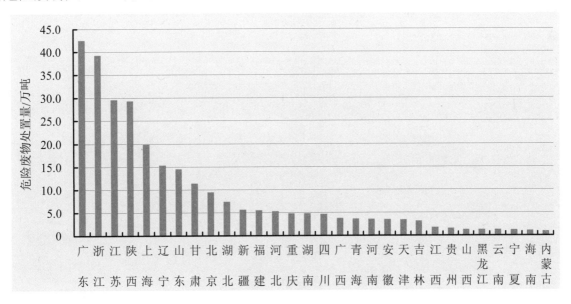

图 1-8　各地区危险废物集中处理（置）量

2

废 水

ANNUAL STATISTIC REPORT ON ENVIRONMENT IN CHINA
2013

2.1 废水及主要污染物排放情况

2.1.1 废水排放情况

2013 年，全国废水排放量 695.4 亿吨，比 2012 年增加 1.5%。

工业废水排放量 209.8 亿吨，比 2012 年减少 5.3%；占废水排放总量的 30.2%，比 2012 年减少 2.1 个百分点。

城镇生活污水排放量 485.1 亿吨，比 2012 年增加 4.8%；占废水排放总量的 69.8%，比 2012 年增加 2.2 个百分点。

集中式污染治理设施废水（不含城镇污水处理厂，下同）排放量 0.5 亿吨。

表 2-1　全国废水及其主要污染物排放情况

年份	排放量＼排放源	合计	工业源	农业源	城镇生活源	集中式
2011	废水/亿吨	659.2	230.9	—	427.9	0.4
	化学需氧量/万吨	2 499.9	354.8	1 186.1	938.8	20.1
	氨氮/万吨	260.4	28.1	82.7	147.7	2.0
2012	废水/亿吨	684.8	221.6	—	462.7	0.5
	化学需氧量/万吨	2 423.7	338.5	1 153.8	912.8	18.7
	氨氮/万吨	253.6	26.4	80.6	144.6	1.9
2013	废水/亿吨	695.4	209.8	—	485.1	0.5
	化学需氧量/万吨	2 352.7	319.5	1 125.8	889.8	17.7
	氨氮/万吨	245.7	24.6	77.9	141.4	1.8
变化率/%	废水	1.5	−5.3	—	4.8	—
	化学需氧量	−2.9	−5.6	−2.4	−2.5	—
	氨氮	−3.1	−6.8	−3.3	−2.2	—

注：①自 2011 年起环境统计中增加农业源的污染排放统计，农业源包括种植业、水产养殖业和畜禽养殖业排放的污染物；②集中式污染治理设施排放量指生活垃圾处理厂（场）和危险废物（医疗废物）集中处理（置）厂（场）垃圾渗滤液/废水及其污染物的排放量；③变化率表示与 2012 年相比指标的变化情况，下同。

2.1.2 化学需氧量排放情况

2013 年，全国废水中化学需氧量排放量 2 352.7 万吨，比 2012 年减少 2.9%。

工业废水中化学需氧量排放量 319.5 万吨，比 2012 年减少 5.6%；占化学需氧量排放总量的 13.6%，比 2012 年减少 0.4 个百分点。

农业源化学需氧量排放量 1 125.8 万吨，比 2012 年减少 2.4%。其中畜禽养殖业排放 1 071.7 万吨，比 2012 年减少 2.5%，水产养殖业排放 54.0 万吨，比 2012 年减少 1.5%。农业源化学需氧量排放占排放总量的 47.9%，比 2012 年增加 0.3 个百分点。

城镇生活污水中化学需氧量排放量 889.8 万吨，比 2012 年减少 2.5%；占化学需氧量排放总量的 37.8%，比 2012 年增加 0.2 个百分点。

集中式污染治理设施废水中化学需氧量排放量 17.7 万吨，其中生活垃圾处理厂（场）17.6 万吨，危险（医疗）废物集中处理（置）厂（场）953.4 吨。

2.1.3 氨氮排放情况

2013 年，全国废水中氨氮排放量 245.7 万吨，比 2012 年减少 3.1%。

工业废水氨氮排放量 24.6 万吨，比 2012 年减少 6.8%；占氨氮排放总量的 10.0%，比 2012 年减少 0.4 个百分点。

农业源氨氮排放量 77.9 万吨，比 2012 年减少 3.3%，其中种植业排放 15.2 万吨，与 2012 年持平，畜禽养殖业排放 60.4 万吨，比 2012 年减少 4.3%，水产养殖业排放 2.3 万吨，与 2012 年持平；农业源氨氮排放量占排放总量的 31.7%，与 2012 年持平。

城镇生活污水中氨氮排放量 141.4 万吨，比 2012 年减少 2.2%；占氨氮排放总量的 57.5%，比 2012 年增加 0.5 个百分点。

集中式污染治理设施废水中氨氮排放量 1.8 万吨，其中生活垃圾处理厂（场）1.8 万吨，危险（医疗）废物集中处理（置）厂（场）159.5 吨。

2.1.4 废水中其他主要污染物排放情况

2013 年，全国工业废水中石油类排放量 1.7 万吨，比 2012 年增加 0.4%；挥发酚排放量 1 259.1 吨，比 2012 年减少 15.0%；氰化物排放量 162.0 吨，比 2012 年减少 5.7%。

工业废水中重金属汞、镉、六价铬、总铬、铅及砷排放量分别为 0.8 吨、17.9 吨、58.1 吨、161.9 吨、74.1 吨和 111.6 吨，重金属污染物分别比 2012 年减少 27.3%、33.0%、17.5%、14.2%、23.7%和 12.6%。

表 2-2　全国工业废水中重金属及其他污染物排放量

单位：吨

年份 \ 污染物	石油类	挥发酚	氰化物	汞	镉	六价铬	总铬	铅	砷
2011	20 589.1	2 410.5	215.4	1.2	35.1	106.2	290.3	150.8	145.2
2012	17 327.2	1 481.4	171.8	1.1	26.7	70.4	188.6	97.1	127.7
2013	17 389.2	1 259.1	162.0	0.8	17.9	58.1	161.9	74.1	111.6
变化率/%	0.4	−15.0	−5.7	−27.3	−33.0	−17.5	−14.2	−23.7	−12.6

2.2　各地区废水及主要污染物排放情况

2.2.1　各地区废水排放情况

2013 年，废水排放量大于 30 亿吨的省份共 8 个，依次为广东、江苏、山东、浙江、河南、河北、四川、湖南。8 个省份废水排放总量为 370.9 亿吨，占全国废水排放量的 53.3%。工业废水排放量前 3 位的是江苏、山东和广东，分别占全国工业废水排放量的 10.5%、8.6%和 8.1%。城镇生活污水排放量前 3 位依次是广东、江苏、山东，分别占全国城镇生活污水排放量的 14.3%、7.7%和 6.5%。

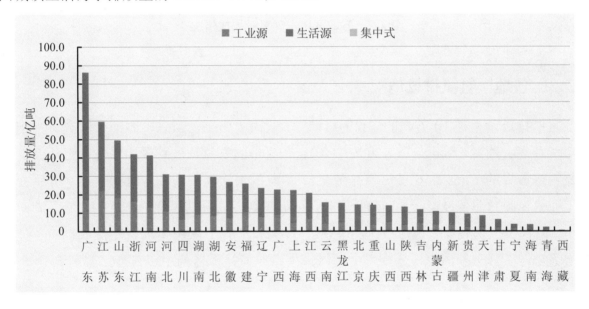

图 2-1　各地区废水排放情况

2.2.2　各地区化学需氧量排放情况

化学需氧量排放量大于 100 万吨的省份有 10 个，依次为山东、广东、黑龙江、河南、河北、辽宁、湖南、四川、江苏和湖北。10 个省份的化学需氧量排放总量为 1 363.2 万吨，占全国化学需氧量排放量的 57.9%。工业化学需氧量排放量前 3 位的依次是广东、江苏和新疆，分别占全国工业化学需氧量排放量的 7.3%、6.5%和 5.8%；农业化学需氧量排放量前 3 位的依次是山东、黑龙江和河北，分别占全国农业化学需氧量排放量的 11.5%、9.2%和 7.9%；城镇生活化学需氧量排放量前 3 位的依次是广东、四川和江苏，分别占全国城镇生活化学需氧量排放量的 10.2%、6.6%和 6.3%。

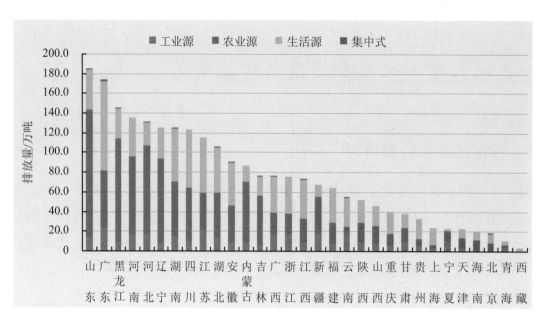

图 2-2　各地区化学需氧量排放情况

2.2.3　各地区氨氮排放情况

氨氮排放量大于 10 万吨的省份有 11 个，依次为广东、山东、湖南、江苏、河南、四川、湖北、浙江、河北、辽宁和安徽。11 个省份的氨氮排放总量为 151.0 万吨，占全国氨氮排放量的 61.5%。工业氨氮排放量前 3 位的依次为湖南、广东和江苏，分别占全国工业氨氮排放量的 9.4%、5.9% 和 5.9%；农业氨氮排放量前 3 位的依次为山东、湖南和河南，分别占全国农业氨氮排放量的 9.1%、7.9% 和 7.8%；城镇生活氨氮排放量前 3 位的依次为广东、江苏和山东，分别占全国城镇生活氨氮排放量的 10.3%、6.7% 和 5.6%。

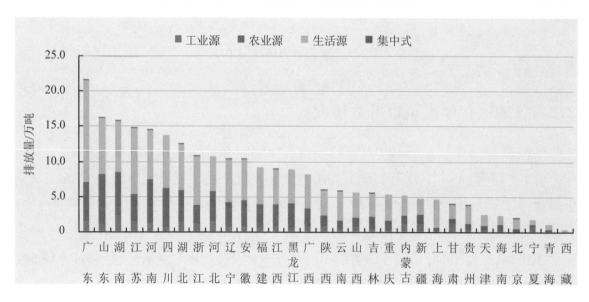

图 2-3　各地区氨氮排放情况

2.2.4 各地区工业石油类、挥发酚和氰化物排放情况

工业废水中石油类排放量大于700吨的省份有11个，依次为江苏、河南、湖北、山西、辽宁、内蒙古、河北、安徽、浙江、江西和新疆。11个省份的石油类排放量为10 251.9吨，占全国工业废水石油类排放量的59.0%。

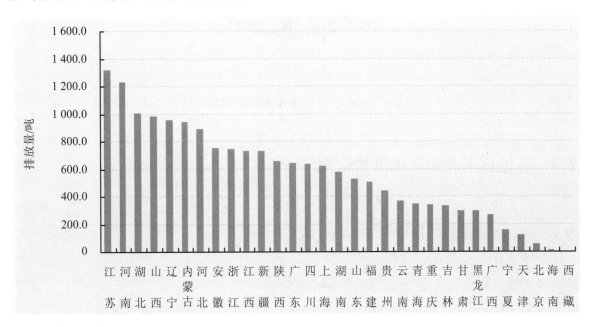

图2-4 各地区工业废水石油类排放情况

工业废水中挥发酚排放量较大的省份有3个，依次为山西、内蒙古和河南。3个省份的挥发酚排放量为1 000.5吨，占全国挥发酚排放量的79.5%。

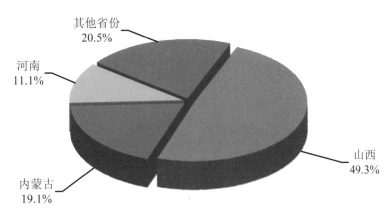

图2-5 各地区工业废水挥发酚排放情况

工业废水中氰化物排放量大于9吨的省份有5个，依次为山西、河南、江苏、湖南和广东。5个省份的氰化物排放量为80.6吨，占全国氰化物排放量的49.8%。

17

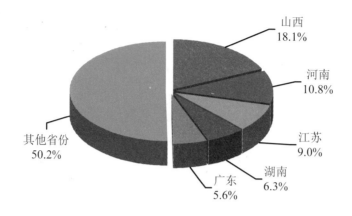

图 2-6　各地区工业废水氰化物排放情况

2.2.5　各地区工业重金属排放情况

工业废水中汞排放量前 3 位省份依次为湖南、广西和甘肃，3 个省份工业废水汞排放量为 0.4 吨，占全国废水汞排放量的 54.0%。

工业废水中镉排放量前 3 位的省份依次为湖南、江西和甘肃，3 个省份工业废水镉排放量为 9.9 吨，占全国工业废水镉排放量的 55.6%。

工业废水中六价铬排放量前 3 位的省份依次为江西、湖北和浙江，3 个省份工业废水六价铬排放量为 35.0 吨，占全国废水六价铬排放量的 60.2%。

工业废水中总铬排放量前 3 位的省份依次为河南、广东和浙江，3 个省份工业废水总铬排放量为 70.7 吨，占全国废水总铬排放量的 43.7%。

工业废水中铅排放量前 3 位的省份依次为湖南、甘肃和广西，3 个省份工业废水铅排放量为 38.5 吨，占全国废水总铬排放量的 52.0%。

工业废水中砷排放量前 3 位的省份依次为湖南、江西和西藏，3 个省份工业废水砷排放量为 60.8 吨，占全国废水砷排放量的 54.4%。

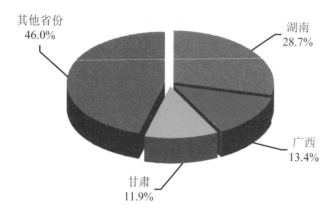

图 2-7　各地区工业废水汞排放情况

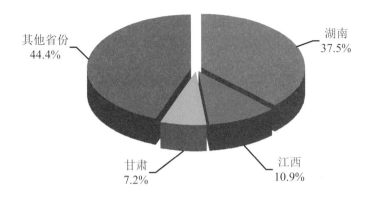

图 2-8　各地区工业废水镉排放情况

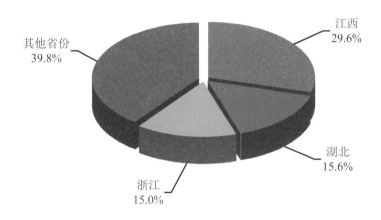

图 2-9　各地区工业废水六价铬排放情况

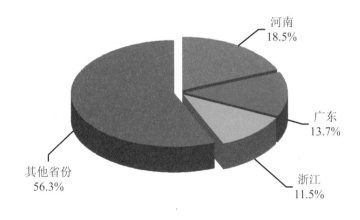

图 2-10　各地区工业废水总铬排放情况

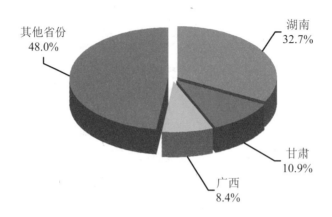

图 2-11　各地区工业废水铅排放情况

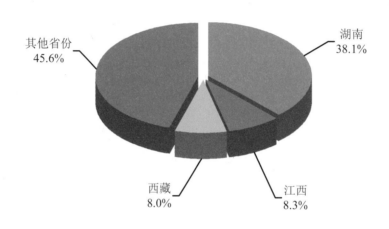

图 2-12　各地区工业废水砷排放情况

2.3　工业行业废水及主要污染物排放情况

2.3.1　行业废水排放情况

2013 年，在调查统计的 41 个工业行业中，废水排放量位于前 4 位的行业依次为造纸和纸制品业、化学原料及化学制品制造业、纺织业、煤炭开采和洗选业。4 个行业的废水排放量为 90.8 亿吨，占重点调查工业企业废水排放总量的 47.5%。

表 2-3　重点行业废水排放情况　　　　　　　　　　　　　　　　　　单位：亿吨

年份 \ 行业	合计	造纸和纸制品业	化学原料及化学制品制造业	纺织业	煤炭开采和洗选业
2011	105.4	38.2	28.8	24.1	14.3
2012	99.6	34.3	27.4	23.7	14.2
2013	90.8	28.5	26.6	21.5	14.3
变化率/%	−8.8	−16.8	−3.1	−9.5	0.6

注：自 2011 年起，环境统计按《国民经济行业分类》（GB/T 4754—2011）标准执行分类统计，下同。

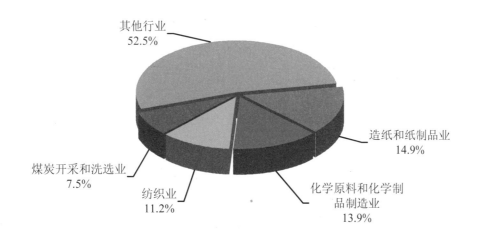

图 2-13　重点行业废水排放情况

造纸和纸制品业废水排放量前 5 位的省份依次是浙江、山东、广东、湖南和广西。5 个省份造纸和纸制品业废水排放量为 13.5 亿吨，占该行业重点调查工业企业废水排放量的 47.3%。

化学原料和化学制品制造业废水排放量前 5 位的省份依次是江苏、山东、湖北、河南和浙江。5 个省份化学原料和化学制品制造业废水排放量为 11.9 亿吨，占该行业重点调查工业企业废水排放量的 45.0%。

纺织业废水排放量前 5 位的省份依次是浙江、江苏、广东、山东和福建。5 个省份纺织业废水排放量为 17.4 亿吨，占该行业重点调查工业企业废水排放量的 81.1%。

煤炭开采和洗选业废水排放量前 5 位的省份依次是河南、山东、山西、贵州和黑龙江。5 个省份煤炭开采和洗选业废水排放量为 7.7 亿吨，占该行业重点调查工业企业废水排放量的 53.6%。

2.3.2　行业化学需氧量排放情况

2013 年，在调查统计的 41 个工业行业中，化学需氧量排放量位于前 4 位的行业依次为造纸和纸制品业、农副食品加工业、化学原料及化学制品制造业、纺织业。4 个行业的化学需氧量排放量为 158.0 万吨，占重点调查工业企业排放总量的 55.4%，较 2012 年下降 1.7 个百分点。

表 2-4　重点行业化学需氧量排放情况　　　　　　　　　　单位：万吨

年份 \ 行业	合计	造纸和纸制品业	农副食品加工业	化学原料和化学制品制造业	纺织业
2011	191.5	74.2	55.3	32.8	29.2
2012	173.6	62.3	51	32.5	27.7
2013	158.0	53.3	47.1	32.2	25.4
变化率/%	−9.0	−14.4	−7.6	−0.9	−8.3

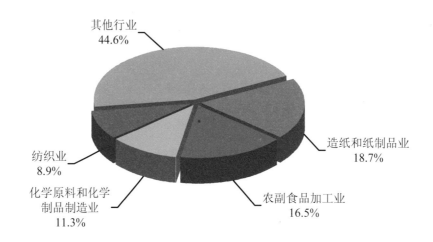

图 2-14　重点行业化学需氧量排放情况

造纸和纸制品业化学需氧量排放量前 5 位的省份依次是广东、广西、宁夏、湖南和河北。5 个省份造纸和纸制品业化学需氧量排放量为 22.4 万吨，占该行业重点调查工业企业化学需氧量排放量的 42.0%。

农副食品加工业化学需氧量排放量前 5 位的省份依次是云南、广西、甘肃、黑龙江和河北。5 个省份农副食品加工业化学需氧量排放量为 21.2 万吨，占该行业重点调查工业企业化学需氧量排放量的 45.0%。

化学原料和化学制品制造业化学需氧量排放量前 5 位的省份依次是江苏、湖北、湖南、河南和山东。5 个省份化学原料和化学制品制造业化学需氧量排放量为 12.1 万吨，占该行业重点调查工业企业化学需氧量排放量的 37.6%。

纺织业化学需氧量排放量较大的省份依次是浙江、广东和江苏。3 个省份纺织业化学需氧量排放量为 15.6 万吨，占该行业重点调查工业企业化学需氧量排放量的 61.4%。

2.3.3 行业氨氮排放情况

2013 年，在调查统计的 41 个工业行业中，氨氮排放量位于前 4 位的行业依次为化学原料和化学制品制造业、农副食品加工业、纺织业、造纸和纸制品业。4 个行业的氨氮排放量 13.1 万吨，占重点调查工业企业排放总量的 58.4%，较 2012 年下降 0.9 个百分点。

表 2-5　重点行业氨氮排放情况　　　　　　　　　　　　　单位：万吨

年份＼行业	合计	化学原料和化学制品制造业	农副食品加工业	纺织业	造纸和纸制品业
2011	15.9	9.3	2.1	2.0	2.5
2012	14.4	8.4	1.9	1.9	2.1
2013	13.1	7.6	1.9	1.8	1.8
变化率/%	−9.0	−9.5	0.0	−5.3	−14.3

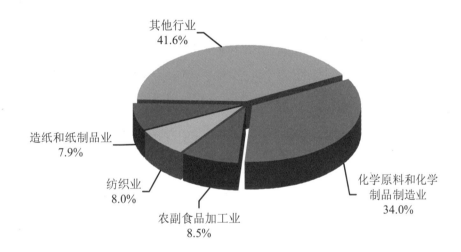

图 2-15　工业行业氨氮排放情况

化学原料和化学制品制造业氨氮排放量前 5 位的省份依次是湖南、新疆、湖北、安徽和甘肃。5 个省份化学原料和化学制品制造业氨氮排放量为 3.8 万吨，占该行业重点调查工业企业氨氮排放量的 49.2%。

农副食品加工业氨氮排放量前 5 位的省份依次是河北、内蒙古、河南、广东和山东。5 个省份农副食品加工业氨氮排放量为 0.6 万吨，占该行业重点调查工业企业氨氮排放量的 33.2%。

纺织业氨氮排放量较大的省份依次是浙江、广东和江苏。3 个省份纺织业氨氮排放量为 1.2 万吨，占该行业重点调查工业企业氨氮排放量的 64.9%。

造纸和纸制品业氨氮排放量前 5 位的省份依次是山东、河北、湖南、广东和广西。5 个省份造纸和纸制品业氨氮排放量为 0.8 万吨，占该行业重点调查工业企业氨氮排放量的 44.8%。

2.3.4 行业石油类排放情况

2013 年，石油类排放量位于前 4 位的行业依次是黑色金属冶炼和压延加工业，化学原料和化学制品制造业，煤炭开采和洗选业，石油加工、炼焦和核燃料加工业。4 个行业石油类排放量为 9 731.1 吨，占重点调查工业企业石油类排放量的 56.0%。

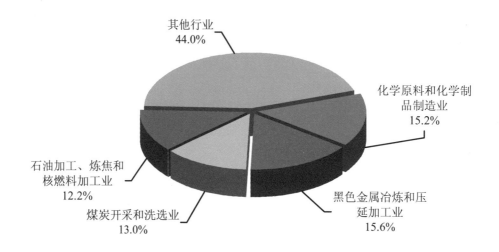

图 2-16　工业行业石油类污染物排放情况

黑色金属冶炼和压延加工业石油类排放量前 5 位的省份依次是内蒙古、江苏、山西、江西和河北。5 个省份黑色金属冶炼和压延加工业石油类排放量为 1 397.9 吨，占该行业重点调查工业企业石油类排放量的 51.7%。

化学原料和化学制品制造业石油类排放量前 5 位的省份依次是江苏、福建、浙江、河南和湖南。5 个省份化学原料和化学制品制造业石油类排放量为 1 027.3 吨，占该行业重点调查工业企业石油类排放量的 38.9%。

煤炭开采和洗选业石油类排放量前 5 位的省份依次是河南、贵州、安徽、山西和内蒙古。5 个省份煤炭开采和洗选业石油类排放量为 1 026.1 吨，占该行业重点调查工业企业石油类排放量的 45.3%。

石油加工、炼焦和核燃料加工业石油类排放量前 5 位的省份依次是山西、辽宁、河南、内蒙古和山东。5 个省份石油加工、炼焦和核燃料加工业石油类排放量为 1 276.5 吨，占该行业重点调查工业企业石油类排放量的 60.0%。

2.3.5 行业挥发酚排放情况

2013 年，挥发酚排放量最大的行业为石油加工、炼焦和核燃料加工业，挥发酚排放

量为 1 031.6 吨，占重点调查工业企业挥发酚排放量的 81.9%；其次为化学原料和化学制
品制造业，挥发酚排放量为 79.6 吨，占重点调查工业企业挥发酚排放量的 6.3%。

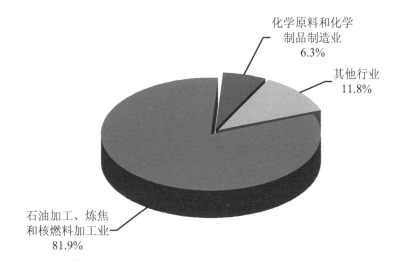

图 2-17　工业行业挥发酚排放情况

石油加工、炼焦和核燃料加工业挥发酚排放量较大的省份为山西和内蒙古，其中山
西石油加工、炼焦和核燃料加工业挥发酚排放量为 595.1 吨，内蒙古为 228.9 吨，分别占
该行业重点调查工业企业挥发酚排放量的 57.7% 和 22.2%。

化学原料和化学制品制造业挥发酚排放量最大的省份为江苏，排放量为 24.4 吨，占
该行业重点调查工业企业挥发酚排放量的 30.6%。

2.3.6　行业氰化物排放情况

2013 年，氰化物排放量位于前 4 位的行业依次为化学原料和化学制品制造业，石油
加工、炼焦和核燃料加工业，黑色金属冶炼和压延加工业，金属制品业。4 个行业石油类
排放量为 148.8 吨，占该行业重点调查工业企业氰化物排放量的 91.8%。

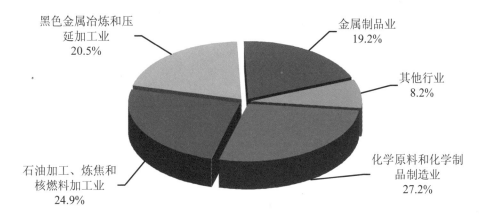

图 2-18　工业行业氰化物排放情况

化学原料和化学制品制造业氰化物排放量前5位的省份依次为湖南、河南、江苏、湖北和安徽。这5个省份化学原料和化学制品制造业氰化物排放量为24.3吨，占该行业重点调查工业企业氰化物排放量的55.2%。

石油加工、炼焦和核燃料加工业氰化物排放量较大的省份依次为山西、河南和河北，其中山西该行业氰化物排放量为18.1吨，河南为6.0吨，河北为4.9吨，分别占该行业重点调查工业企业氰化物排放量的44.8%、14.9%和12.1%。

黑色金属冶炼和压延加工业氰化物排放量前5位的省份依次是山西、河南、新疆、河北和吉林。5个省份黑色金属冶炼和压延加工业石氰化物放量为19.2吨，占该行业重点调查工业企业氰化物排放量的57.7%。

金属制品业氰化物排放量前5位的省份依次是广东、浙江、江西、江苏和河南。这5个省份金属制品业氰化物排放量为25.4吨，占该行业重点调查工业企业氰化物排放量的81.6%。

2.3.7 行业重金属污染物排放情况

2013年，重金属（汞、镉、六价铬、总铬、铅、砷）排放量位于前4位的行业依次为金属制品业，有色金属冶炼和压延加工业，皮革、毛皮、羽毛及其制品和制鞋业，有色金属矿采选业。4个行业重金属排放量为307.2吨，占重点调查工业企业重金属排放量的72.3%。

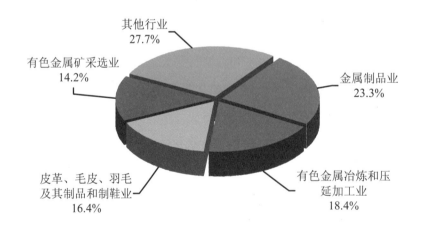

其他行业
27.7%

有色金属矿采选业
14.2%

金属制品业
23.3%

皮革、毛皮、羽毛及其制品和制鞋业
16.4%

有色金属冶炼和压延加工业
18.4%

图 2-19 工业行业重金属排放情况

汞排放量前3位的行业依次是有色金属矿采选业、有色金属冶炼和压延加工业和化学原料和化学制品制造业，这3个行业汞排放量为0.698吨，占重点调查工业企业汞排放量的89.2%。有色金属矿采选业汞排放量前3位的省份依次是广西、内蒙古和江西，这3个省份有色金属矿采选业汞排放量为181.7千克，占该行业重点调查工业企业汞排放量的

71.5%；有色金属冶炼和压延加工业汞排放量前 3 位的省份依次是湖南、甘肃和江西，这 3 个省份有色金属冶炼和压延加工业汞排放量为 170.5 千克，占该行业重点调查工业企业汞排放量的 75.5%；化学原料和化学制品制造业汞排放量前 3 位的省份依次是湖南、甘肃和内蒙古，这 3 个省份化学原料和化学制品制造业汞排放量为 161.9 千克，占该行业重点调查工业企业汞排放量的 74.1%。

镉排放量前 3 位的行业依次是有色金属冶炼和压延加工业，有色金属矿采选业，化学原料和化学制品制造业，这 3 个行业镉排放量为 16.331 吨，占重点调查工业企业镉排放量的 91.4%。有色金属冶炼和压延加工业镉排放量前 3 位的省份依次是湖南、江西和云南，这 3 个省份有色金属冶炼和压延加工业的镉排放量为 7.234 吨，占该行业重点调查工业企业镉排放量的 58.2%；有色金属矿采选业镉排放量前 3 位的省份依次是江西、湖南和广西，这 3 个省份有色金属矿采选业镉排放量为 1.425 吨，占该行业重点调查工业企业镉排放量的 63.5%；化学原料和化学制品制造业镉排放量前 3 位的省份依次是湖南、甘肃和湖北，这 3 个省份化学原料和化学制品制造业镉排放量为 1.434 吨，占该行业重点调查工业企业镉排放量的 86.5%。

铅排放量前 3 位的行业依次是有色金属冶炼和压延加工业，有色金属矿采选业，化学原料和化学制品制造业，这 3 个行业铅排放量为 66.074 吨，占重点调查工业企业铅排放量的 89.2%。有色金属冶炼和压延加工业铅排放量前 3 位的省份依次是湖南、云南和江西，这 3 个省份有色金属冶炼和压延加工业铅排放量为 17.928 吨，占该行业重点调查工业企业铅排放量的 55.4%；有色金属矿采选业铅排放量前 3 位的省份依次是湖南、广西和内蒙古，这 3 个省份有色金属矿采选业铅排放量为 14.520 吨，占该行业重点调查工业企业铅排放量的 57.4%；化学原料和化学制品制造业铅排放量为湖南、甘肃和福建，这 3 个省份化学原料和化学制品制造业行业铅排放量为 7.067 吨，占该行业重点调查工业企业铅排放量的 83.8%。

砷排放量前 3 位的行业依次是化学原料和化学制品制造业、有色金属矿采选业、有色金属冶炼和压延加工业，这 3 个行业砷排放量为 95.165 吨，占重点调查工业企业砷排放的 85.3%。化学原料和化学制品制造业砷排放量前 3 位的省份依次是湖南、湖北和安徽，3 个省份化学原料和化学制品制造业砷排放量为 36.387 吨，占该行业重点调查工业企业砷排放量的 94.8%；有色金属矿采选业砷排放量前 3 位的省份依次是广西、内蒙古和云南，3 个省份有色金属矿采选业砷排放量为 17.291 吨，占该行业重点调查工业企业砷排放量的 55.0%；有色金属冶炼和压延加工业砷排放量前 3 位的省份依次是江西、湖南和云南，3 个省份有色金属冶炼和压延加工业砷排放量为 13.277 吨，占该行业重点调查工业企业砷排放量的 52.4%。

总铬排放量较大的行业依次是皮革、毛皮、羽毛及其制品和制鞋业，金属制品业。

六价铬排放量较大的行业是金属制品业。皮革、毛皮、羽毛及其制品和制鞋业总铬排放量最高的省份为河南，其次为湖南，分别占 42.8%和 15.2%。金属制品业总铬、六价铬排放量为 60.786 吨和 36.601 吨，分别占重点调查工业企业总铬、六价铬排放量的 37.6%和 63.0%，主要分布在广东、江西和浙江 3 个省份，这 3 个省份金属制品业总铬、六价铬排放量为 39.813 吨和 26.661 吨，分别占该行业重点调查工业企业总铬、六价铬排放量的 65.5%和 72.8%。

表 2-6　工业行业废水重金属污染物排放情况

污染物	排放量总计/吨	主要行业及所占比例	排放量大的地区及所占比例
汞	0.8	有色金属矿采选业 32.4%，有色金属冶炼和压延加工业 28.8%，化学原料和化学制品制造业 27.9%	湖南 28.7%，广西 13.4%，甘肃 11.9%
镉	17.9	有色金属冶炼和压延加工业 69.5%，有色金属矿采选业 12.6%，化学原料和化学制品制造业 9.3%	湖南 37.5%，江西 10.9%，甘肃 7.2%
铅	74.1	有色金属冶炼和压延加工业 43.7%，有色金属矿采选业 34.18%，化学原料和化学制品制造业 11.4%	湖南 32.7%，甘肃 10.9%，广西 8.4%
砷	111.6	化学原料和化学制品制造业 34.4%，有色金属矿采选业 28.2%，有色金属冶炼和压延加工业 22.7%	湖南 38.1%，江西 8.3%，西藏 8.0%
六价铬	58.1	金属制品业 63.0%，汽车制造业 14.3%，有色金属冶炼和压延加工业 4.6%	江西 29.6%，湖北 15.6%，浙江 15.1%
总铬	161.9	皮革、毛皮、羽毛及其制品和制鞋业 41.2%，金属制品业 37.6%，汽车制造业 5.6%	河南 18.6%，广东 13.7%，浙江 11.5%

2.4　重点流域废水及主要污染物排放与治理情况

根据《重点流域水污染防治"十二五"规划》中的流域分区，松花江、辽河、海河、黄河中上游、淮河、长江中下游、太湖、巢湖、滇池、三峡库区及其上游、丹江口库区及其上游等重点流域总体排放情况如下：

表 2-7　重点流域废水及废水中污染物总体排放情况

污染物 流域	废水/ 亿吨	化学需氧量/ 万吨	氨氮/ 万吨	工业石油类/ 吨	工业挥发酚/ 吨	工业氰化物/ 吨	工业重金属/ 吨
松花江	22.8	194.9	12.3	491.5	13.4	1.8	0.4
辽河	18.5	122.2	9.4	501.6	3.7	2.8	3.1
海河	81.2	275.3	23.9	1 800.9	359.4	22.5	22.7
黄河中上游	40.8	168.0	16.9	2 429.4	556.2	32.3	38.9
淮河	66.1	255.1	28.0	1 808.6	154.9	19.7	17.5
长江中下游	126.6	375.6	47.6	3 999.7	65.7	37.7	166.5
太湖	34.5	32.9	5.2	380.2	7.1	6.9	9.1
巢湖	4.9	12.8	1.3	43.0	0.0	0.0	0.1
滇池	4.1	0.9	0.4	46.4	0.1	0.1	2.6
三峡库区	59.5	197.7	23.6	1 525.5	13.2	4.2	14.7
丹江口库区	4.8	20.7	2.8	151.3	0.1	0.1	16.7

注：本年报中重点流域是根据《重点流域水污染防治"十二五"规划》中流域分区汇总得出，下同。

　2013 年，重点流域的废水排放总量为 463.7 亿吨，较 2012 年上升了 2.2%，占全国废水排放总量的 66.7%。其中，松花江、辽河、海河、黄河中上游、淮河、长江中下游、太湖、巢湖、滇池、三峡库区及其上游、丹江口库区及其上游流域的废水排放量分别为 22.8 亿吨、18.5 亿吨、81.2 亿吨、40.8 亿吨、66.1 亿吨、126.6 亿吨、34.5 亿吨、4.9 亿吨、4.1 亿吨、59.5 亿吨和 4.8 亿吨，分别占重点流域排放总量的 4.9%、4.0%、17.5%、8.8%、14.3%、27.3%、7.4%、1.1%、0.9%、12.8%和 1.0%。

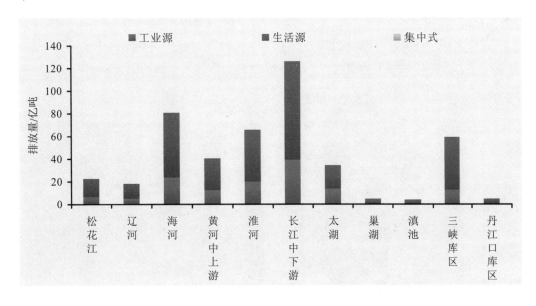

图 2-20　重点流域废水排放情况

　重点流域的化学需氧量排放总量为 1 656.1 万吨，较 2012 年下降了 2.7%，占全国化学需氧量排放总量的 70.4%。其中，松花江、辽河、海河、黄河中上游、淮河、长江中下游、太湖、巢湖、滇池、三峡库区及其上游、丹江口库区及其上游流域的化学需氧量排放量分别为 194.9 万吨、122.2 万吨、275.3 万吨、168.0 万吨、255.1 万吨、375.6 万吨、32.9 万吨、12.8 万吨、0.9 万吨、197.7 万吨和 20.7 万吨，分别占重点流域排放总量的 11.8%、7.4%、16.6%、10.1%、15.4%、22.7%、2.0%、0.8%、0.1%、11.9%和 1.2%。

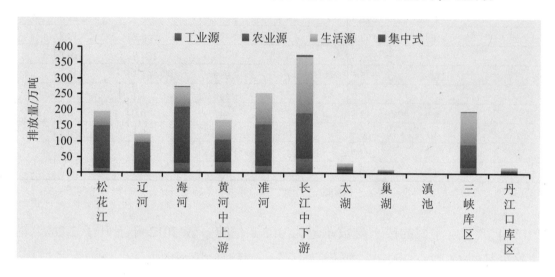

图 2-21 重点流域化学需氧量排放情况

重点流域的氨氮排放总量为 171.4 万吨，较 2012 年下降了 3.0%，占全国氨氮排放总量的 69.8%。其中，松花江、辽河、海河、黄河中上游、淮河、长江中下游、太湖、巢湖、滇池、三峡库区及其上游、丹江口库区及其上游流域的氨氮排放量分别为 12.3 万吨、9.4 万吨、23.9 万吨、16.9 万吨、28.0 万吨、47.6 万吨、5.2 万吨、1.3 万吨、0.4 万吨、23.6 万吨和 2.8 万吨，分别占重点流域排放总量的 7.2%、5.5%、14.0%、9.9%、16.3%、27.8%、3.0%、0.7%、0.2%、13.8%和 1.6%。

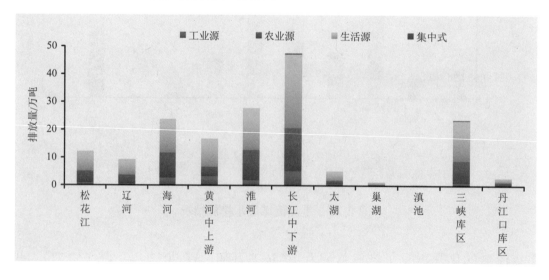

图 2-22 重点流域氨氮排放情况

2.4.1 松花江流域

（1）废水及污染物排放情况

《重点流域水污染防治"十二五"规划》中松花江流域含吉林、黑龙江和内蒙古自治区 3 个省份的 26 个地市和 173 个区县。2013 年，重点调查了工业企业 2 752 家，规模化畜禽养殖场 4 815 家，规模化畜禽养殖小区 624 家。

松花江流域共排放废水 22.8 亿吨，其中，工业废水 6.8 亿吨，城镇生活污水 16.0 亿吨。化学需氧量排放量为 194.9 万吨，其中，工业化学需氧量为 13.2 万吨，农业化学需氧量 136.9 万吨，城镇生活化学需氧量 43.7 万吨。氨氮排放量为 12.3 万吨，其中，工业氨氮 1.0 万吨，农业氨氮 4.3 万吨，城镇生活氨氮 7.0 万吨。

松花江流域工业石油类排放量为 491.5 吨，工业挥发酚排放量为 13.4 吨，工业氰化物排放量为 1.8 吨，工业废水中 6 种重金属（包括铅、镉、汞、六价铬、总铬及砷）排放总量为 0.4 吨。

表 2-8　松花江流域废水及主要污染物排放情况

污染物　　年份	废水/亿吨			化学需氧量/万吨				氨氮/万吨			
	工业源	生活源	集中式	工业源	农业源	生活源	集中式	工业源	农业源	生活源	集中式
2011	6.56	15.88	0.01	14.31	143.93	49.41	1.17	0.98	4.48	7.68	0.13
2012	8.03	15.84	0.01	13.95	138.94	47.40	1.13	0.96	4.28	7.40	0.12
2013	6.80	15.95	0.01	13.19	136.86	43.68	1.12	0.96	4.25	6.98	0.12
变化率/%	−15.4	0.7	—	−5.4	−1.5	−7.8	—	−0.7	−0.8	−5.8	—

（2）废水及主要污染物在各地区的分布

2013 年，松花江流域废水、化学需氧量和氨氮排放量最大的均是黑龙江省，分别占该流域各类污染物排放量的 58.8%、62.3% 和 60.1%。其中，工业废水、化学需氧量和氨氮排放量最大的均是黑龙江省，分别占该流域工业排放总量的 58.6%、58.9% 和 56.7%。农业化学需氧量和氨氮排放量最大的均是黑龙江省，分别占该流域农业排放总量的 63.3% 和 60.8%。城镇生活废水、化学需氧量和氨氮排放量最大的均是黑龙江省，分别占该流域生活排放总量的 58.9%、61.6% 和 60.8%。

31

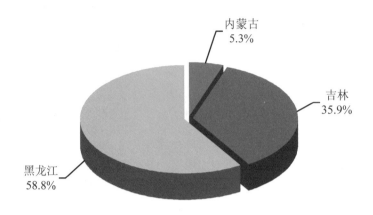

图 2-23 松花江流域废水排放区域构成

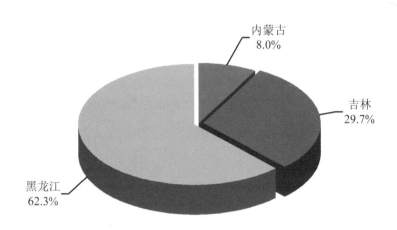

图 2-24 松花江流域化学需氧量排放区域构成

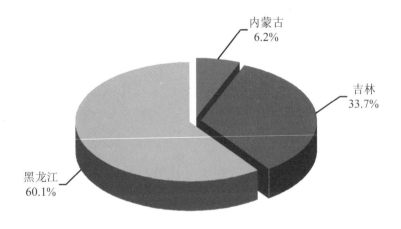

图 2-25 松花江流域氨氮排放区域构成

（3）废水及主要污染物在行业的分布

2013 年，在调查统计的 41 个工业行业中，松花江流域废水排放量位于前 4 位的行业依次为农副食品加工业，煤炭开采和洗选业，化学原料及化学制品制造业，石油加工、炼焦和核燃料加工业。4 个行业的废水排放量为 3.5 亿吨，占重点调查工业企业废水排放总量的 54.4%。

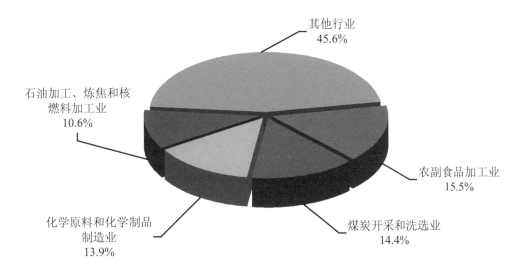

图 2-26　松花江流域工业废水排放量行业构成

2013 年，在调查统计的 41 个工业行业中，松花江流域化学需氧量排放量位于前 4 位的行业依次为农副食品加工业，造纸和纸制品业，酒、饮料和精制茶制造业，煤炭开采和洗选业。4 个行业的化学需氧量排放量为 8.0 万吨，占重点调查工业企业化学需氧量排放总量的 68.6%。

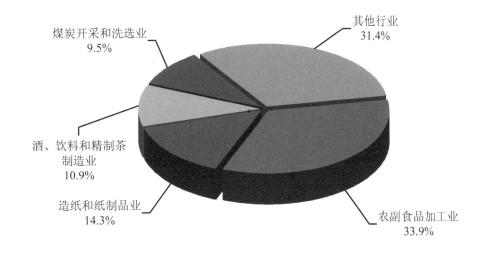

图 2-27　松花江流域工业化学需氧量排放量行业构成

2013 年，在调查统计的 41 个工业行业中，松花江流域氨氮排放量位于前 4 位的行业依次为农副食品加工业，石油加工、炼焦和核燃料加工业，化学原料和化学制品制造业，食品制造业。4 个行业的氨氮排放量为 0.54 万吨，占重点调查工业企业氨氮排放总量的 64.7%。

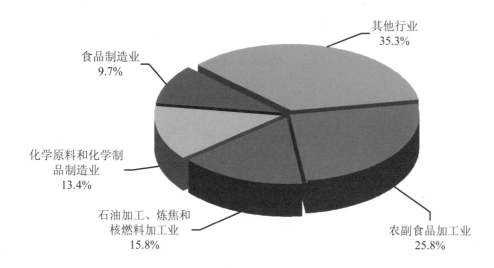

图 2-28　松花江流域工业氨氮排放量行业构成

（4）废水及污染物治理情况

2013 年，松花江流域纳入统计的污水处理厂 128 座，形成了 581 万吨/日的处理能力，年运行费用达 12.3 亿元，共处理污水 13.1 亿吨，其中生活污水 12.0 亿吨。去除化学需氧量 32.4 万吨、氨氮 3.0 万吨、油类 0.2 万吨、总氮 2.5 万吨、总磷 0.3 万吨。

松花江流域重点调查工业企业共有废水治理设施 1 513 套，形成了 809 万吨/日的废水处理能力，年运行费用达 33.6 亿元，共处理了 12.6 亿吨工业废水。去除工业化学需氧量 125.2 万吨、氨氮 8.1 万吨、石油类 4.4 万吨、挥发酚 5 190.9 吨、氰化物 56.6 吨。

2.4.2　辽河流域

（1）废水及污染物排放情况

《重点流域水污染防治"十二五"规划》中辽河流域含辽宁、吉林和内蒙古自治区 3 个省份的 16 个地市和 106 个区县。2013 年，重点调查了工业企业 5 211 家，规模化畜禽养殖场 3 785 家，规模化畜禽养殖小区 1 193 家。

辽河流域共排放废水 18.5 亿吨，其中，工业废水 5.3 亿吨，城镇生活污水 13.2 亿吨。化学需氧量排放量为 122.2 万吨，其中，工业化学需氧量为 7.7 万吨，农业化学需氧量 89.8 万吨，城镇生活化学需氧量 24.1 万吨。氨氮排放量为 9.4 万吨，其中，工业氨氮为 0.5 万吨，农业氨氮 3.2 万吨，城镇生活氨氮 5.6 万吨。

辽河流域工业石油类排放量为 501.6 吨，工业挥发酚排放量为 3.7 吨，工业氰化物排放量为 2.8 吨，工业废水中 6 种重金属（包括铅、镉、汞、六价铬、总铬及砷）排放总量为 3.1 吨。

表 2-9　辽河流域废水及主要污染物排放情况

年份	废水/亿吨			化学需氧量/万吨				氨氮/万吨			
污染物	工业源	生活源	集中式	工业源	农业源	生活源	集中式	工业源	农业源	生活源	集中式
2011	5.57	11.72	0.01	9.53	94.85	27.51	0.80	0.66	3.49	5.88	0.11
2012	5.35	12.76	0.01	8.79	90.27	26.23	0.66	0.60	3.36	5.76	0.09
2013	5.34	13.16	0.01	7.68	89.77	24.07	0.66	0.54	3.24	5.57	0.09
变化率/%	−0.01	3.1	—	−12.6	−0.6	−8.2	—	−8.8	−3.6	−3.3	—

（2）废水及主要污染物在各地区的分布

2013 年，辽河流域工业废水、化学需氧量和氨氮排放量最大的均是辽宁省，分别占该流域各类污染物排放量的 80.5%、76.7% 和 79.3%。其中，工业废水、化学需氧量和氨氮排放量最大的均是辽宁省，分别占该流域工业排放总量的 76.3%、60.2% 和 48.2%。农业化学需氧量和氨氮排放量最大的均是辽宁省，分别占该流域农业排放总量的 76.1% 和 80.8%。城镇生活废水、化学需氧量和氨氮排放量最大的均是辽宁省，分别占该流域生活排放总量的 82.2%、84.5% 和 81.3%。

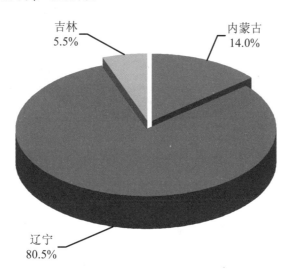

吉林 5.5%
内蒙古 14.0%
辽宁 80.5%

图 2-29　辽河流域废水排放区域构成

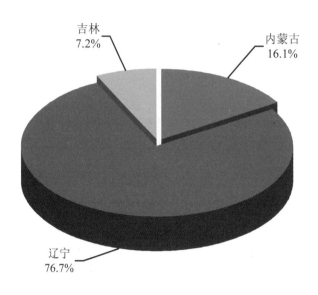

图 2-30　辽河流域化学需氧量排放区域构成

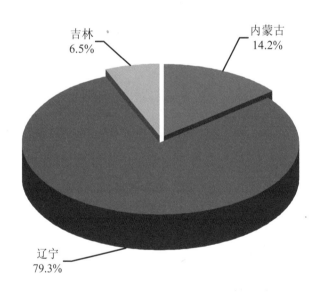

图 2-31　辽河流域氨氮排放区域构成

（3）废水及主要污染物在行业的分布

2013 年，在调查统计的 41 个工业行业中，辽河流域废水排放量位于前 4 位的行业依次为化学原料和化学制品制造业、农副食品加工业、黑色金属矿采选业、煤炭开采和洗选业，4 个行业的废水排放量为 2.0 亿吨，占重点调查工业企业废水排放总量的39.5%。

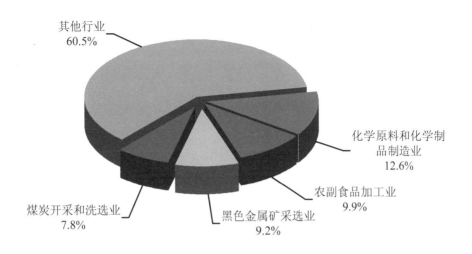

图 2-32　辽河流域工业废水排放量行业构成

2013 年，在调查统计的 41 个工业行业中，辽河流域化学需氧量位于前 4 位的行业依次为农副食品加工业，化学原料和化学制品制造业，酒、饮料和精制茶制造业，造纸和纸制品业。4 个行业的化学需氧量排放量为 4.1 万吨，占重点调查工业企业化学需氧量排放总量的 59.5%。

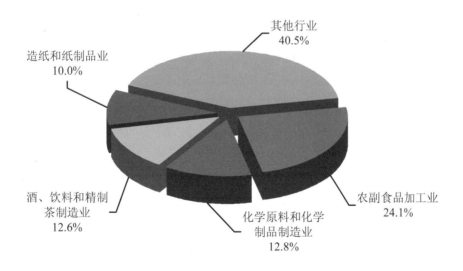

图 2-33　辽河流域工业化学需氧量排放量行业构成

2013 年，在调查统计的 41 个工业行业中，辽河流域氨氮排放量位于前 4 位的行业依次为农副食品加工业，化学原料和化学制品制造业，食品制造业，酒、饮料和精制茶制造业。4 个行业的氨氮排放量为 0.32 万吨，占重点调查工业企业氨氮排放总量的 64.9%。

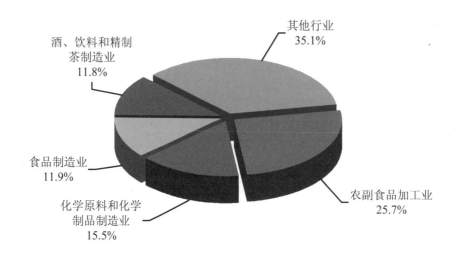

图 2-34 辽河流域工业氨氮排放量行业构成

（4）废水及污染物治理情况

2013 年，辽河流域纳入统计的污水处理厂 139 座，形成了 674 万吨/日的处理能力，年运行费用达 10.9 亿元，共处理污水 14.7 亿吨，其中生活污水 12.9 亿吨。去除化学需氧量 37.2 万吨、氨氮 2.6 万吨、油类 0.2 万吨、总氮 1.5 万吨、总磷 0.2 万吨。

辽河流域重点调查工业企业共有废水治理设施 1 640 套，形成了 1 122 万吨/日的废水处理能力，年运行费用达 19.8 亿元，处理了 20.5 亿吨工业废水。去除工业化学需氧量 31.3 万吨、氨氮 1.4 万吨、石油类 1.4 万吨、挥发酚 1 374.1 吨、氰化物 159.5 吨。

2.4.3 海河流域

（1）废水及污染物排放情况

《重点流域水污染防治"十二五"规划》中海河流域含北京、天津、河北、山西、内蒙古、山东、河南 7 个省（市）的 64 个地市和 335 个区县。2013 年，重点调查了工业企业 20 237 家，规模化畜禽养殖场 18 296 家，规模化畜禽养殖小区 1 948 家。

海河流域共排放废水 81.2 亿吨，其中，工业废水 24.1 亿吨，城镇生活污水 57.0 亿吨。化学需氧量排放量为 275.3 万吨，其中，工业化学需氧量 31.9 万吨，农业化学需氧量 178.5 万吨，城镇生活化学需氧量 63.0 万吨。氨氮排放量为 23.9 万吨，其中，工业氨氮为 2.7 万吨，农业氨氮 9.1 万吨，城镇生活氨氮 11.9 万吨。

海河流域工业石油类排放量为 1 800.9 吨，工业挥发酚排放量为 359.4 吨，工业氰化物排放量为 22.5 吨，工业废水中 6 种重金属（包括铅、镉、汞、六价铬、总铬及砷）排放总量为 22.7 吨。

表 2-10　海河流域废水及主要污染物排放情况

年份	废水/亿吨			化学需氧量/万吨				氨氮/万吨			
污染物	工业源	生活源	集中式	工业源	农业源	生活源	集中式	工业源	农业源	生活源	集中式
2011	26.02	48.20	0.04	35.75	188.83	69.89	1.79	3.25	9.82	12.80	0.16
2012	25.76	53.34	0.04	34.78	183.01	65.28	1.85	2.98	9.48	12.32	0.16
2013	24.11	57.04	0.04	31.86	178.53	63.00	1.95	2.73	9.11	11.94	0.16
变化率/%	−6.4	6.9	—	−8.4	−2.4	−3.5	—	−8.2	−3.9	−3.1	—

（2）废水及主要污染物在各地区的分布

2013 年，海河流域废水、化学需氧量和氨氮排放量最大的均是河北省，分别占该流域各类污染物排放量的 36.1%、45.3% 和 43.0%。其中，工业废水、化学需氧量和氨氮排放量最大的均是河北省，分别占该流域工业排放总量的 41.3%、50.7% 和 48.7%。农业化学需氧量和氨氮排放量最大的均是河北省，分别占该流域农业排放总量的 47.8% 和45.6%。城镇生活废水、化学需氧量和氨氮排放量最大的均是河北省，分别占该流域生活排放总量的 33.9%、35.6% 和 39.7%。

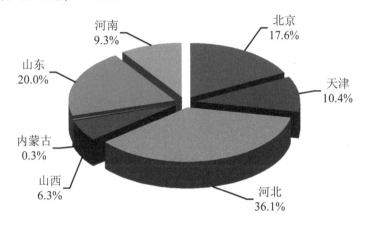

图 2-35　海河流域废水排放区域构成

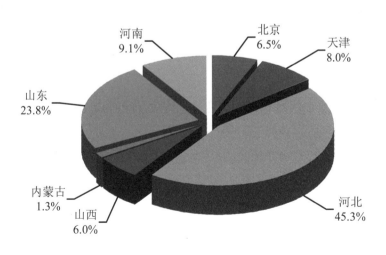

图 2-36　海河流域化学需氧量排放区域构成

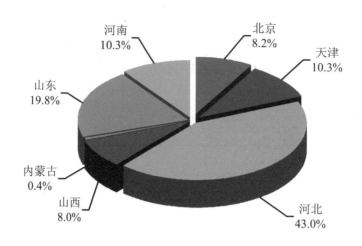

图 2-37　海河流域氨氮排放区域构成

（3）废水及主要污染物在行业的分布

2013 年，在调查统计的 41 个工业行业中，海河流域废水排放量位于前 4 位的行业依次为造纸和纸制品业、化学原料和化学制品制造业、农副食品加工业、纺织业，4 个行业的废水排放量为 11.3 亿吨，占重点调查工业企业废水排放总量的 51.9%。

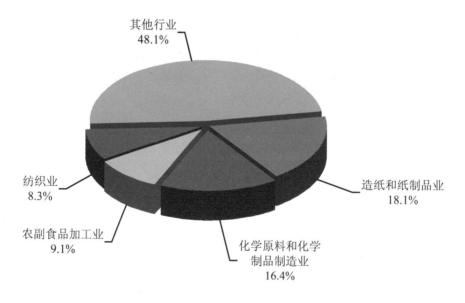

图 2-38　海河流域工业废水排放量行业构成

2013 年，在调查统计的 41 个工业行业中，海河流域化学需氧量位于前 4 位的行业依次为造纸和纸制品业、化学原料和化学制品制造业、农副食品加工业、纺织业，4 个行业的化学需氧量排放量为 16.2 万吨，占重点调查工业企业化学需氧量排放总量的 57.5%。

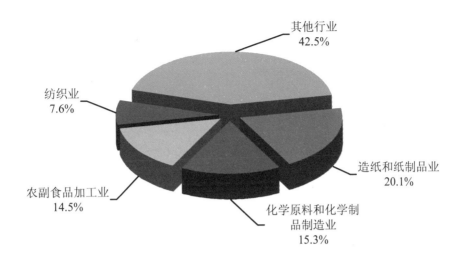

图 2-39 海河流域工业化学需氧量排放量行业构成

2013 年，在调查统计的 41 个工业行业中，海河流域氨氮排放量位于前 4 位的行业依次为化学原料和化学制品制造业、造纸和纸制品业、农副食品加工业、纺织业，4 个行业的氨氮排放量为 1.5 万吨，占重点调查工业企业氨氮排放总量的 60.2%。

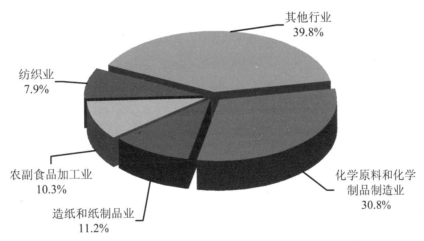

图 2-40 海河流域工业氨氮排放量行业构成

（4）废水及污染物治理情况

2013 年，海河流域纳入统计的污水处理厂 651 座，形成了 2 188 万吨/日的处理能力，年运行费用达 53.6 亿元，共处理污水 61.3 亿吨，其中生活污水 50.4 亿吨。去除化学需氧量 181.3 万吨、氨氮 19.8 万吨、油类 1.1 万吨、总氮 15.7 万吨、总磷 1.8 万吨。

海河流域重点调查工业企业共有废水治理设施 9 456 套，形成了 5 144 万吨/日的废水处理能力，年运行费用达 98.4 亿元，处理了 93.0 亿吨工业废水。去除工业化学需氧量 305.0 万吨、氨氮 14.8 万吨、石油类 4.4 万吨、挥发酚 1.6 万吨、氰化物 603.2 吨。

2.4.4　黄河中上游流域

（1）废水及污染物排放情况

《重点流域水污染防治"十二五"规划》中黄河中上游流域含山西、内蒙古、河南、陕西、甘肃、青海、宁夏 7 个省份的 52 个地市和 339 个区县。2013 年，重点调查了工业企业 12 400 家，规模化畜禽养殖场 8 190 家，规模化畜禽养殖小区 1 221 家。

黄河中上游流域共排放废水 40.8 亿吨，其中，工业废水 12.7 亿吨，城镇生活污水 28.1 亿吨。化学需氧量排放量为 168.0 万吨，其中，工业化学需氧量为 35.1 万吨，农业化学需氧量 72.0 万吨，城镇生活化学需氧量 60.1 万吨。氨氮排放量为 16.9 万吨，其中，工业氨氮 3.4 万吨，农业氨氮 3.4 万吨，城镇生活氨氮 10.0 万吨。

黄河中上游流域工业石油类排放量为 2 429.4 吨，工业挥发酚排放量为 556.2 吨，工业氰化物排放量为 32.3 吨，工业废水中 6 种重金属（包括铅、镉、汞、六价铬、总铬及砷）排放总量为 38.9 吨。

表 2-11　黄河中上游流域废水及主要污染物排放情况

年份 \ 污染物	废水/亿吨			化学需氧量/万吨				氨氮/万吨			
	工业源	生活源	集中式	工业源	农业源	生活源	集中式	工业源	农业源	生活源	集中式
2011	13.35	24.12	0.01	38.65	82.92	63.15	1.00	3.34	3.70	10.67	0.12
2012	13.24	26.61	0.01	36.67	74.00	61.74	1.01	3.54	3.58	10.21	0.12
2013	12.70	28.08	0.02	35.09	71.96	60.11	0.87	3.35	3.41	10.04	0.09
变化率/%	−4.1	5.5	—	−4.3	−2.8	−2.6	—	−5.4	−4.6	−1.7	—

（2）废水及主要污染物在各地区的分布

2013 年，黄河中上游流域废水、化学需氧量和氨氮排放量在各地区的分布相对比较平均。废水、化学需氧量和氨氮排放量最大的均是陕西省，分别占该流域各类污染物排放量的 24.6%、22.4% 和 23.5%。其中，工业废水排放量最大的是河南省、工业化学需氧量和氨氮排放量最大的是宁夏回族自治区，分别占该流域工业排放总量的 24.2%、28.7% 和 24.7%。农业化学需氧量排放量最大的是内蒙古自治区，农业氨氮排放量最大的是河南省，分别占该流域农业排放总量的 21.8% 和 25.9%。城镇生活废水、化学需氧量和氨氮排放量最大的均是陕西省，分别占该流域生活排放总量的 26.9%、26.7% 和 26.4%。

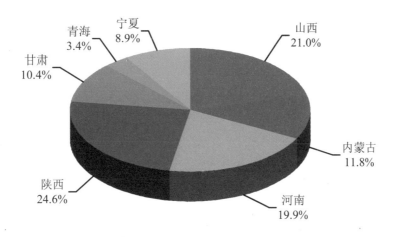

图 2-41 黄河中上游流域废水排放区域构成

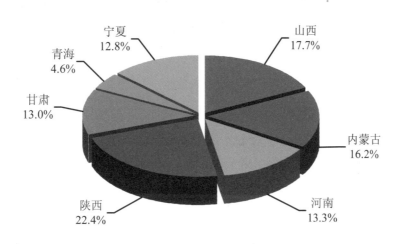

图 2-42 黄河中上游流域化学需氧量排放区域构成

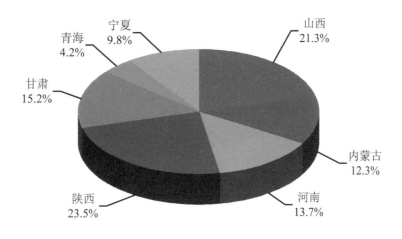

图 2-43 黄河中上游流域氨氮排放区域构成

（3）废水及主要污染物在行业的分布

2013 年，在调查统计的 41 个工业行业中，黄河中上游流域废水排放量位于前 4 位的行业依次为化学原料和化学制品制造业、煤炭开采和洗选业、造纸和纸制品业、黑色金属冶炼和压延加工业，4 个行业的废水排放量为 6.2 亿吨，占重点调查工业企业废水排放总量的 53.0%。

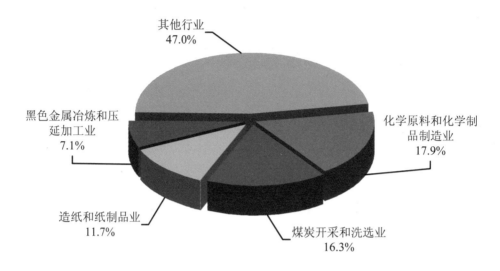

图 2-44　黄河中上游流域工业废水排放量行业构成

2013 年，在调查统计的 41 个工业行业中，黄河中上游流域化学需氧量位于前 5 位的行业依次为造纸和纸制品业，农副食品加工业，化学原料和化学制品制造业，食品制造业，酒、饮料和精制茶制造业。5 个行业的化学需氧量排放量为 21.6 万吨，占重点调查工业企业化学需氧量排放总量的 67.6%。

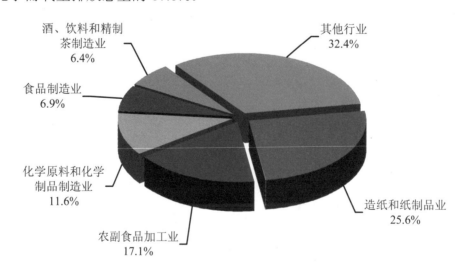

图 2-45　黄河中上游流域工业化学需氧量排放量行业构成

2013 年，在调查统计的 41 个工业行业中，黄河中上游流域氨氮排放量位于前 4 位的行业依次为有色金属冶炼和压延加工业，化学原料和化学制品制造业，石油加工、炼焦和核燃料加工业，食品制造业。4 个行业的氨氮排放量为 2.1 万吨，占重点调查工业企业氨氮排放总量的 69.2%。

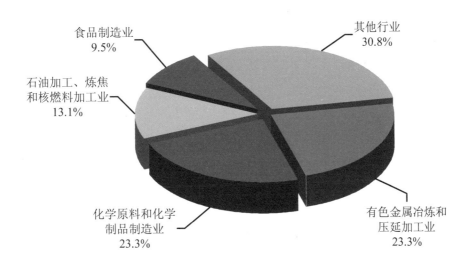

图 2-46　黄河中上游流域工业氨氮排放量行业构成

（4）废水及污染物治理情况

2013 年，黄河中上游流域纳入统计的污水处理厂 436 座，形成了 1 031 万吨/日的处理能力，年运行费用达 23.1 亿元，共处理污水 24.3 亿吨，其中生活污水 22.0 亿吨。去除化学需氧量 70.6 万吨、氨氮 7.0 万吨、油类 0.4 万吨、总氮 7.2 万吨、总磷 0.9 万吨。

黄河中上游流域重点调查工业企业共有废水治理设施 5 652 套，形成了 1 853 万吨/日的废水处理能力，年运行费用达 43.9 亿元，处理了 31.7 亿吨工业废水。去除工业化学需氧量 132.0 万吨、氨氮 16.0 万吨、石油类 2.8 万吨、挥发酚 1.6 万吨、氰化物603.2 吨。

2.4.5　淮河流域

（1）废水及污染物排放情况

《重点流域水污染防治"十二五"规划》中淮河流域含江苏、安徽、山东、河南 4 个省份的 35 个地市和 218 个区县。2013 年，重点调查了工业企业 12 198 家，规模化畜禽养殖场 25 726 家，规模化畜禽养殖小区 919 家。

淮河流域共排放废水 66.1 亿吨，其中，工业废水 20.1 亿吨，城镇生活污水 45.9 亿吨。化学需氧量排放量为 255.1 万吨，其中，工业化学需氧量为 23.6 万吨，农业化学需氧量131.4 万吨，城镇生活化学需氧量 98.5 万吨。氨氮排放量为 28.0 万吨，其中，工业氨氮为 1.9 万吨，农业氨氮 11.0 万吨，城镇生活氨氮 14.9 万吨。

淮河流域工业石油类排放量为 1 808.6 吨，工业挥发酚排放量为 154.9 吨，工业氰化物排放量为 19.7 吨，工业废水中 6 种重金属（包括铅、镉、汞、六价铬、总铬及砷）排放总量为 17.5 吨。

表 2-12　淮河流域废水及主要污染物排放情况

年份 污染物	废水/亿吨			化学需氧量/万吨				氨氮/万吨			
	工业源	生活源	集中式	工业源	农业源	生活源	集中式	工业源	农业源	生活源	集中式
2011	22.16	41.76	0.03	27.44	140.94	100.59	1.83	2.27	11.60	15.24	0.20
2012	21.60	44.37	0.03	25.87	136.08	99.16	1.82	2.15	11.28	15.09	0.20
2013	20.14	45.94	0.03	23.64	131.39	98.49	1.58	1.95	10.97	14.88	0.17
变化率/%	-6.8	3.5	—	-8.6	-3.4	-0.7	—	-9.3	-2.8	-1.4	—

（2）废水及主要污染物在各地区的分布

2013 年，淮河流域废水、化学需氧量和氨氮排放量在各地区的分布相对比较平均。废水、化学需氧量和氨氮排放量最大的均是河南省，分别占该流域各类污染物排放量的 32.5%、31.3% 和 30.4%。其中，工业废水排放量最大的是河南省，工业化学需氧量和氨氮排放量最大的均是江苏省，分别占该流域工业排放总量的 31.7%、37.6% 和 34.6%。农业化学需氧量和氨氮排放量最大的均是河南省，分别占该流域农业排放总量的 35.6% 和 34.6%。城镇生活废水排放量最大的是河南省，生活化学需氧量和氨氮排放量最大的是江苏省，分别占该流域生活排放总量的 32.8%、35.3% 和 33.1%。

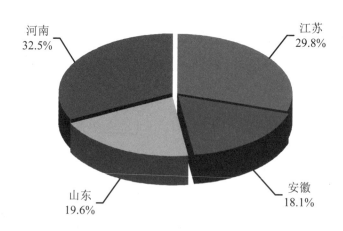

图 2-47　淮河流域废水排放区域构成

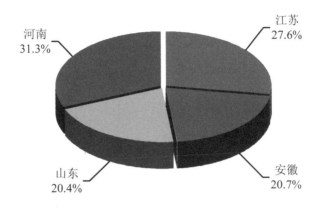

图 2-48　淮河流域化学需氧量排放区域构成

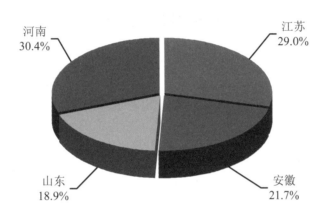

图 2-49　淮河流域氨氮排放区域构成

（3）废水及主要污染物在行业的分布

2013 年，在调查统计的 41 个工业行业中，淮河流域废水排放量位于前 4 位的行业依次为煤炭开采和洗选业、化学原料和化学制品制造业、造纸和纸制品业、农副食品加工业，4 个行业的废水排放量为 11.3 亿吨，占重点调查工业企业废水排放总量的 61.2%。

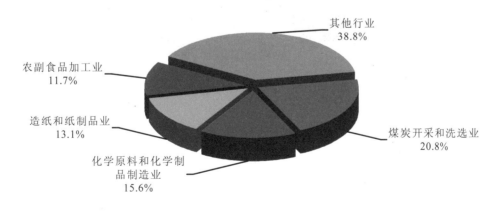

图 2-50　淮河流域工业废水排放量行业构成

2013 年，在调查统计的 41 个工业行业中，淮河流域化学需氧量位于前 4 位的行业依次为造纸和纸制品业、化学原料和化学制品制造业、农副食品加工业、煤炭开采和洗选业，4 个行业的化学需氧量排放量为 12.8 万吨，占重点调查工业企业化学需氧量排放总量的 60.0%。

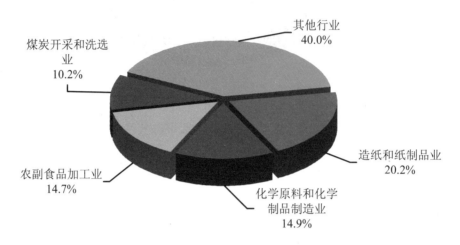

图 2-51 淮河流域工业化学需氧量排放量行业构成

2013 年，在调查统计的 41 个工业行业中，淮河流域氨氮排放量位于前 4 位的行业依次为化学原料和化学制品制造业，农副食品加工业，造纸和纸制品业，酒、饮料和精制茶制造业。4 个行业的氨氮排放量为 1.2 万吨，占重点调查工业企业氨氮排放总量的 67.5%。

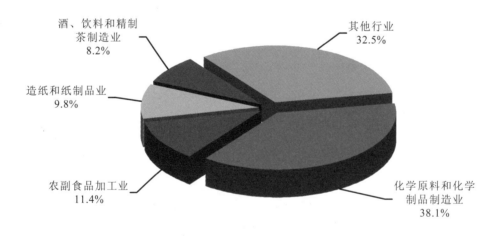

图 2-52 淮河流域工业氨氮排放量行业构成

（4）废水及污染物治理情况

2013 年，淮河流域纳入统计的污水处理厂 485 座，形成了 1 302 万吨/日的处理能力，年运行费用达 23.7 亿元，共处理污水 36.7 亿吨，其中生活污水 33.5 亿吨。去除化学需氧量 74.1 万吨、氨氮 7.4 万吨、油类 0.2 万吨、总氮 5.8 万吨、总磷 0.8 万吨。

淮河流域重点调查工业企业共有废水治理设施 5 256 套，形成了 1 584 万吨/日的废水处理能力，年运行费用达 39.0 亿元，处理了 26.1 亿吨工业废水。去除工业化学需氧量 179.4 万吨、氨氮 7.7 万吨、石油类 1.0 万吨、挥发酚 2 785.5 吨、氰化物 137.9 吨。

2.4.6 长江中下游流域

（1）废水及污染物排放情况

《重点流域水污染防治"十二五"规划》中长江中下游流域含上海、江苏、安徽、江西、河南、湖北、湖南、广西 8 个省（市）的 55 个地市和 408 个区县，包括长江干流控制区 109 个区县、长江口控制区 45 个区县、汉江中下游控制区 35 个区县、洞庭湖控制区 130 个区县和鄱阳湖控制区 89 个区县。2013 年，重点调查了工业企业 22 362 家，规模化畜禽养殖场 31 817 家，规模化畜禽养殖小区 612 家。

长江中下游流域共排放废水 126.6 亿吨，其中，工业废水 39.4 亿吨，城镇生活污水 87.0 亿吨。化学需氧量排放量为 375.6 万吨，其中，工业化学需氧量为 47.3 万吨，农业化学需氧量 143.9 万吨，城镇生活化学需氧量 179.5 万吨。氨氮排放量为 47.6 万吨，其中，工业氨氮为 5.3 万吨，农业氨氮 15.5 万吨，城镇生活氨氮 26.4 万吨。

长江中下游流域工业石油类排放量为 3 999.7 吨，工业挥发酚排放量为 65.7 吨，工业氰化物排放量为 37.7 吨，工业废水中 6 种重金属（包括铅、镉、汞、六价铬、总铬及砷）排放总量为 166.6 吨。

表 2-13 长江中下游流域废水及主要污染物排放情况

年份	污染物	废水/亿吨			化学需氧量/万吨				氨氮/万吨			
		工业源	生活源	集中式	工业源	农业源	生活源	集中式	工业源	农业源	生活源	集中式
2011	长江中下游	41.90	79.03	0.14	55.01	147.50	181.37	5.82	6.42	16.11	26.90	0.52
	其中：洞庭湖	9.77	18.58	0.04	17.37	59.44	54.92	1.68	2.80	6.59	7.36	0.15
	其中：鄱阳湖	6.81	11.73	0.02	10.91	23.60	36.40	1.07	0.99	2.94	4.59	0.09
2012	长江中下游	40.25	83.77	0.17	49.89	145.35	180.37	5.46	5.79	15.73	26.81	0.48
	其中：洞庭湖	9.78	21.28	0.04	15.28	57.09	54.94	1.61	2.60	6.38	7.39	0.15
	其中：鄱阳湖	6.44	12.53	0.02	9.41	22.63	37.24	1.00	0.90	2.80	4.72	0.08
2013	长江中下游	39.42	87.02	0.17	47.27	143.94	179.52	4.85	5.28	15.52	26.41	0.42
	其中：洞庭湖	9.37	21.96	0.04	14.24	56.02	55.92	1.25	2.32	6.24	7.48	0.12
	其中：鄱阳湖	6.49	13.06	0.03	8.79	22.01	37.24	0.93	0.86	2.70	4.71	0.08
变化率/%	长江中下游	−2.1	3.9	—	−5.2	−1.0	−0.5	—	−8.9	−1.4	−1.5	—
	其中：洞庭湖	−4.2	3.2	—	−6.8	−1.9	1.8	—	−10.8	−2.2	1.2	—
	其中：鄱阳湖	0.8	4.2	—	−6.6	−2.7	0.0	—	−4.4	−3.6	−0.2	—

表 2-14　洞庭湖及鄱阳湖其他污染物排放情况

年份	污染物	工业石油类/吨	工业挥发酚/吨	工业氰化物/吨	工业重金属/吨
2011	长江中下游	4 926.07	229.55	46.22	272.39
	其中：洞庭湖	907.44	103.22	12.83	148.52
	其中：鄱阳湖	663.90	74.64	10.20	62.23
2012	长江中下游	4 101.43	62.89	37.69	221.69
	其中：洞庭湖	810.45	16.80	11.30	125.78
	其中：鄱阳湖	536.51	7.73	7.69	51.76
2013	长江中下游	3 999.66	65.72	37.70	166.55
	其中：洞庭湖	606.45	15.64	10.45	85.93
	其中：鄱阳湖	708.30	13.76	7.07	51.27
变化率/%	长江中下游	−2.5	4.5	0.0	−24.9
	其中：洞庭湖	−25.2	−6.9	−7.5	−31.7
	其中：鄱阳湖	32.0	78.0	−8.1	−1.0

（2）废水及主要污染物在各地区的分布

2013 年，长江中下游流域废水、化学需氧量和氨氮排放量主要集中在湖南、湖北和江西 3 个省。废水、化学需氧量和氨氮排放量最大的均是湖南省，分别占该流域各类污染物排放量的 23.8%、32.4% 和 32.5%。其中，工业废水排放量最大的是上海市、化学需氧量和氨氮排放量最大的均是湖南省，分别占该流域工业排放总量的 23.0%、28.6% 和 43.2%。农业化学需氧量和氨氮排放量最大的均是湖南省，分别占该流域农业排放总量的 37.9% 和 38.7%。城镇生活废水、化学需氧量和氨氮排放量最大的均是湖南省，分别占该流域生活排放总量的 24.2%、29.3% 和 26.8%。

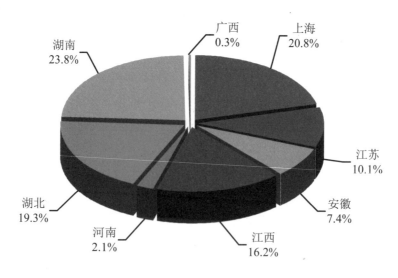

图 2-53　长江中下游流域废水排放区域构成

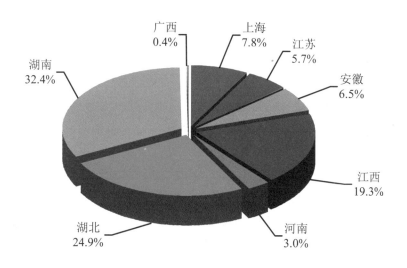

图 2-54　长江中下游流域化学需氧量排放区域构成

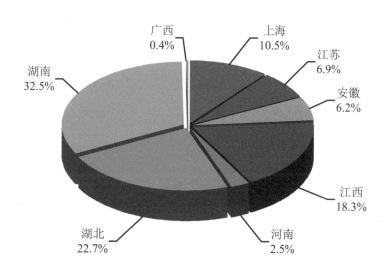

图 2-55　长江中下游流域氨氮排放区域构成

（3）废水及主要污染物在行业的分布

2013 年，在调查统计的 41 个工业行业中，长江中下游流域废水排放量位于前 4 位的行业依次为化学原料和化学制品制造业，造纸和纸制品业，黑色金属冶炼和压延加工业，石油加工、炼焦和核燃料加工业。4 个行业的废水排放量为 17.9 亿吨，占重点调查工业企业废水排放总量的 50.3%。

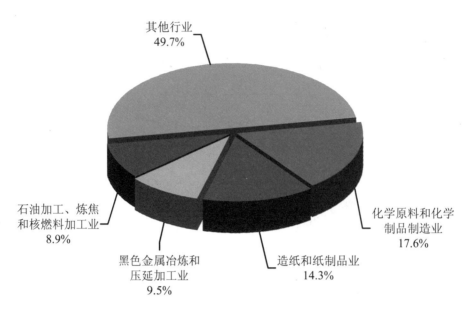

图 2-56　长江中下游流域工业废水排放量行业构成

2013 年，在调查统计的 41 个工业行业中，长江中下游流域化学需氧量位于前 4 位的行业依次为造纸和纸制品业，化学原料和化学制品制造业，农副食品加工业，石油加工、炼焦和核燃料加工业。4 个行业的化学需氧量排放量为 21.8 万吨，占重点调查工业企业化学需氧量排放总量的 51.3%。

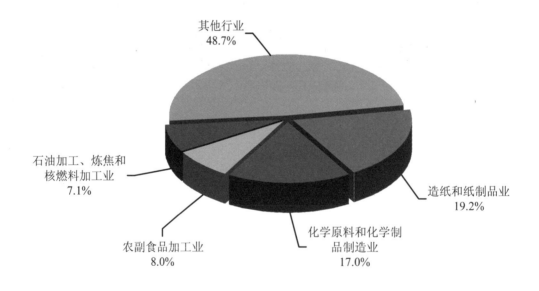

图 2-57　长江中下游流域工业化学需氧量排放量行业构成

2013 年，在调查统计的 41 个工业行业中，长江中下游流域氨氮排放量位于前 4 位的行业依次为化学原料和化学制品制造业，造纸和纸制品业，有色金属冶炼和压延加工业，石油加工、炼焦和核燃料加工业。4 个行业的氨氮排放量为 3.2 万吨，占重点调查工业企业氨氮排放总量的 67.2%。

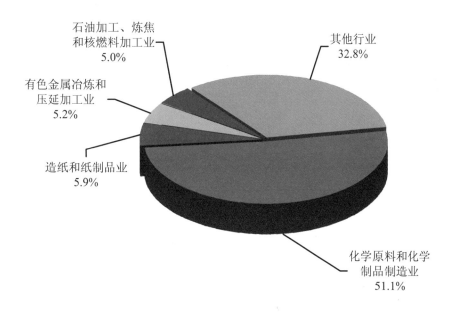

图 2-58　长江中下游流域工业氨氮排放量行业构成

（4）废水及污染物治理情况

2013 年，长江中下游流域纳入统计的污水处理厂 755 座，形成了 3 477 万吨/日的处理能力，年运行费用达 73.9 亿元，共处理污水 98.6 亿吨，其中生活污水 84.5 亿吨。去除化学需氧量 216.1 万吨、氨氮 17.5 万吨、油类 1.2 万吨、总氮 14.2 万吨、总磷 2.4 万吨。

长江中下游流域重点调查工业企业共有废水治理设施 13 535 套，形成了 4 908 万吨/日的废水处理能力，年运行费用达 119.7 亿元，处理了 101.1 亿吨工业废水。共去除工业化学需氧量 239.6 万吨、氨氮 11.2 万吨、石油类 3.1 万吨、挥发酚 1.0 万吨、氰化物 1 050.5 吨。

2.4.7　太湖、巢湖及滇池

（1）废水及污染物排放情况

《重点流域水污染防治"十二五"规划》中太湖流域含上海、江苏、浙江 3 个省（市）的 8 个地市和 51 个区县。巢湖流域含安徽省的 3 个地市和 13 个区县。滇池流域含云南省的 1 个地市和 6 个区县。

2013 年，重点调查了太湖流域工业企业 7 555 家，规模化畜禽养殖场 2 190 家，规模化畜禽养殖小区 94 家。巢湖流域工业企业 1 298 家，规模化畜禽养殖场 650 家，规模化畜禽养殖小区 26 家。滇池流域工业企业 169 家，规模化畜禽养殖场 72 家，规模化畜禽养殖小区 16 家。

表 2-15　太湖、巢湖及滇池流域废水及主要污染物排放情况

年份	污染物	废水/亿吨			化学需氧量/万吨				氨氮/万吨			
		工业源	生活源	集中式	工业源	农业源	生活源	集中式	工业源	农业源	生活源	集中式
2011	太湖	15.72	17.15	0.02	10.71	11.50	14.36	0.10	0.81	1.34	3.45	0.02
	巢湖	0.67	3.67	0.00	1.23	4.69	6.83	0.14	0.07	0.37	0.97	0.01
	滇池	0.09	2.86	0.00	0.12	0.36	0.47	0.08	0.01	0.05	0.26	0.01
2012	太湖	14.99	19.52	0.02	10.11	11.28	13.39	0.13	0.78	1.31	3.33	0.02
	巢湖	0.62	3.69	0.00	1.15	5.03	6.02	0.14	0.06	0.44	0.88	0.01
	滇池	0.08	3.82	0.00	0.13	0.42	0.15	0.08	0.01	0.06	0.25	0.01
2013	太湖	14.04	20.43	0.03	9.69	10.86	12.20	0.11	0.72	1.24	3.19	0.01
	巢湖	0.57	4.28	0.00	0.96	5.82	5.86	0.12	0.04	0.41	0.83	0.01
	滇池	0.08	4.01	0.00	0.13	0.37	0.35	0.08	0.01	0.05	0.33	0.01
变化率/%	太湖	−6.4	4.7	—	−4.2	−3.7	−8.9	—	−7.0	−5.6	−4.1	—
	巢湖	−7.4	16.0	—	−16.2	15.7	−2.7	—	−33.9	−7.8	−6.1	—
	滇池	0.8	5.0	—	−0.7	−11.1	136.1	—	−32.3	−13.6	32.9	—

太湖、巢湖、滇池流域共排放废水 43.3 亿吨。其中，工业废水 14.7 亿吨，城镇生活污水 28.8 亿吨。化学需氧量排放总量为 46.6 万吨。其中，工业化学需氧量为 10.8 万吨，农业化学需氧量 17.1 万吨，城镇生活化学需氧量 18.4 万吨。氨氮排放总量为 6.8 万吨。其中，工业氨氮为 0.8 万吨，农业氨氮 1.7 万吨，城镇生活氨氮 4.4 万吨。

太湖、巢湖、滇池流域工业石油类排放总量为 469.7 吨，工业挥发酚排放总量为 7.2 吨，工业氰化物排放总量为 7.1 吨，工业废水中 6 种重金属（包括铅、镉、汞、六价铬、总铬及砷）排放总量为 11.8 吨。

（2）废水及主要污染物在各地区的分布

2013 年，太湖流域废水、化学需氧量和氨氮排放量最大的均是江苏省，分别占该流域各类污染物排放量的 68.5%、52.7% 和 49.8%。其中，工业废水、化学需氧量和氨氮排放量最大的均是江苏省，分别占该流域工业排放总量的 72.5%、65.0% 和 54.0%。农业化学需氧量和氨氮排放量最大的均是浙江省，分别占该流域农业排放总量的 53.9% 和56.8%。城镇生活废水、化学需氧量和氨氮排放量最大的均是江苏省，分别占该流域生活排放总量的 65.9%、50.5% 和 51.8%。

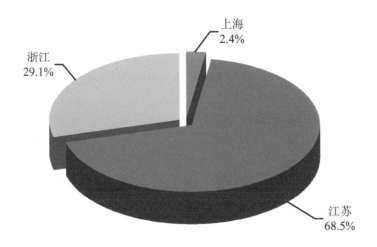

图 2-59　太湖流域废水排放区域构成

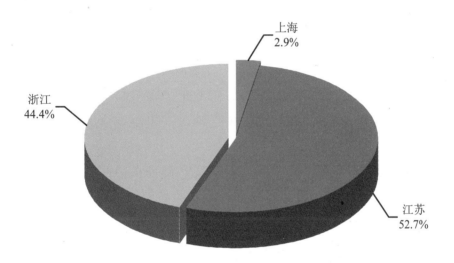

图 2-60　太湖流域化学需氧量排放区域构成

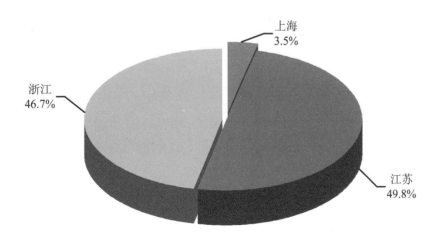

图 2-61　太湖流域氨氮排放区域构成

（3）废水及主要污染物在行业的分布

①太湖流域

2013 年，在调查统计的 41 个工业行业中，太湖流域废水排放量位于前 4 位的行业依次为纺织业，化学原料和化学制品制造业，造纸和纸制品业，计算机、通信和其他电子设备制造业。4 个行业的废水排放量为 9.2 亿吨，占重点调查工业企业废水排放总量的 73.3%。

太湖流域化学需氧量排放量位于前 4 位的行业依次为纺织业，化学原料和化学制品制造业，造纸和纸制品业，计算机、通信和其他电子设备制造业。4 个行业的化学需氧量排放量为 6.5 万吨，占重点调查工业企业化学需氧量排放总量的 74.5%。

太湖流域氨氮排放量位于前 4 位的行业依次为纺织业，化学原料和化学制品制造业，造纸和纸制品业，计算机、通信和其他电子设备制造业。4 个行业的氨氮排放量为 0.55 万吨，占重点调查工业企业氨氮排放总量的 79.6%。

②巢湖流域

2013 年，在调查统计的 41 个工业行业中，巢湖流域废水排放量位于前 4 位的行业依次为化学纤维制造业、黑色金属冶炼和压延加工业、化学原料和化学制品制造业、造纸和纸制品业。4 个行业的废水排放量为 0.33 亿吨，占重点调查工业企业废水排放总量的 64.4%。

巢湖流域化学需氧量排放量位于前 4 位的行业依次为黑色金属冶炼和压延加工业、造纸和纸制品业、化学纤维制造业、化学原料和化学制品制造业。4 个行业的化学需氧量排放量为 0.65 万吨，占重点调查工业企业化学需氧量排放总量的 72.5%。

巢湖流域氨氮排放量位于前 4 位的行业依次为酒、饮料和精制茶制造业，黑色金属冶炼和压延加工业，化学纤维制造业，化学原料和化学制品制造业。4 个行业的氨氮排放量为 0.03 万吨，占重点调查工业企业氨氮排放总量的 68.4%。

③滇池流域

2013 年，在调查统计的 41 个工业行业中，滇池流域废水排放量位于前 4 位的行业依次为化学原料和化学制品制造业，农副食品加工业，酒、饮料和精制茶制造业，医药制造业。4 个行业的废水排放量为 0.05 亿吨，占重点调查工业企业废水排放总量的 71.9%。

滇池流域化学需氧量排放量位于前 4 位的行业依次为化学原料和化学制品制造业、农副食品加工业、烟草制品业、医药制造业。4 个行业的化学需氧量排放量为 0.07 万吨，占重点调查工业企业化学需氧量排放总量的 64.9%。

滇池流域氨氮排放量位于前 4 位的行业依次为化学原料和化学制品制造业，酒、饮料和精制茶制造业，农副食品加工业，造纸和纸制品业。4 个行业的氨氮排放量为 59 吨，占重点调查工业企业氨氮排放总量的 89.4%。

（4）废水及污染物治理情况

2013 年，太湖、巢湖、滇池流域纳入统计的污水处理厂分别为 341 座、21 座、10 座，形成了 1 032 万吨/日、111 万吨/日、84 万吨/日的处理能力，年运行费用达 34.0 亿元、1.9 亿元、2.4 亿元，共处理污水 28.2 亿吨、4.1 亿吨、2.9 亿吨，其中生活污水 18.3 亿吨、3.9 亿吨、2.9 亿吨。分别去除化学需氧量 81.9 万吨、6.6 万吨、7.7 万吨，氨氮 6.2 万吨、0.7 万吨、0.7 万吨，油类 0.3 万吨、0.05 万吨、0.1 万吨，总氮 4.4 万吨、0.6 万吨、0.8 万吨，总磷 0.9 万吨、0.1 万吨、0.1 万吨。

太湖、巢湖、滇池流域重点调查工业企业分别有废水治理设施 5 348 套、299 套、163 套，形成了 1 087 万吨/日、163 万吨/日、17 万吨/日的废水处理能力，年运行费用达 45.0 亿元、1.7 亿元、0.9 亿元，处理了 20.5 亿吨、2.5 亿吨、0.2 亿吨工业废水。分别去除工业化学需氧量 87.5 万吨、8.1 万吨、0.8 万吨，氨氮 2.1 万吨、0.4 万吨、0.1 万吨，石油类 0.4 万吨、0.1 万吨、0.003 万吨，挥发酚 331.2 吨、0.1 吨、189.7 吨，氰化物 115.8 吨、3.3 吨、1.6 吨。

2.4.8 三峡库区及其上游

（1）废水及污染物排放情况

《重点流域水污染防治"十二五"规划》中三峡库区及其上游流域含湖北、重庆、四川、贵州、云南 5 个省份的 42 个地市，含库区、影响区及上游区共 320 个区县。2013 年，重点调查了工业企业 13 896 家，规模化畜禽养殖场 11 160 家，规模化畜禽养殖小区 824 家。

三峡库区及其上游流域共排放废水 59.5 亿吨，其中，工业废水 13.1 亿吨，城镇生活污水 46.3 亿吨。化学需氧量排放量为 197.7 万吨，其中，工业化学需氧量为 21.7 万吨，农业化学需氧量 72.1 万吨，城镇生活化学需氧量 102.4 万吨。氨氮排放量为 23.6 万吨。其中，工业氨氮为 1.2 万吨，农业氨氮 7.9 万吨，城镇生活氨氮 14.4 万吨。

三峡库区及其上游流域工业石油类排放量为 1 525.5 吨，工业挥发酚排放量为 13.2 吨，工业氰化物排放量为 4.2 吨，工业废水中 6 种重金属（包括铅、镉、汞、六价铬、总铬及砷）排放总量为 14.7 吨。

表 2-16　三峡库区及其上游流域主要污染物排放情况

污染物 年份	废水/亿吨			化学需氧量/万吨				氨氮/万吨			
	工业源	生活源	集中式	工业源	农业源	生活源	集中式	工业源	农业源	生活源	集中式
2011	14.52	38.80	0.04	25.13	75.03	106.81	2.30	1.22	8.29	14.93	0.20
2012	13.05	42.44	0.05	22.93	73.88	104.34	1.48	1.23	8.11	14.57	0.19
2013	13.11	46.32	0.05	21.74	72.15	102.37	1.46	1.20	7.88	14.38	0.19
变化率/%	0.5	9.1	—	−5.2	−2.3	−1.9	—	−2.2	−2.8	−1.3	—

（2）废水及主要污染物在各地区的分布

2013 年，三峡库区及其上游流域废水、化学需氧量和氨氮排放量最大的均是四川省，分别占该流域各类污染物排放量的 51.7%、62.3%和 58.0%。其中，工业废水、化学需氧量和氨氮排放量最大的均是四川省，分别占该流域工业排放总量的 49.5%、49.0%和 41.4%。农业化学需氧量和氨氮排放量最大的均是四川省，分别占该流域农业排放总量的 73.3%和 71.0%。城镇生活废水、化学需氧量和氨氮排放量最大的均是四川省，分别占该流域生活排放总量的 52.4%、57.8%和 52.5%。

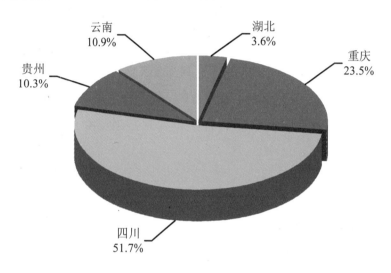

图 2-62 三峡库区及其上游流域废水排放区域构成

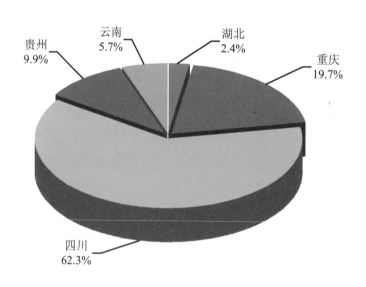

图 2-63 三峡库区及其上游流域化学需氧量排放区域构成

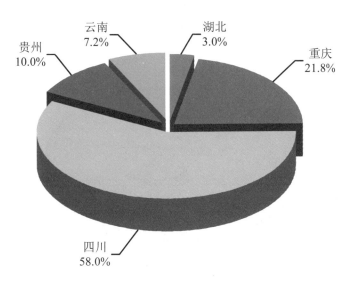

图 2-64 三峡库区及其上游流域氨氮排放区域构成

（3）废水及主要污染物在行业的分布

2013 年，在调查统计的 41 个工业行业中，三峡库区及其上游流域废水排放量位于前 4 位的行业依次为化学原料和化学制品制造业，煤炭开采和洗选业，造纸和纸制品业，酒、饮料和精制茶制造业。4 个行业的废水排放量为 7.3 亿吨，占重点调查工业企业废水排放总量的 60.8%。

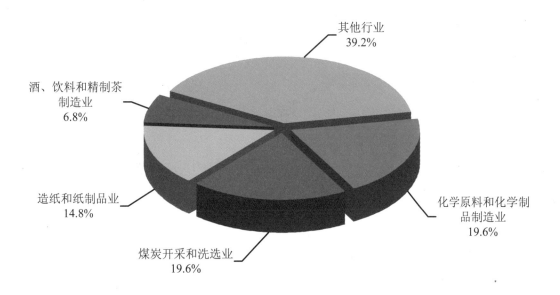

图 2-65 三峡库区及其上游流域工业废水排放量行业构成

2013 年，在调查统计的 41 个工业行业中，三峡库区及其上游流域化学需氧量位于前 5 位的行业依次为造纸和纸制品业，酒、饮料和精制茶制造业，农副食品加工业，煤炭开采和洗选业，化学原料和化学制品制造业。5 个行业的化学需氧量排放量为 14.2 万吨，占重点调查工业企业化学需氧量排放总量的 74.3%。

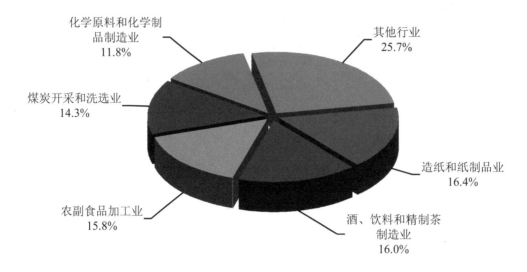

图 2-66　三峡库区及其上游流域工业化学需氧量排放量行业构成

2013 年，在调查统计的 41 个工业行业中，三峡库区及其上游流域氨氮排放量位于前 4 位的行业依次为化学原料和化学制品制造业，农副食品加工业，酒、饮料和精制茶制造业，造纸和纸制品业。4 个行业的氨氮排放量为 0.7 万吨，占重点调查工业企业氨氮排放总量的 65.6%。

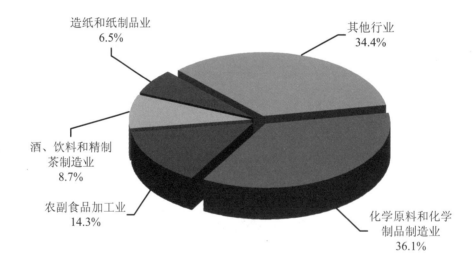

图 2-67　三峡库区及其上游流域工业氨氮排放量行业构成

（4）废水及污染物治理情况

2013 年，三峡库区及其上游流域纳入统计的污水处理厂 625 座，形成了 1 156 万吨/日的处理能力，年运行费用达 32.3 亿元，共处理污水 35.2 亿吨，其中生活污水 34.7 亿吨。去除化学需氧量 76.4 万吨、氨氮 7.7 万吨、油类 0.4 万吨、总氮 8.0 万吨、总磷 0.8 万吨。

三峡库区及其上游流域重点调查工业企业共有废水治理设施 7 371 套,形成了 1 664 万吨/日的废水处理能力,年运行费用达 34.0 亿元,处理了 28.4 亿吨工业废水。去除工业化学需氧量 97.0 万吨、氨氮 13.3 万吨、石油类 0.9 万吨、挥发酚 3 466.0 吨、氰化物 260.6 吨。

2.4.9 丹江口库区及其上游

(1)废水及污染物排放情况

《重点流域水污染防治"十二五"规划》中丹江口库区及其上游流域含河南、湖北、陕西 3 个省份的 8 个地市和 43 个区县。2013 年,重点调查了工业企业 1 697 家,规模化畜禽养殖场 1 270 家,规模化畜禽养殖小区 51 家。

丹江口库区及其上游流域共排放废水 4.8 亿吨,其中,工业废水 1.1 亿吨,城镇生活污水 3.7 亿吨。化学需氧量排放量为 20.7 万吨,其中,工业化学需氧量为 2.8 万吨,农业化学需氧量 8.3 万吨,城镇生活化学需氧量 9.1 万吨。氨氮排放量为 2.8 万吨,其中,工业氨氮为 0.3 万吨,农业氨氮 1.0 万吨,城镇生活氨氮 1.4 万吨。

丹江口库区及其上游流域工业石油类排放量为 151.3 吨,工业挥发酚排放量为 0.1 吨,工业氰化物排放量为 0.1 吨,工业废水中 6 种重金属(包括铅、镉、汞、六价铬、总铬及砷)排放总量为 16.7 吨。

表 2-17 丹江口库区及其上游流域主要污染物排放情况

年份 \ 污染物	废水/亿吨			化学需氧量/万吨				氨氮/万吨			
	工业源	生活源	集中式	工业源	农业源	生活源	集中式	工业源	农业源	生活源	集中式
2011	1.13	3.21	0.00	2.77	8.36	9.80	0.43	0.35	1.03	1.47	0.04
2012	1.08	3.44	0.01	2.79	8.82	9.83	0.42	0.34	1.05	1.48	0.04
2013	1.08	3.71	0.01	2.81	8.32	9.14	0.42	0.34	0.98	1.39	0.04
变化率/%	—	7.9	—	0.7	−5.6	−7.0	—	−0.6	−6.3	−6.0	—

(2)废水及主要污染物在各地区的分布

2013 年,丹江口库区及其上游流域废水、化学需氧量和氨氮排放量最大的均是陕西省,分别占该流域各类污染物排放量的 44.6%、55.9%和 61.1%。其中,工业废水、化学需氧量和氨氮排放量最大的均是陕西省,分别占该流域工业排放总量的 45.3%、50.2%和 69.1%。农业化学需氧量和氨氮排放量最大的均是陕西省,分别占该流域农业排放总量的 54.3%和 67.4%。城镇生活废水、化学需氧量和氨氮排放量最大的均是陕西省,分别占该流域生活排放总量的 44.3%、58.5%和 54.6%。

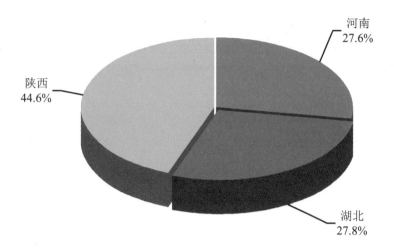

图 2-68　丹江口库区及其上游流域废水排放区域构成

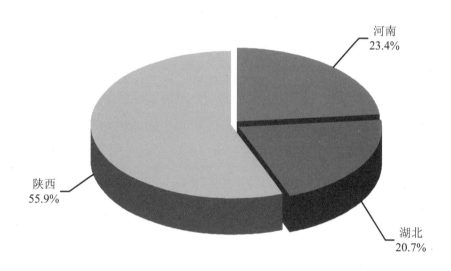

图 2-69　丹江口库区及其上游流域化学需氧量排放区域构成

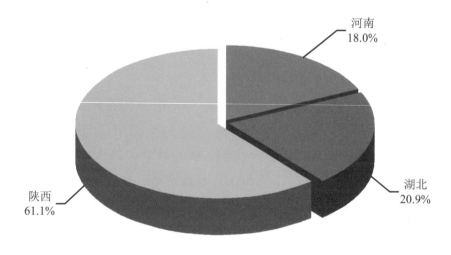

图 2-70　丹江口库区及其上游流域氨氮排放区域构成

（3）废水及主要污染物在行业的分布

2013 年，在调查统计的 41 个工业行业中，丹江口库区及其上游流域废水排放量位于前 3 位的行业依次为有色金属矿采选业、化学原料和化学制品制造业、造纸和纸制品业，3 个行业的废水排放量为 0.5 亿吨，占重点调查工业企业废水排放总量的 53.5%。

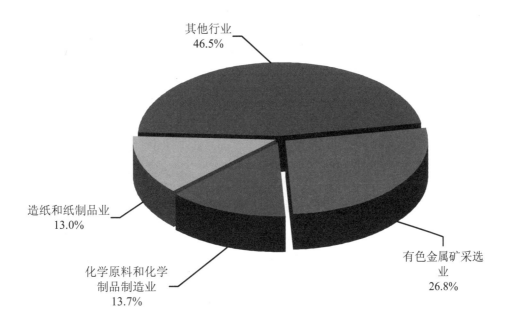

图 2-71　丹江口库区及其上游流域工业废水排放量行业构成

2013 年，在调查统计的 41 个工业行业中，丹江口库区及其上游流域化学需氧量位于前 5 位的行业依次为医药制造业、农副食品加工业、有色金属矿采选业、化学原料和化学制品制造业、造纸和纸制品业，5 个行业的化学需氧量排放量为 1.8 万吨，占重点调查工业企业化学需氧量排放总量的 71.4%。

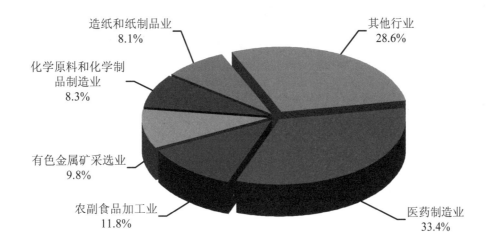

图 2-72　丹江口库区及其上游流域工业化学需氧量排放量行业构成

2013 年，在调查统计的 41 个工业行业中，丹江口库区及其上游流域氨氮排放量位于前 3 位的行业依次为有色金属矿采选业、化学原料和化学制品制造业、医药制造业，3 个行业的氨氮排放量为 0.26 万吨，占重点调查工业企业氨氮排放总量的 80.7%。

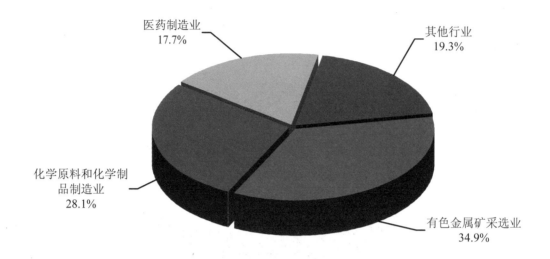

图 2-73　丹江口库区及其上游流域工业氨氮排放量行业构成

（4）废水及污染物治理情况

2013 年，丹江口库区及其上游流域纳入统计的污水处理厂 48 座，形成了 112 万吨/日的处理能力，年运行费用达 2.1 亿元，共处理污水 2.6 亿吨，其中生活污水 2.4 亿吨。去除化学需氧量 5.4 万吨、氨氮 0.4 万吨、总氮 0.3 万吨、总磷 0.04 万吨。

丹江口库区及其上游流域重点调查工业企业共有废水治理设施 846 套，形成了 184 万吨/日的废水处理能力，年运行费用达 3.3 亿元，处理了 2.7 亿吨工业废水。去除工业化学需氧量 6.4 万吨、氨氮 0.5 万吨、石油类 0.05 万吨、挥发酚 2.0 吨、氰化物 1.5 吨。

2.5　沿海地区废水及主要污染物排放情况

2.5.1　废水及污染物排放情况

2013 年，沿海地区的统计范围为天津、河北、辽宁、上海、江苏、浙江、福建、山东、广东、广西、海南 11 个沿海省（市）的 56 个地市和 212 个区县。我国沿海地区重点调查工业企业数为 22 395 家，规模化畜禽养殖场 13 290 家，规模化畜禽养殖小区 444 家。

表 2-18　沿海地区废水及主要污染物接纳情况

年份 \ 污染物	废水/亿吨			化学需氧量/万吨				氨氮/万吨			
	工业源	生活源	集中式	工业源	农业源	生活源	集中式	工业源	农业源	生活源	集中式
2011	45.84	59.75	0.08	42.48	114.23	109.66	2.09	3.22	9.95	18.48	0.18
2012	38.57	63.87	0.11	42.49	108.06	106.08	2.01	3.07	9.51	18.12	0.18
2013	36.07	65.22	0.10	40.21	108.22	102.08	1.89	2.71	9.45	17.40	0.18
变化率/%	−6.5	2.0	—	−5.4	0.1	−3.8	—	−12.9	0.0	−3.9	—

表 2-19　沿海地区工业其他污染物排放情况

年份 \ 污染物	工业石油类/吨	工业挥发酚/吨	工业氰化物/吨	工业重金属/吨
2011	2 481.61	66.18	16.97	60.15
2012	2 492.19	62.82	13.04	48.30
2013	2 266.93	35.44	12.59	44.01
变化率/%	−9.0	−43.6	−3.1	−8.9

沿海地区废水排放总量为 101.4 亿吨。其中，工业废水排放量为 36.1 亿吨，城镇生活污水排放量为 65.2 亿吨。其中，东海沿海地区工业和城镇生活的废水排放量均居四大海域之首，分别占沿海地区的 47.9%和 39.8%。

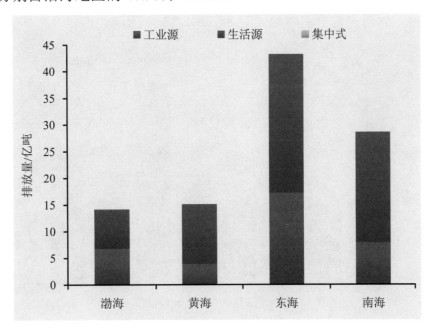

图 2-74　四大海域废水排放情况

沿海地区化学需氧量排放总量为 252.4 万吨。其中，工业废水化学需氧量为 40.2 万吨，农业源化学需氧量 108.2 万吨，城镇生活污水化学需氧量 102.1 万吨。其中，工业化学需氧量排放量最大的是东海，占整个沿海地区的 36.9%，农业源化学需氧量较大的是渤海和黄海，分别占沿海地区的 30.3%和 28.7%，城镇生活化学需氧量最大的是东海，占沿海地区的 41.7%。

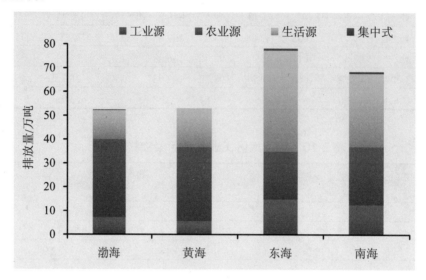

图 2-75　四大海域化学需氧量排放情况

沿海地区氨氮排放总量为 29.7 万吨。其中，工业废水氨氮为 2.7 万吨，农业源氨氮为 9.5 万吨，城镇生活污水氨氮为 17.4 万吨。其中，工业、农业、城镇生活氨氮排放量最大的均为东海，分别占整个沿海地区的 34.1%、33.7%和 44.0%。

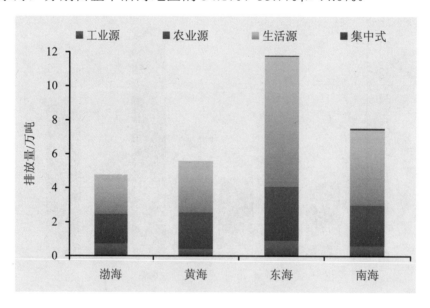

图 2-76　四大海域氨氮排放情况

沿海地区工业石油类排放量为 2 266.9 吨，工业挥发酚排放量为 35.4 吨，工业氰化物排放量为 12.6 吨，工业废水中 6 种重金属（包括铅、镉、汞、六价铬、总铬及砷）排放总量为 44.0 吨。

2.5.2 废水及主要污染物在各地区的分布

2013年，沿海地区废水、化学需氧量和氨氮排放量主要来自广东、山东、福建和浙江4个省。废水、化学需氧量和氨氮排放量最大的均是广东省，分别占该流域各类污染物排放量的26.0%、21.0%和21.1%。其中，工业废水排放量最大的是福建省，工业化学需氧量和氨氮排放量最大的是广东省，分别占该流域工业排放总量的20.3%、26.6%和21.4%。农业化学需氧量排放量最大的是山东省，农业氨氮排放量最大的是福建省，分别占该流域农业排放总量的24.1%和18.7%。城镇生活废水、化学需氧量和氨氮排放量最大的均是广东省，分别占该流域生活排放总量的29.1%、25.1%和22.3%。

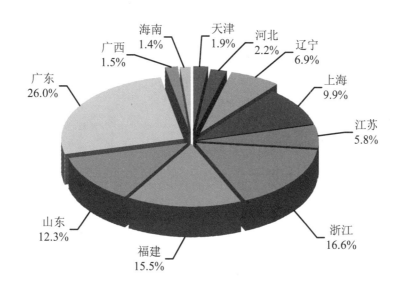

图 2-77 沿海地区废水排放区域构成

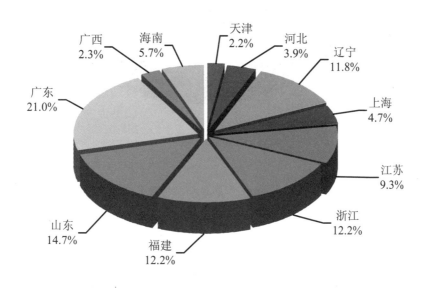

图 2-78 沿海地区化学需氧量排放区域构成

67

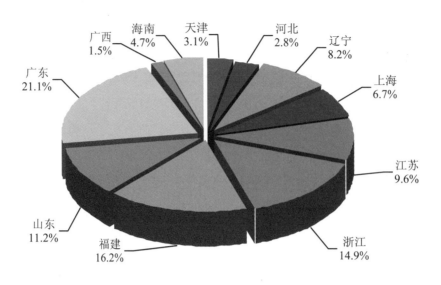

图 2-79　沿海地区氨氮排放区域构成

2.5.3　主要污染物在行业的分布

2013 年，在调查统计的 41 个工业行业中，沿海地区废水排放量位于前 4 位的行业依次为纺织业，电力、热力生产和供应业，造纸和纸制品业，石油加工、炼焦和核燃料加工业。4 个行业的废水排放量为 19.1 亿吨，占重点调查工业企业废水排放总量的 58.6%。

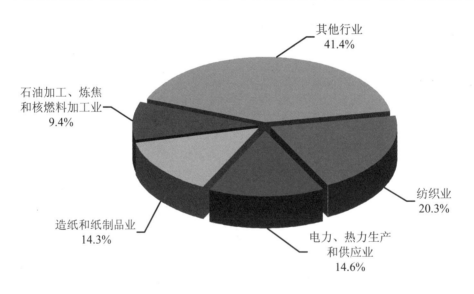

图 2-80　沿海地区工业废水排放量行业构成

2013 年，在调查统计的 41 个工业行业中，沿海地区化学需氧量位于前 4 位的行业依次为纺织业、造纸和纸制品业、农副食品加工业、化学原料和化学制品制造业。4 个行业的化学需氧量排放量为 22.7 万吨，占重点调查工业企业化学需氧量排放总量的 65.7%。

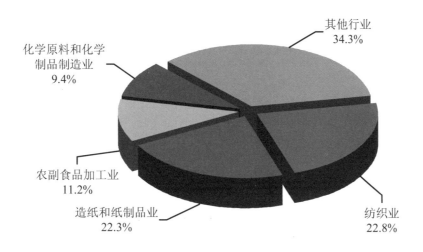

图 2-81　沿海地区工业化学需氧量排放量行业构成

2013 年，在调查统计的 41 个工业行业中，沿海地区氨氮排放量位于前 5 位的行业依次为纺织业，造纸和纸制品业，化学原料和化学制品制造业，石油加工、炼焦和核燃料加工业，农副食品加工业。5 个行业的氨氮排放量为 1.7 万吨，占重点调查工业企业氨氮排放总量的 68.9%。

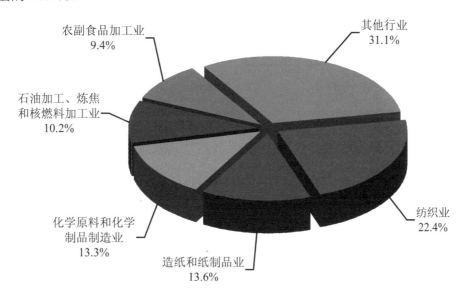

图 2-82　沿海地区工业氨氮排放量行业构成

2.5.4　废水及主要污染物治理情况

2013 年，沿海地区纳入统计的集中式污水处理厂 769 座，形成了 3 072 万吨/日的处理能力，年运行费用达 74.6 亿元，共处理污水 86.0 亿吨，其中生活污水 72.0 亿吨。去除化学需氧量 227.5 万吨、氨氮 18.3 万吨、油类 0.8 万吨、总氮 13.9 万吨、总磷

2.6 万吨。

沿海地区重点调查工业企业共有废水治理设施 14 196 套，形成了 3 053 万吨/日的处理能力，年运行费用达 112.7 亿元，处理了 57.3 亿吨工业废水。去除工业化学需氧量 297.9 万吨，氨氮 12.7 万吨，石油类 5.0 万吨，挥发酚 3 234.2 吨，氰化物 1 087.1 吨。

3

废气

ANNUAL STATISTIC REPORT ON ENVIRONMENT IN CHINA
2013

3.1 废气及废气中主要污染物排放情况

3.1.1 二氧化硫排放情况

2013 年，全国工业废气排放量 669 361 亿米^3（标态），比 2012 年增加 5.3%。

全国二氧化硫排放量 2 043.9 万吨，比 2012 年减少 3.5%。

工业二氧化硫排放量 1 835.2 万吨，比 2012 年减少 4.0%，占全国二氧化硫排放总量的 89.8%。

城镇生活二氧化硫排放量 208.5 万吨，比 2012 年增加 1.4%，占全国二氧化硫排放总量的 10.2%。

集中式污染治理设施二氧化硫排放量 0.2 万吨。

表 3-1　全国二氧化硫排放量　　　　　　　　　　　单位：万吨

年份 \ 排放源	合计	工业源	城镇生活源	集中式
2011	2 217.9	2 017.2	200.4	0.3
2012	2 117.6	1 911.7	205.7	0.3
2013	2 043.9	1 835.2	208.5	0.2
变化率/%	−3.5	−4.0	1.4	−33.3

注：①集中式污染治理设施包括生活垃圾处理厂（场）和危险废物（医疗废物）集中处理（置）厂焚烧废气中排放的污染物，下同；

②变化率表示与2012 年相比指标的变化情况，下同。

3.1.2 氮氧化物排放情况

2013 年，全国氮氧化物排放量 2 227.4 万吨，比 2012 年减少 4.7%。

工业氮氧化物排放量 1 545.6 万吨，比 2012 年减少 6.8%，占全国氮氧化物排放总量的 69.4%。

城镇生活氮氧化物排放量 40.7 万吨，比 2012 年增加 3.6%，占全国氮氧化物排放总量的 1.8%。

机动车氮氧化物排放量 640.6 万吨，比 2012 年增加 0.1%，占全国氮氧化物排放总量的 28.8%。

集中式污染治理设施氮氧化物排放量 0.4 万吨。

表 3-2　全国氮氧化物排放量　　　　　　　　　单位：万吨

年份＼排放源	合计	工业源	城镇生活源	机动车	集中式
2011	2 404.3	1 729.7	36.6	637.6	0.3
2012	2 337.8	1 658.1	39.3	640.0	0.4
2013	2 227.4	1 545.6	40.7	640.6	0.4
变化率/%	−4.7	−6.8	3.6	0.1	—

注：自 2011 年起机动车排气污染物排放情况与生活源分开单独统计。

3.1.3　烟（粉）尘排放情况

2013 年，全国烟（粉）尘排放量 1 278.1 万吨，比 2012 年增加 3.5%。

工业烟（粉）尘排放量 1 094.6 万吨，比 2012 年增加 6.3%，占全国烟（粉）尘排放总量的 85.6%。

城镇生活烟（粉）尘排放量 123.9 万吨，比 2012 年减少 13.2%，占全国烟（粉）尘排放总量的 9.7%。

机动车颗粒物排放量 59.4 万吨，比 2012 年减少 4.3%，占全国烟（粉）尘排放总量的 4.6%。

集中式污染治理设施烟（粉）尘排放量 0.2 万吨。

表 3-3　全国烟（粉）尘排放量　　　　　　　　　单位：万吨

年份＼排放源	合计	工业源	城镇生活源	机动车	集中式
2011	1 278.8	1 100.9	114.8	62.9	0.2
2012	1 234.3	1 029.3	142.7	62.1	0.2
2013	1 278.1	1 094.6	123.9	59.4	0.2
变化率/%	3.5	6.3	−13.2	−4.3	—

注：① 自 2011 年起不再单独统计烟尘和粉尘，统一以烟（粉）尘进行统计；
　　② 机动车的烟（粉）尘排放量指机动车的颗粒物排放量。

3.2　各地区废气中主要污染物排放情况

3.2.1　二氧化硫排放情况

2013 年，二氧化硫排放量超过 100 万吨的省份依次为山东、内蒙古、河北、山西、河南和辽宁，6 个省份的二氧化硫排放量占全国排放总量的 38.3%。各地区中，工业二氧化硫排放量最大的是山东，城镇生活二氧化硫排放量最大的是贵州，集中式污染治理设施二氧化硫排放量最大的是广东。

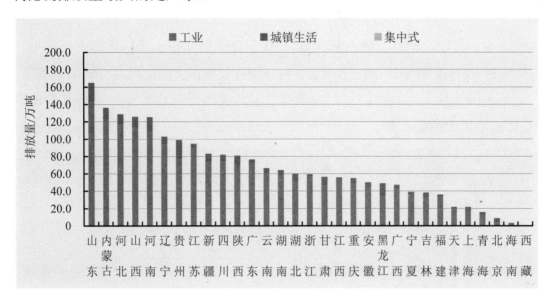

图 3-1　各地区二氧化硫排放情况

3.2.2　氮氧化物排放情况

2013 年，氮氧化物排放量超过 100 万吨的省份依次为河北、山东、河南、内蒙古、江苏、广东和山西，7 个省份氮氧化物排放量占全国氮氧化物排放总量的 44.7%。工业氮氧化物排放量最大的是山东，城镇生活氮氧化物排放量最大的是黑龙江，机动车氮氧化物排放量最大的是河北，集中式污染治理设施氮氧化物排放量最大的是广东。

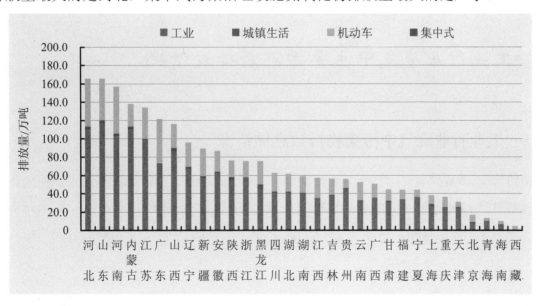

图 3-2　各地区氮氧化物排放情况

3.2.3 烟（粉）尘排放情况

2013 年，烟（粉）尘排放量超过 50 万吨的省份依次为河北、山西、内蒙古、新疆、黑龙江、山东、辽宁、河南、陕西和江苏，10 个省份烟（粉）尘排放量占全国烟（粉）尘排放总量的 60.1%。各地区中，工业烟（粉）尘排放量最大的是河北，城镇生活烟（粉）尘排放量最大的是黑龙江，机动车颗粒物排放量最大的是河南，集中式污染治理设施烟（粉）尘排放量最大的是广东。

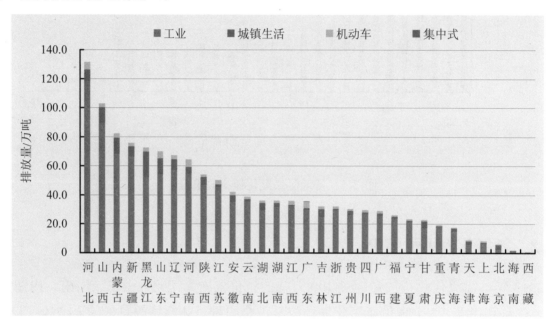

图 3-3　各地区烟（粉）尘排放情况

3.3　工业行业废气中主要污染物排放情况

3.3.1　工业行业废气中污染物排放总体情况

2013 年，调查统计的 41 个工业行业中，二氧化硫排放量位于前 3 位的行业依次为电力、热力生产和供应业，黑色金属冶炼及压延加工业，非金属矿物制品业。3 个行业共排放二氧化硫 1 151.8 万吨，占重点调查工业企业二氧化硫排放总量的 68.2%。

表 3-4　重点行业二氧化硫排放情况　　　　　　　　　　　　　单位：万吨

年份　　　　　行业	合计	电力、热力生产和供应业	黑色金属冶炼及压延加工业	非金属矿物制品业
2011	1 354.3	901.2	251.4	201.7
2012	1 237.4	797.0	240.6	199.8
2013	1 151.8	720.6	235.1	196.0
变化率/%	−6.9	−9.6	−2.3	−1.9

2013 年，调查统计的工业行业中，氮氧化物排放量位于前 3 位的行业依次为电力、热力生产和供应业，非金属矿物制品业，黑色金属冶炼及压延加工业。3 个行业共排放氮氧化物 1 268.3 万吨，占重点调查工业企业氮氧化物排放总量的 86.6%。

表 3-5　重点行业氮氧化物排放情况　　　　　　　　　　　　　单位：万吨

年份　　　　　行业	合计	电力、热力生产和供应业	非金属矿物制品业	黑色金属冶炼及压延加工业
2011	1 471.3	1 106.8	269.4	95.1
2012	1 390.1	1 018.7	274.2	97.2
2013	1 268.3	896.9	271.6	99.7
变化率/%	−8.8	−12.0	−1.0	2.6

2013 年，调查统计工业行业中，烟（粉）尘排放量位于前 3 位的行业依次为电力、热力生产和供应业，非金属矿物制品业，黑色金属冶炼及压延加工业。3 个行业共排放烟（粉）尘 722.6 万吨，占重点调查工业企业烟（粉）尘排放量的 70.7%。

表 3-6　重点行业烟（粉）尘排放情况　　　　　　　　　　　　　单位：万吨

年份　　　　　行业	合计	电力、热力生产和供应业	非金属矿物制品业	黑色金属冶炼及压延加工业
2011	700.9	215.6	279.1	206.2
2012	659.3	222.8	255.2	181.3
2013	722.6	270.3	258.8	193.5
变化率/%	9.6	21.3	1.4	6.7

3.3.2　电力、热力生产和供应业废气污染物排放及处理情况

（1）电力、热力生产和供应业总体情况

2013 年，电力、热力生产和供应业重点调查工业企业 5 830 家，占重点调查工业企业的 3.9%；全年工业废气排放量为 225 446.6 亿米³（标态），占重点调查工业企业废气排

放量的 33.7%；二氧化硫排放量为 720.6 万吨，占重点调查工业企业的 42.7%；氮氧化物排放量为 896.9 万吨，占重点调查工业企业的 61.2%；烟（粉）尘排放量为 270.3 万吨，占重点调查工业企业的 26.4%。

电力、热力生产和供应业拥有废气治理设施 23 852 套，占重点调查工业企业废气治理设施总数的 10.2%。其中，脱硫设施 5 916 套，占重点调查工业企业脱硫设施总数的 27.3%；脱硝设施 1 168 套，占重点调查工业企业脱硝设施总数的 54.3%；除尘设施 16 247 套，占重点调查工业企业除尘设施总数的 9.0%。二氧化硫去除量为 2 620.1 万吨，去除率为 78.4%，较重点调查工业企业平均水平（72.2%）高出 6.2 个百分点；氮氧化物去除量为 301.8 万吨，去除率为 25.2%，较重点调查工业企业平均水平（19.2%）高出 6.0 个百分点；烟（粉）尘去除量为 38 238.0 万吨，去除率为 99.3%，较重点调查工业企业平均水平（98.6%）高出 0.7 个百分点。

电力、热力生产和供应业二氧化硫排放量居全国前 4 位的省份依次为内蒙古、山东、山西和贵州，其二氧化硫排放量占电力、热力生产和供应业总排放量的 34.2%；氮氧化物排放量居前 4 位的省份依次为内蒙古、山东、河南和江苏，其氮氧化物排放量占电力、热力生产和供应业总排放量的 30.8%；烟（粉）尘排放量居前 4 位的省份依次为内蒙古、黑龙江、辽宁和山西，其烟（粉）尘排放量占电力、热力生产和供应业总排放量的 39.4%。

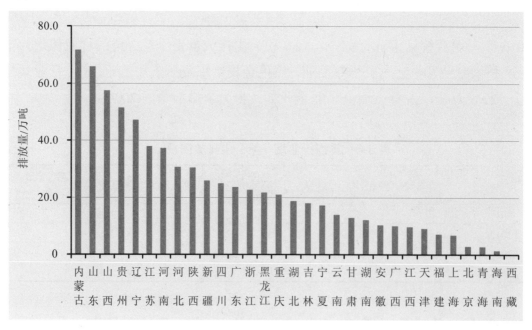

图 3-4　各地区电力、热力生产和供应业二氧化硫排放情况

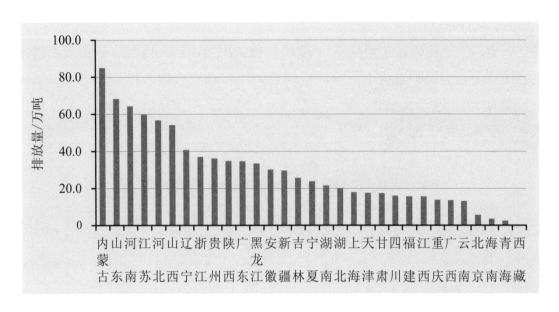

图 3-5　各地区电力、热力生产和供应业氮氧化物排放情况

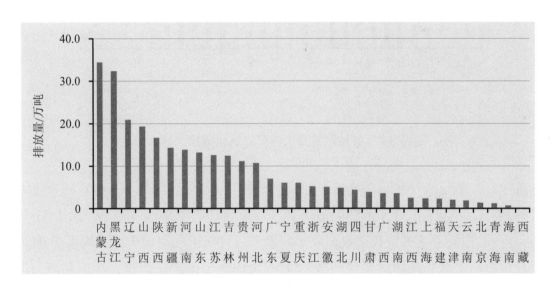

图 3-6　各地区电力、热力生产和供应业烟（粉）尘排放情况

（2）火电厂废气污染物排放及治理情况

2013 年，纳入重点调查统计范围的火电厂共 3 102 家，占重点调查工业企业数量的 2.1%。其中，独立火电厂 1 853 家，拥有 4 825 台机组，共有脱硫设施 3 547 套，脱硝设施 1 076 套，除尘设施 5 140 套；自备电厂 1 249 家，拥有 2 690 台机组，其中 1 106 台机组有脱硫设施，162 台有脱硝设施，1 569 台有除尘设施。

独立火电厂二氧化硫排放量为 634.1 万吨，比 2012 年减少 10.2%，占重点调查工业企业排放量的 37.5%，占电力、热力生产和供应业排放量的 88.0%；共去除二氧化硫 2 584.7 万吨，二氧化硫去除率达 80.3%，比 2012 年提高 3.1 个百分点，比重点调查工业企业平均水平高 8.1 个百分点。

独立火电厂二氧化硫排放量居前 4 位的省份依次为山东、内蒙古、山西和贵州，其二氧化硫排放量占全国独立火电厂排放量的 35.5%。独立火电厂单位发电量二氧化硫平均排放强度为 1.63 克/千瓦·时，比上年减少 14.2%，其中重庆、四川和贵州 3 个省份的排放强度大于 4.0 克/千瓦·时。

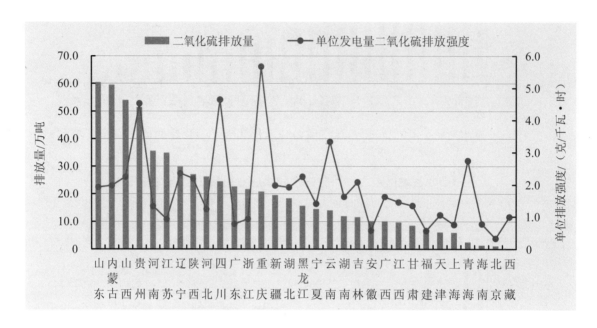

图 3-7　各地区独立火电厂二氧化硫排放情况

自备电厂二氧化硫排放量为 148.6 万吨，比 2012 年减少 2.6%，占重点调查工业企业排放量的 8.8%；共去除二氧化硫 141.6 万吨，二氧化硫去除率达 48.8%，比 2012 年提高 3.5 个百分点，但低于重点调查工业企业平均水平，且低于独立火电厂平均水平 31.5 个百分点。自备电厂二氧化硫排放量居前 4 位的省份依次为山东、新疆、江苏和河南，其二氧化硫排放量占全国自备电厂排放量的 36.1%。

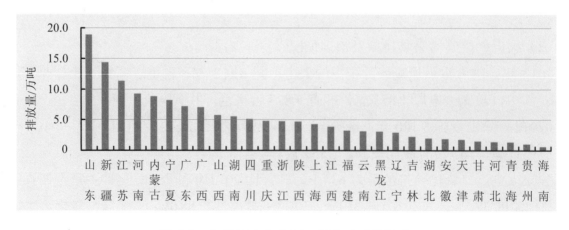

图 3-8　各地区自备电厂二氧化硫排放情况

独立火电厂氮氧化物排放量为 861.8 万吨，比 2012 年减少 12.2%，占重点调查工业企业排放量的 58.8%，占电力、热力生产和供应业排放量的 96.1%；共去除氮氧化物 300.3 万吨，氮氧化物去除率为 25.8%，比 2012 年提高 15.6 个百分点，比重点调查工业企业平均水平高出 6.6 个百分点。

独立火电厂氮氧化物排放量居前 4 位的省份依次为内蒙古、山东、河南和江苏，其氮氧化物排放量占独立火电厂排放量的 31.0%。独立火电厂单位发电量氮氧化物平均排放强度为 2.2 克/千瓦·时，比 2012 年减少 15.4%，其中黑龙江和吉林 2 个省份的排放强度大于 4.0 克/千瓦·时。

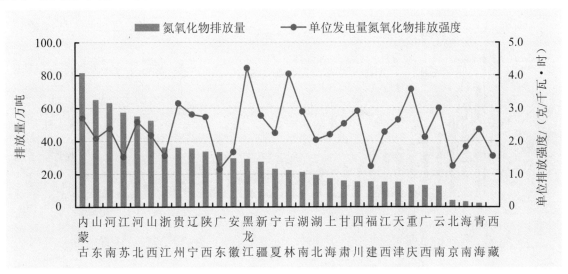

图 3-9 各地区独立火电厂氮氧化物排放情况

自备电厂氮氧化物排放量为 102.8 万吨，占重点调查工业企业排放量的 7.0%。共去除氮氧化物 3.4 万吨，氮氧化物去除率为 3.2%，比 2012 年降低 0.2 个百分点，低于重点调查工业企业平均水平，且低于独立火电厂平均水平 22.6 个百分点。自备电厂氮氧化物排放量居前 4 位的省份依次为江苏、新疆、山东和内蒙古，其氮氧化物排放量占自备电厂排放量的 37.5%。

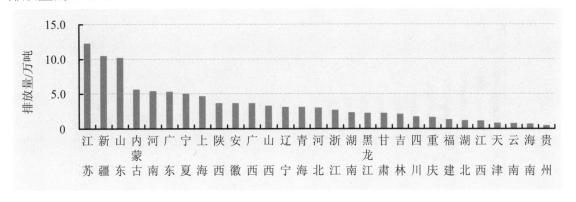

图 3-10 各地区自备电厂氮氧化物排放情况

独立火电厂烟（粉）尘排放量为 183.9 万吨，比 2012 年增加 27.5%，占重点调查工业企业排放量的 18.0%，占电力、热力生产和供应业排放量的 68.0%；共去除烟（粉）尘 3.6 亿吨，烟（粉）尘去除率达 99.5%，比 2012 年降低 0.1 个百分点，比重点调查工业企业平均水平高出 0.9 个百分点。

独立火电厂烟（粉）尘排放量居前 4 位的省份依次为内蒙古、黑龙江、山西和河南，其烟（粉）尘排放量占独立火电厂排放量的 38.5%。独立火电厂单位发电量烟（粉）尘平均排放强度为 0.5 克/千瓦·时，比 2012 年提高 25.0%，其中黑龙江、重庆和吉林 3 个省份的排放强度大于 1.0 克/千瓦·时。

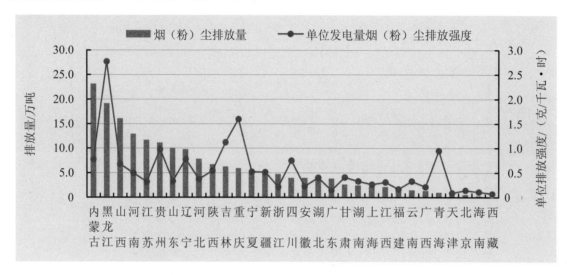

图 3-11　各地区独立火电厂烟（粉）尘排放情况

自备电厂烟（粉）尘排放量为 34.9 万吨，占重点调查工业企业排放量的 3.4%。共去除烟（粉）尘 2 893.8 万吨，烟（粉）尘去除率达 98.8%，与 2012 年持平，比重点调查工业企业平均水平高 0.2 个百分点，但低于独立火电厂平均水平 0.7 个百分点。自备电厂烟（粉）尘排放量居前 4 位的省份依次为内蒙古、广西、江苏和黑龙江，其烟（粉）尘排放量占自备电厂排放量的 32.7%。

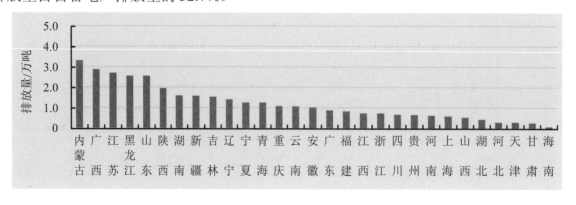

图 3-12　各地区自备电厂烟（粉）尘排放情况

3.3.3 非金属矿物制品业废气污染物排放及处理情况

（1）非金属矿物制品业总体情况

2013 年，非金属矿物制品业重点调查工业企业 32 532 家，占重点调查工业企业总数的 22.0%；全年工业废气排放量为 120 337.2 亿米³（标态），占重点调查统计工业企业废气排放量的 18.0%；二氧化硫排放量为 196.0 万吨，占重点调查统计工业企业排放量的 11.6%；氮氧化物排放量为 271.6 万吨，占全重点调查统计工业企业排放量的 18.5%；烟（粉）尘排放量为 258.8 万吨，占重点调查统计工业企业排放量的 25.3%。

非金属矿物制品业拥有废气治理设施 66 461 套，占重点调查工业企业废气治理设施总数的 28.4%。其中，脱硫设施数 2 051 套，占全国脱硫设施总数的 9.5%；脱硝设施数 568 套，占重点调查工业企业脱硝设施总数的 26.4%；除尘设施数 61 393 套，占重点调查工业企业除尘设施总数的 34.1%。二氧化硫去除量为 34.1 万吨，去除率为 14.8%；氮氧化物去除量为 29.4 万吨，去除率为 9.8%；烟（粉）尘去除量为 22 361.6 万吨，去除率为 98.9%。

非金属矿物制品业二氧化硫排放量居全国前 4 位的省份依次为河南、安徽、山东和江西，其二氧化硫排放量占非金属矿物制品业排放量的 32.2%；氮氧化物排放量居前 4 位的省份依次为安徽、山东、广东和河南，其氮氧化物排放量占非金属矿物制品业排放量的 28.9%；烟（粉）尘排放量居前 4 位的省份依次为河北、河南、云南和安徽，其烟（粉）尘排放量占非金属矿物制品业排放量的 28.0%。

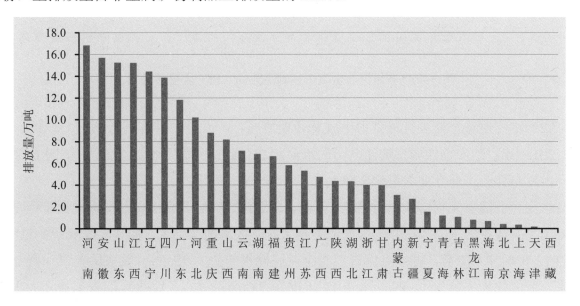

图 3-13　各地区非金属矿物制品业二氧化硫排放情况

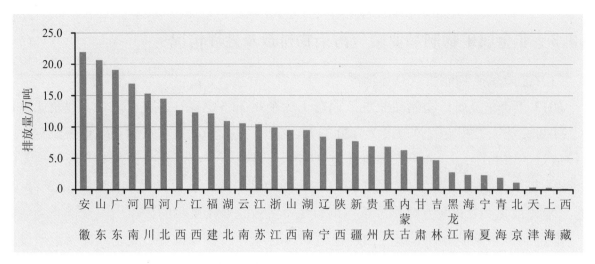

图 3-14 各地区非金属矿物制品业氮氧化物排放情况

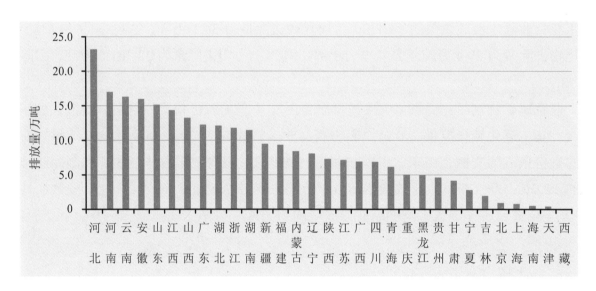

图 3-15 各地区非金属矿物制品业烟（粉）尘排放情况

（2）水泥制造企业废气污染物排放及处理情况

2013 年，纳入重点调查统计范围的水泥制造企业（简称"水泥企业"，下同）3 679 家，占重点调查工业企业数量的 2.5%。

水泥企业氮氧化物排放量为 196.9 万吨，比 2012 年减少 0.5%，占重点调查工业企业氮氧化物排放量的 13.4%，占非金属矿物制品业排放量的 72.5%。拥有 538 套脱硝设施，比 2012 年增加 454 套，占重点调查工业企业脱硝设施数的 25.0%；共去除氮氧化物 25.7 万吨，氮氧化物去除率 11.5%，比 2012 年提高 9.6 个百分点，比重点调查工业企业平均水平低 7.7 个百分点。

水泥企业氮氧化物排放量居前 4 位的省份依次为安徽、山东、河南和广西，其氮氧化物排放量占全国水泥企业排放量的 27.2%。

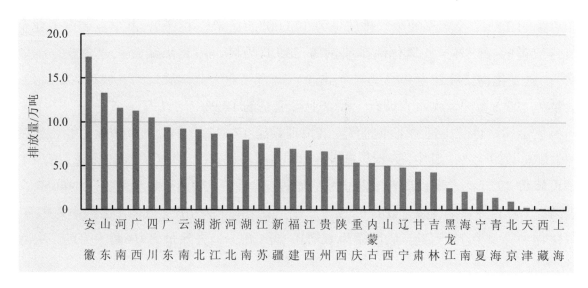

图 3-16　各地区水泥企业氮氧化物排放及处理情况

水泥企业烟（粉）尘排放量为 64.9 万吨，比 2012 年减少 3.3%，占重点调查工业企业烟（粉）尘排放量的 7.0%。拥有除尘设施 46 592 套，比 2012 年减少 1 747 套，占重点调查工业企业除尘设施数的 25.9%；共去除烟（粉）尘 17 906.1 万吨，烟（粉）尘去除率为 99.6%，与 2012 年持平，比重点调查工业企业平均水平高 1.0 个百分点。

水泥企业烟（粉）尘排放量居前 4 位的省份依次为内蒙古、湖南、河北和新疆，其烟（粉）尘排放量占全国水泥企业排放量的 27.2%。

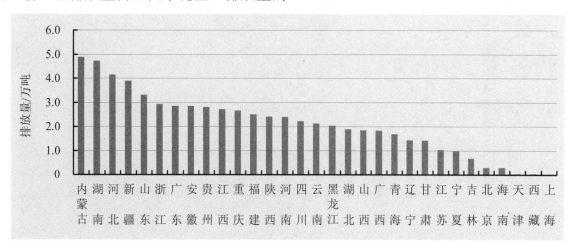

图 3-17　各地区水泥企业烟（粉）尘排放及处理情况

3.3.4　黑色金属冶炼及压延加工业废气污染物排放及处理情况

（1）黑色金属冶炼及压延加工业总体情况

2013 年，黑色金属冶炼及压延加工业重点调查工业企业 4 202 家，占重点调查工业企业总数的 2.8%；全年工业废气排放量为 173 002.5 亿米³（标态），占重点调查工业企业废气排放量的 25.8%；二氧化硫排放量为 235.1 万吨，占重点调查工业企业排放量的 13.9%；氮氧化物排放量为 99.7 万吨，占重点调查工业企业排放量的 6.8%；烟（粉）尘排放量为 193.5 万吨，占重点调查工业企业排放量的 18.9%。

黑色金属冶炼及压延加工业拥有废气治理设施 17 017 套，占重点调查工业企业废气治理设施总数的 7.3%。其中，脱硫设施 894 套，占重点调查工业企业脱硫设施总数的 4.1%；脱硝设施数 49 套，占重点调查工业企业脱硝设施总数的 2.3%；除尘设施 15 346 套，占重点调查工业企业除尘设施总数的 8.5%。二氧化硫去除量为 107.7 万吨，去除率为 31.4%；氮氧化物去除量为 7.4 万吨，去除率为 6.8%；烟（粉）尘去除量为 6 645.5 万吨，去除率为 97.2%。

黑色金属冶炼及压延加工业二氧化硫排放量居全国前 4 位的省份依次为河北、四川、山西和内蒙古，其二氧化硫排放量占黑色金属冶炼及压延加工业排放量的 38.0%；氮氧化物排放量居前 4 位的省份依次为河北、辽宁、江苏和山东，其氮氧化物排放量占黑色金属冶炼及压延加工业排放量的 46.7%；烟（粉）尘排放量居前 4 位的省份依次为河北、山西、山东和辽宁，其烟（粉）尘排放量占黑色金属冶炼及压延加工业排放量的 54.9%。

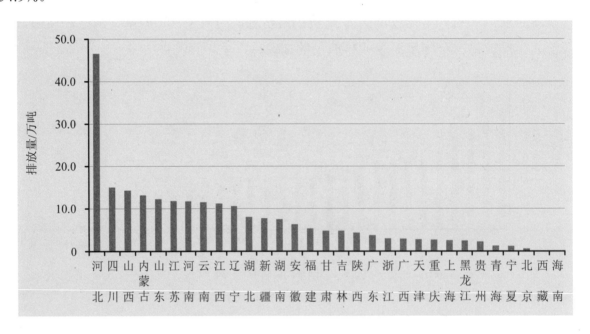

图 3-18　各地区黑色金属冶炼及压延加工业二氧化硫排放情况

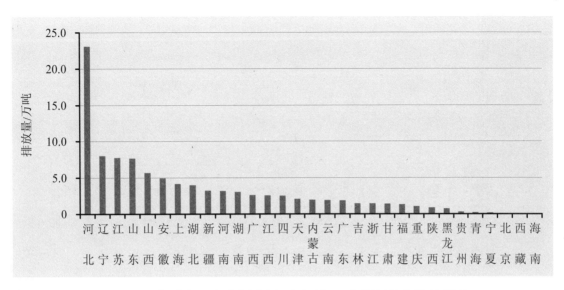

图 3-19　各地区黑色金属冶炼及压延加工业氮氧化物排放情况

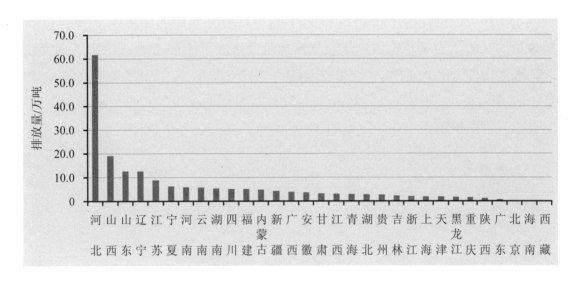

图 3-20　各地区黑色金属冶炼及压延加工业烟（粉）尘排放情况

（2）钢铁冶炼企业废气污染物排放及处理情况

2013 年，纳入重点调查统计范围的、有烧结机或球团设备的钢铁冶炼企业 740 家，占重点调查工业企业总数的 0.5%。共拥有烧结机数 1 258 台，其中 447 台有脱硫设施，837 台有除尘设施；有球团设备数 598 套，其中 52 套有脱硫设施，365 套有除尘设施。

钢铁冶炼企业二氧化硫排放量为 199.3 万吨，比 2012 年降低 0.4%，占重点调查工业企业排放量的 11.8%，占黑色金属冶炼及压延加工业排放量的 84.8%。二氧化硫去除量为 75.8 万吨，二氧化硫去除率为 27.6%，比 2012 年提高 7.2 个百分点，比重点调查工业企业平均水平低 44.6 个百分点。钢铁冶炼企业二氧化硫排放量居前 4 位的省份依次为河北、山西、四川和内蒙古，占全国钢铁企业排放量的 38.8%。

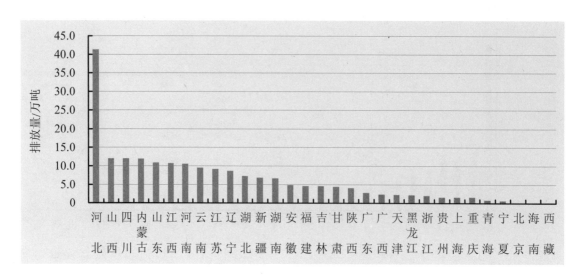

图 3-21　各地区钢铁冶炼企业二氧化硫排放及处理情况

钢铁冶炼企业氮氧化物排放量为 55.5 万吨，比 2012 年增加 2.2%，占重点调查工业企业排放量的 3.8%，占黑色金属冶炼及压延加工业排放量的 55.6%。氮氧化物去除量 5.5 万吨，氮氧化物去除率为 9.1%，比 2012 年提高 7.9 个百分点，比重点调查工业企业平均水平低 10.1 个百分点。钢铁冶炼企业氮氧化物排放量居前 4 位的省份依次为河北、江苏、山东和辽宁，占全国钢铁企业排放量的 52.9%。

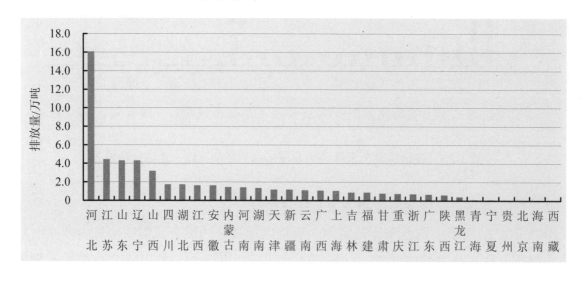

图 3-22　各地区钢铁冶炼企业氮氧化物排放及处理情况

钢铁冶炼企业烟（粉）尘排放量为 61.9 万吨，占重点调查工业企业排放量的 6.1%。烟（粉）尘去除量 2 222.0 万吨，烟（粉）尘去除率为 97.3%，与 2012 年持平，比重点调查工业企业平均水平低 1.3 个百分点。钢铁冶炼企业烟（粉）尘放量居前 4 位的省份依次为河北、山西、辽宁和山东，占全国钢铁企业排放量的 61.1%。

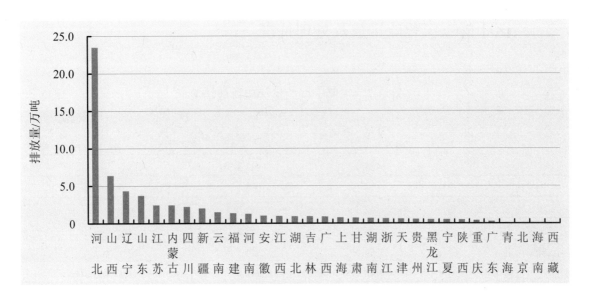

图 3-23　各地区钢铁冶炼企业烟（粉）尘排放及处理情况

3.4　大气污染防治重点区域（三区十群）废气污染物排放及处理情况

3.4.1　三区十群总体情况

大气污染防治重点区域（以下简称三区十群）调查统计范围包括京津冀、长三角、珠三角地区，辽宁中部、山东、武汉及其周边、长株潭、成渝、海峡西岸、山西中北部、陕西关中、甘宁、新疆乌鲁木齐城市群，共涉及 19 个省份（含 114 个地级及以上城市、3 个县级城市、1 个副省级开发区、1 个正厅级实验区），区域总面积 132.56 万千米^2。

2013 年，三区十群工业废气排放量为 361 126 亿米^3（标态），占全国工业废气排放量的 54.0%。三区十群二氧化硫排放量 932.8 万吨，占全国二氧化硫排放量的 45.6%。氮氧化物排放量为 1 076.6 万吨，占全国氮氧化物排放量的 48.3%。烟（粉）尘排放量为 528.1 万吨，占全国烟（粉）尘排放量的 41.3%。

3.4.2 三区十群二氧化硫排放及处理情况

表 3-7 三区十群二氧化硫排放情况 单位：万吨

排放源 区域	合计	工业源	城镇生活源	单位面积排放强度 （吨/千米^2）
京津冀	158.9	143.3	15.5	7.3
长三角	175.1	166.1	8.9	8.3
珠三角	42.3	41.8	0.4	7.7
辽宁中部城市群	54.0	49.8	4.2	8.3
山东城市群	164.5	144.5	19.9	10.5
武汉及其周边城市群	35.1	31.8	3.4	5.9
长株潭城市群	11.5	10.9	0.6	4.1
成渝城市群	118.7	107.8	10.9	5.4
海峡西岸城市群	36.1	34.2	1.9	2.9
山西中北部城市群	42.2	36.1	6.1	7.4
陕西关中城市群	49.4	43.1	6.3	9.0
甘宁城市群	28.2	26.7	1.5	6.5
新疆乌鲁木齐城市群	16.8	15.5	1.3	5.3
总　计	932.8	851.7	81.0	7.0

2013 年，三区十群二氧化硫排放量为 932.8 万吨，其中，工业二氧化硫排放量为 851.7 万吨，占三区十群二氧化硫排放总量的 91.3%，城镇生活二氧化硫排放量为 81.0 万吨，占三区十群二氧化硫排放总量的 8.7%，集中式二氧化硫排放量为 0.1 万吨。

三区十群共有工业脱硫设施 14 185 套，占全国工业脱硫设施数的 65.4%，二氧化硫去除率为 71.0%。全年工业二氧化硫去除量为 2 039.3 万吨，占全国工业二氧化硫去除量的 45.5%。

三区十群中，二氧化硫排放量居前 4 位的区域依次为长三角地区、山东城市群、京津冀地区和成渝城市群，其二氧化硫排放量占三区十群排放量的 66.2%。

三区十群单位面积二氧化硫排放强度为 7.0 吨/千米^2，是全国平均排放强度（2.1 吨/千米^2）的 3.3 倍。三区十群中各区域的单位面积二氧化硫排放强度均高于全国平均水平。其中，单位面积二氧化硫排放强度较大的区域为山东城市群和陕西关中城市群，其污染物排放强度分别为 10.5 吨/千米^2 和 9.0 吨/千米^2。二氧化硫排放量和单位面积二氧化硫排放强度均较大的区域为长三角地区、山东城市群和京津冀地区。

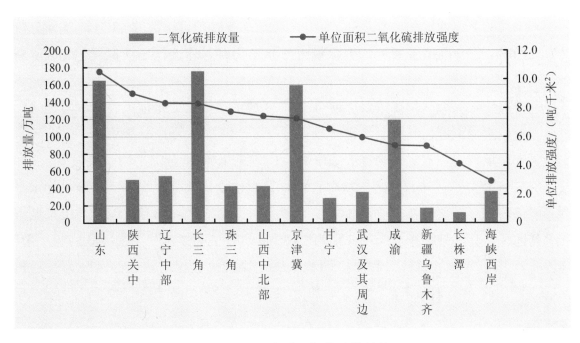

图 3-24　三区十群二氧化硫排放情况

3.4.3　三区十群氮氧化物排放及处理情况

2013 年，三区十群氮氧化物排放量 1 076.6 万吨。其中，工业氮氧化物排放量 758.6 万吨，占区域氮氧化物排放量的 70.5%；生活氮氧化物排放量 17.3 万吨，占区域氮氧化物排放量的 1.6%；机动车氮氧化物排放量 300.4 万吨，占区域氮氧化物排放量的 27.9%；集中式氮氧化物排放量 0.3 万吨。

表 3-8　三区十群氮氧化物排放量　　　　　　　　　　　　　　单位：万吨

排放源 区域	合计	工业源	城镇生活源	机动车	单位面积排放强度 （吨/千米²）
京津冀	213.1	143.2	4.2	65.6	9.7
长三角	247.1	182.1	3.3	61.7	11.7
珠三角	75.5	43.3	0.2	31.9	13.8
辽宁中部城市群	53.6	39.4	1.1	13.1	8.3
山东城市群	165.1	117.2	2.6	45.3	10.5
武汉及其周边城市群	36.1	25.5	0.7	9.9	6.1
长株潭城市群	12.7	7.9	0.1	4.7	4.5
成渝城市群	87.5	59.3	1.3	26.9	4.0
海峡西岸城市群	43.8	33.1	0.2	10.5	3.5
山西中北部城市群	40.6	31.3	1.2	8.1	7.1
陕西关中城市群	48.0	36.1	1.7	10.2	8.7
甘宁城市群	29.5	21.9	0.4	7.2	6.8
新疆乌鲁木齐城市群	23.9	18.4	0.3	5.2	7.6
总　计	1 076.6	758.6	17.3	300.4	8.1

三区十群共有工业脱硝设施 1 367 套，占全国工业脱硝设施数的 63.5%，氮氧化物去除率为 21.6%。全年工业氮氧化物去除量为 208.7 万吨，占全国工业氮氧化物去除量的 58.2%。

三区十群中，氮氧化物排放量居前 4 位的区域依次为长三角地区、京津冀地区、山东城市群和成渝城市群，其氮氧化物排放量占三区十群排放量的 66.2%。

三区十群单位面积氮氧化物排放强度为 8.1 吨/千米^2，是全国平均排放强度（2.3 吨/千米^2）的 3.5 倍。三区十群中各区域单位面积氮氧化物排放强度均高于全国平均水平。其中，单位面积氮氧化物排放强度较大的区域为珠三角地区、长三角地区、山东城市群和京津冀地区，其污染物排放强度分别为 13.8 吨/千米^2、11.7 吨/千米^2、10.5 吨/千米^2 和 9.7 吨/千米^2。氮氧化物排放量和单位面积氮氧化物排放强度均较大的区域为长三角地区、京津冀地区和山东城市群。

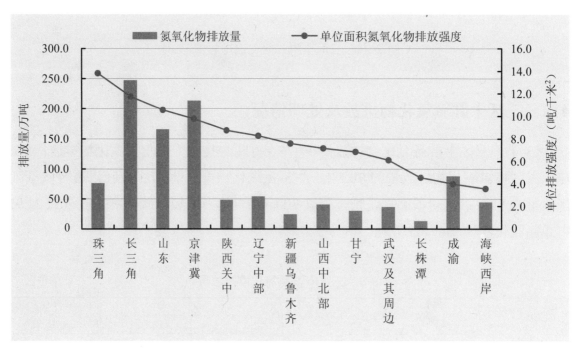

图 3-25　三区十群氮氧化物排放情况

3.4.4　三区十群烟（粉）尘排放及处理情况

2013 年，三区十群烟（粉）尘排放量为 528.1 万吨。其中，工业烟（粉）尘排放量 455.9 万吨，占区域烟（粉）尘排放量的 86.3%；生活烟（粉）尘排放量 45.8 万吨，占区域烟（粉）尘排放量的 8.7%；机动车烟（粉）尘排放量 26.2 万吨，占区域烟（粉）尘排放量的 5.0%；集中式烟（粉）尘排放量 0.1 万吨。

表 3-9　三区十群烟（粉）尘排放情况　　　　　　　　　　　单位：万吨

区域＼排放源	合计	工业源	城镇生活源	机动车	单位面积排放强度（吨/千米^2）
京津冀	146.0	127.7	12.4	5.9	6.7
长三角	90.1	81.9	3.0	5.0	4.3
珠三角	20.2	16.9	0.1	3.1	3.7
辽宁中部城市群	42.6	37.3	4.0	1.3	6.5
山东城市群	69.7	54.2	10.8	4.6	4.4
武汉及其周边城市群	17.8	14.7	2.2	0.9	3.0
长株潭城市群	6.0	5.1	0.5	0.3	2.1
成渝城市群	37.6	34.5	1.2	1.9	1.7
海峡西岸城市群	25.9	24.1	1.0	0.9	2.1
山西中北部城市群	27.9	22.0	5.2	0.7	4.9
陕西关中城市群	26.8	22.5	3.7	0.5	4.9
甘宁城市群	9.6	8.3	0.6	0.6	2.2
新疆乌鲁木齐城市群	8.0	6.6	1.0	0.4	2.5
总　计	528.1	455.9	45.8	26.2	4.0

三区十群共有工业除尘设施 98 326 套，占全国工业除尘设施数的 54.6%，烟（粉）尘去除率为 98.7%。全年工业烟（粉）尘去除量为 35 568.5 万吨，占全国工业烟（粉）尘去除量的 47.5%。

三区十群中，烟（粉）尘排放量居前 4 位的区域依次为京津冀地区、长三角地区、山东城市群和辽宁中部城市群，其烟（粉）尘排放量占三区十群排放量的 66.0%。

三区十群单位面积烟（粉）尘排放强度为 4.0 吨/千米^2，是全国平均排放强度（1.3 吨/千米^2）的 3.1 倍。三区十群中各区域的单位面积烟（粉）尘排放强度均高于全国平均水平。其中，单位面积烟（粉）尘排放强度较大的区域为京津冀地区、辽宁中部城市群、山西中北部城市群和陕西关中城市群，其污染物排放强度分别为 6.7 吨/千米^2、6.5 吨/千米^2、4.9 吨/千米^2 和 4.9 吨/千米^2。烟（粉）尘排放量和单位面积烟（粉）尘排放强度均较大的区域为京津冀地区、长三角地区和山东城市群。

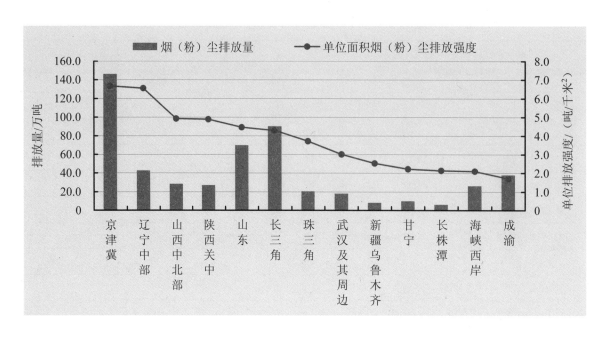

图 3-26　三区十群氮氧化物排放情况

3.5　机动车主要污染物排放情况

2013 年，全国机动车 4 项污染物排放总量为 4 570.9 万吨，比 2012 年减少 0.9%。其中，一氧化碳（CO）3 439.7 万吨，碳氢化合物（HC）431.2 万吨，氮氧化物（NOₓ）640.6 万吨，颗粒物（PM）59.4 万吨。机动车污染物排放量最大的是汽车，其一氧化碳（CO）、碳氢化合物（HC）、氮氧化物（NOₓ）和颗粒物（PM）排放量分别占机动车排放量的 84.7%、80.9%、91.9%、95.5%。

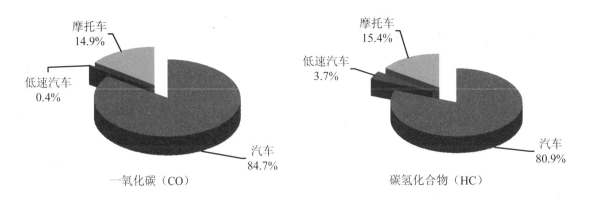

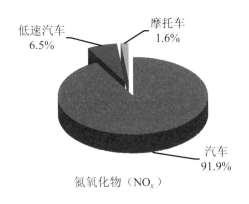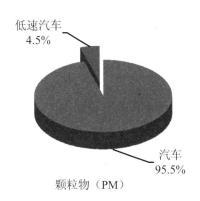

图 3-27　机动车污染物排放分担率

3.5.1　一氧化碳（CO）排放情况

2013 年全国机动车一氧化碳（CO）排放量为 3 439.7 万吨。其中，汽车排放 2 912.1 万吨，占 84.7%；低速汽车排放 14.5 万吨，占 0.4%；摩托车排放 513.1 万吨，占 14.9%。

2013 年全国机动车污染物排放量中，一氧化碳（CO）排放量前 5 位的省份依次为广东、河北、山东、河南、江苏。

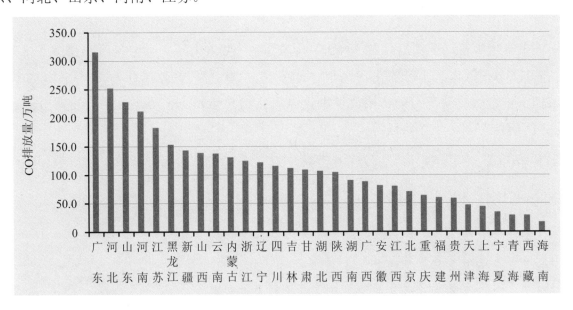

图 3-28　2013 年全国各省（市）机动车一氧化碳（CO）排放量

3.5.2　碳氢化合物（HC）排放情况

2013 年全国机动车碳氢化合物（HC）排放量为 431.2 万吨。其中，汽车排放 349.0 万吨，占 80.9%；低速汽车排放 15.8 万吨，占 3.7%；摩托车排放 66.4 万吨，占 15.4%。

2013 年全国机动车污染物排放量中，碳氢化合物（HC）排放量前 5 位的省份依次为广东、河北、山东、河南、江苏。

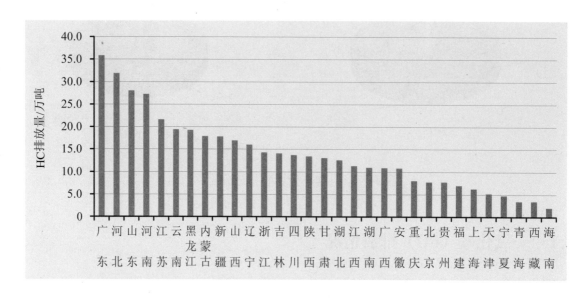

图 3-29　2013 年全国各省（市）机动车碳氢化合物（HC）排放量

3.5.3　氮氧化物（NO_x）排放情况

2013 年全国机动车氮氧化物（NO_x）排放量为 640.6 万吨。其中，汽车排放 588.7 万吨，占 91.9%；低速汽车排放 41.9 万吨，占 6.5%；摩托车排放 10.0 万吨，占 1.6%。

2013 年全国机动车污染物排放量中，氮氧化物（NO_x）排放量前 5 位的省份依次为河北、河南、广东、山东、江苏。

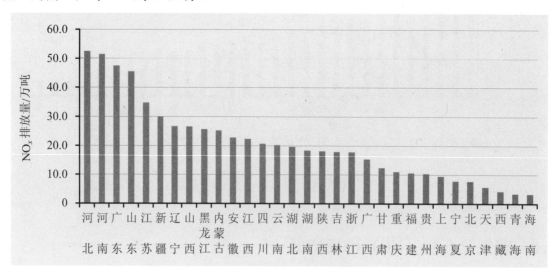

图 3-30　2013 年全国各省（市）机动车氮氧化物（NO_x）排放量

3.5.4 颗粒物（PM）排放情况

2013 年全国机动车颗粒物（PM）排放量为 59.4 万吨。其中，汽车排放 56.7 万吨，占 95.5%；低速汽车排放 2.7 万吨，占 4.5%。

2013 年全国机动车污染物排放量中，颗粒物（PM）排放量前 5 位的省份依次为河南、河北、山东、广东、内蒙古。

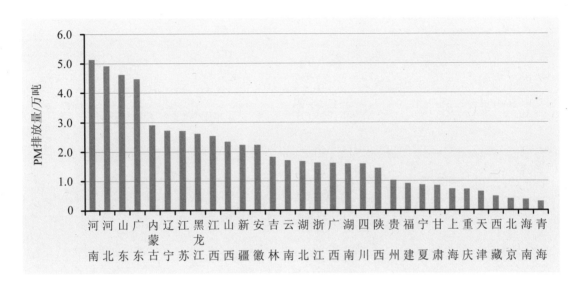

图 3-31 2013 年全国各省（市）机动车颗粒物（PM）排放量

4

工业固体废物

ANNUAL STATISTIC REPORT ON ENVIRONMENT IN CHINA
2013

4.1 一般工业固体废物产生及处理情况

2013 年，全国一般工业固体废物产生量 32.8 亿吨，比 2012 年减少 0.41%，综合利用量为 20.6 亿吨，比 2012 年增加 1.71%，综合利用率为 62.2%，贮存量为 4.3 亿吨，比 2012 年减少 28.7%；处置量为 8.3 亿吨，比 2012 年增加 17.3%；倾倒丢弃量为 129.3 万吨，比 2012 年减少 10.3%。

表 4-1　全国一般工业固体废物产生及处理情况　　　　　　　　　　单位：万吨

年份	产生量	综合利用量	贮存量	处置量	倾倒丢弃量
2011	322 722.3	195 214.6	60 424.3	70 465.3	433.3
2012	329 044.3	202 461.9	59 786.3	70 744.8	144.2
2013	327 701.9	205 916.3	42 634.2	82 969.5	129.3
变化率/%	−0.41	1.71	−28.69	17.28	−10.33

注：①"综合利用量"包括综合利用往年贮存量，"处置量"包括处置往年贮存量；
②工业固体废物综合利用率=工业固体废物综合利用量/（工业固体废物产生量+综合利用往年贮存量）；
③变化率表示与 2012 年相比指标的变化情况，下同。

重点调查工业企业的一般工业固体废物产生量为 31.3 亿吨。其中，尾矿产生量为 10.6 亿吨，占 34.0%，尾矿综合利用量为 3.3 亿吨，综合利用率为 30.7%；粉煤灰产生量 4.6 亿吨，占 14.8%，综合利用量为 4.0 亿吨，综合利用率为 86.2%；煤矸石产生量 3.8 亿吨，占 12.3%，综合利用量为 2.8 亿吨，综合利用率为 71.1%；冶炼废渣产生量 3.7 亿吨，占 11.8%，综合利用量为 3.4 亿吨，综合利用率为 91.8%；炉渣产生量 2.6 亿吨，占 8.5%，综合利用量为 2.4 亿吨，综合利用率为 89.9%。

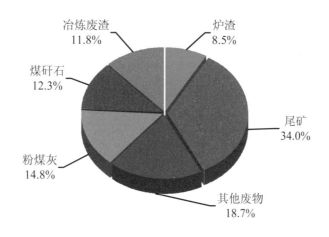

图 4-1　一般工业固体废物构成

尾矿产生量较大的省份依次为河北 2.4 亿吨、辽宁 1.1 亿吨、四川 0.8 亿吨、江西 0.7 亿吨、云南 0.6 亿吨。其中河北、辽宁两省尾矿产生量占全国的 33.3%。

粉煤灰产生量较大的省份依次为山西 4 192.4 万吨、山东 4 159.6 万吨、内蒙古 4 082.6 万吨、河南 3 670.3 万吨和江苏 3 048.8 万吨，这 5 个省份粉煤灰产生量占全国的 41.4%。

煤矸石产生量较大的省份依次为山西 1.4 亿吨、内蒙古 0.5 亿吨、安徽 0.3 亿吨、河南 0.3 亿吨和山东 0.2 亿吨。其中，山西省煤矸石产生量占全国的 35.2%。

冶炼废渣产生量较大的省份依次为河北 7 394.4 万吨、辽宁 3 417.4 万吨、山东 3 047.1 万吨、江苏 2 875.3 万吨和山西 1 969.6 万吨，这 5 个省份冶炼废渣产生量占全国的 50.8%。其中河北省冶炼废渣产生量占 20.1%。

4.1.1 各地区一般工业固体废物产生及处理情况

2013 年，一般工业固体废物产生量较大的省份为河北 4.3 亿吨，占全国工业企业产生量的 13.2%；山西 3.1 亿吨，占全国工业企业产生量的 9.3%；辽宁 2.7 亿吨，占全国工业企业产生量的 8.2%；内蒙古 2.0 亿吨，占全国工业企业产生量的 6.1%；山东 1.8 亿吨，占全国工业企业产生量的 5.5%。

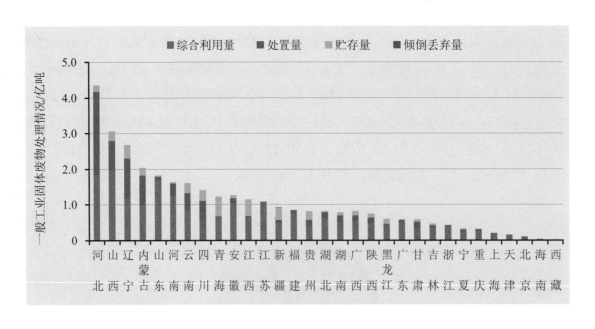

图 4-2 各地区一般工业固体废物综合利用、处置、贮存、倾倒丢弃情况

综合利用量超过 1 亿吨的省（市）有 7 个省，较大的省份为山西 2.0 亿吨，主要为煤矸石，占全省工业企业综合利用量的 44.7%；河北 1.8 亿吨，主要为冶炼废渣，占全省工业企业的 40.1%；山东 1.7 亿吨，主要为粉煤灰、冶炼废渣和尾矿，占全省工业企业的 55.5%；

辽宁 1.2 亿吨，主要为尾矿、冶炼废渣和粉煤灰，占全省工业企业的 53.2%；河南 1.2 亿吨，主要为粉煤灰和煤矸石，占全省工业企业的 51.1%；江苏 1.1 亿吨，主要为冶炼废渣、粉煤灰和炉渣，占全省工业企业的 75.7%；安徽 1.0 亿吨，主要为煤矸石、粉煤灰和尾矿，占全省工业企业的 61.8%。这 7 个省（市）的一般工业固体废物综合利用量占全国工业企业的 48.8%。一般工业固体废物综合利用率较大的省（市）为天津、上海、江苏、浙江和山东，均高于 90%。

处置量较大的省份为河北 2.3 亿吨，主要为尾矿，占全省工业企业的 89.8%；辽宁 1.1 亿吨，主要为尾矿，占全省工业企业处置量的 45.4%；内蒙古 0.8 亿吨，主要为尾矿和煤矸石，占全省工业企业的 82.5%；山西 0.8 亿吨，主要为煤矸石，占全省工业企业的 53.7%；四川 0.5 亿吨，主要为尾矿，占全省工业企业的 87.6%。这 5 个省份的一般工业固体废物处置量占全国工业企业的 68.1%。

贮存量较大的省份为青海 0.6 亿吨，主要为尾矿和其他废物，占全省工业企业的 98.0%；江西 0.5 亿吨，主要为尾矿，占全省工业企业的 92.0%；辽宁 0.4 亿吨，主要为尾矿，占全省工业企业的 87.2%；新疆 0.4 亿吨，主要为尾矿，占全省工业企业贮存量的 59.0%；四川 0.3 亿吨，主要为尾矿，占全省工业企业的 81.0%。这 5 个省份的一般工业固体废物贮存量占全国工业企业的 49.0%。

倾倒丢弃量较大的省份为云南 48.9 万吨，主要为尾矿，占全省一般工业固体废物倾倒丢弃量的 66.2%；新疆 26.5 万吨，主要为尾矿，占全省工业企业的 78.2%；贵州 19.2 万吨，主要为其他废物，占全省工业企业的 61.8%；重庆 11.5 万吨，主要为煤矸石，占全省工业企业的 77.0%；辽宁 9.1 万吨，主要为其他废物，占全省工业企业的 98.7%。这 5 个省份的工业固体废物倾倒丢弃量占全国工业企业的 89.0%。

4.1.2　工业行业固体废物产生及处理情况

2013 年，一般工业固体废物产生量较大的行业依次为黑色金属矿采选业 6.8 亿吨，占重点调查工业企业的 21.7%；电力、热力生产和供应业 6.1 亿吨，占重点调查工业企业的 19.4%；黑色金属冶炼和压延加工业 4.4 亿吨，占重点调查工业企业的 14.1%；煤炭开采和洗选业 3.9 亿吨，占重点调查工业企业的 12.5%；有色金属矿采选业 3.8 亿吨，占重点调查工业企业的 12.1%；化学原料和化学制品制造业 2.8 亿吨，占重点调查工业企业的 8.9%。这 6 个行业一般工业固体废物产生量占重点调查工业企业的 88.7%。

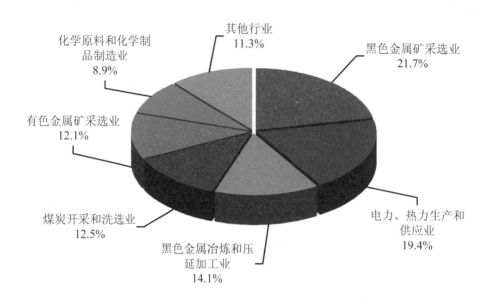

化学原料和化学制品制造业
8.9%

其他行业
11.3%

黑色金属矿采选业
21.7%

有色金属矿采选业
12.1%

煤炭开采和洗选业
12.5%

黑色金属冶炼和压延加工业
14.1%

电力、热力生产和供应业
19.4%

图 4-3　一般工业固体废物产生量行业构成

综合利用量较大的行业依次为电力、热力生产和供应业 5.2 亿吨，占重点调查工业企业的 26.7%；黑色金属冶炼和压延加工业 4.0 亿吨，占重点调查工业企业的 20.4%；煤炭开采和洗选业 2.9 亿吨，占重点调查工业企业的 14.7%；化学原料和化学制品制造业 1.8 亿吨，占重点调查工业企业的 9.2%；黑色金属矿采选业 1.7 亿吨，占全国的 8.7%；有色金属矿采选业 1.4 亿吨，占重点调查工业企业的 7.1%。

处置量较大的行业依次为黑色金属矿采选业 4.15 亿吨，占重点调查工业企业的 51.7%；有色金属矿采选业 1.14 亿吨，占重点调查工业企业的 14.2%；煤炭开采和洗选业 0.90 亿吨，占重点调查工业企业的 11.2%；化学原料和化学制品制造业 0.44 亿吨，占重点调查工业企业的 5.5%；电力、热力生产和供应业 0.42 亿吨，占重点调查工业企业的 5.2%；有色金属冶炼和压延加工业 0.40 亿吨，占重点调查工业企业的 5.0%。

贮存量较大的行业依次为有色金属矿采选业 1.3 亿吨，占重点调查工业企业的 32.8%；黑色金属矿采选业 0.9 亿吨，占重点调查工业企业的 23.0%；化学原料和化学制品制造业 0.6 亿吨，占重点调查工业企业的 14.7%；电力、热力生产和供应业 0.5 亿吨，占重点调查工业企业的 12.0%；煤炭开采和洗选业 0.3 亿吨，占重点调查工业企业的 6.8%；黑色金属冶炼和压延加工业 0.2 亿吨，占重点调查工业企业的 4.5%。

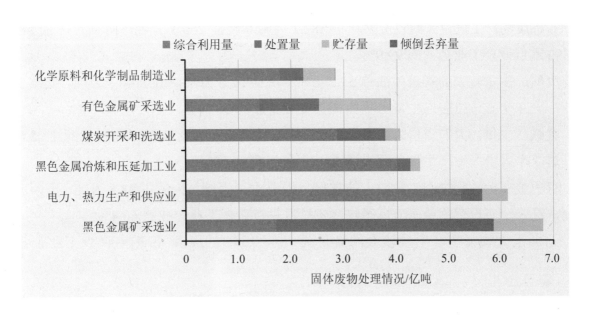

图 4-4　一般工业固体废物处理情况

4.2　工业危险废物产生和处理情况

2013 年，全国工业危险废物产生量为 3 156.9 万吨，比 2012 年减少 8.9%；综合利用量为 1 700.1 万吨，比 2012 年减少 15.2%；处置量为 701.2 万吨，比 2012 年增加 0.43%；贮存量为 810.8 万吨，比 2012 年减少 4.3%。工业危险废物处置利用率为 74.8%，比 2012 年下降 1.3 个百分点。

表 4-2　全国工业危险废物产生及处理情况

年份	产生量/万吨	综合利用量/万吨	处置量/万吨	贮存量/万吨	倾倒丢弃量/吨
2011	3 431.2	1 773.1	916.5	823.7	95.7
2012	3 465.2	2 004.6	698.2	846.9	16.1
2013	3 156.9	1 700.1	701.2	810.9	0.0
变化率/%	−8.90	−15.19	0.43	−4.25	—

注：危险废物处置利用率=（危险废物综合利用量＋处置量）/（危险废物产生量＋综合利用往年贮存量＋处置往年贮存量）。

产生量较大的危险废物种类为石棉废物 651.3 万吨，占重点调查工业企业的 20.6%；废酸 373.8 万吨，占重点调查工业企业的 11.8%；废碱 361.0 万吨，占重点调查工业企业的 11.4%。有色金属冶炼废物 298.3 万吨，占重点调查工业企业的 9.4%；无机氰化物废物 211.8 万吨，占重点调查工业企业的 6.7%。

石棉废物产生量较大的省份为青海 382.2 万吨和新疆 268.8 万吨，两省的石棉废物产生量占重点调查工业企业的 99.9%。

废酸产生量较大的省份为江苏 59.2 万吨，广西 52.6 万吨，山东 46.8 万吨，安徽 30.4 万吨，这 4 个省份废酸产生量占重点调查工业企业的 50.5%。

废碱产生量较大的省份为山东 193.8 万吨，其次分别为湖南 58.2 万吨和浙江 25.0 万吨，这 3 个省份废碱产生量占重点调查工业企业的 76.9%。

有色金属冶炼废物产生量较大的省份为云南 94.4 万吨，内蒙古 48.8 万吨，广西 36.9 万吨，这 3 个省份有色金属冶炼废物产生量占重点调查工业企业的 60.4%。

无机氰化物废物产生量较大的省份为山东 169.3 万吨，占重点调查工业企业的 79.9%。

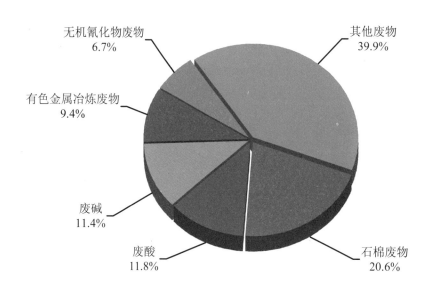

图 4-5　工业危险废物产生量构成

4.2.1　各地区工业危险废物产生和处理情况

2013 年，各地区工业危险废物产生量较大的省份为山东 509.1 万吨，占全国工业企业危险废物产生量的 16.1%；青海 399.9 万吨，占全国工业企业的 12.7%；新疆 302.7 万吨，占全国工业企业的 9.6%；湖南 284.3 万吨，占全国工业企业的 9.0%；江苏 218.1 万吨，占全国工业企业的 6.9%。

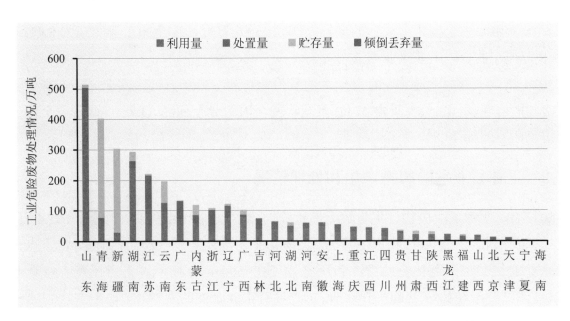

图 4-6 各地区工业危险废物综合利用、处置、贮存、倾倒丢弃情况

2013 年，工业危险废物综合利用量较大的省份为山东 442.4 万吨，占全国工业企业危险废物综合利用量的 26.0%；湖南 242.8 万吨，占全国工业企业的 14.3%；江苏 107.3 万吨，占全国工业企业的 6.3%；云南 98.2 万吨，占全国工业企业的 5.8%；广西 79.7 万吨，占全国工业企业的 4.7%；全国工业危险废物综合利用率为 53.1%。共有 14 个省份工业危险废物综合利用率超过全国平均水平，其中山东、湖南、江西、安徽和广西均超过 80%。

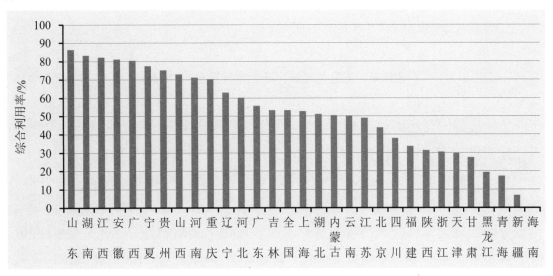

图 4-7 各地区工业危险废物综合利用率

2013 年，工业危险废物处置量较大的省份为江苏 109.2 万吨，占全国工业企业工业危险废物处置量的 15.6%；浙江 71.0 万吨，占全国工业企业工业危险废物处置量的 10.1%；

山东 61.8 万吨，占全国工业企业的 8.8%；广东 58.6 万吨，占全国工业企业的 8.4%；辽宁 39.8 万吨，占全国工业企业的 5.7%。

2013 年，工业危险废物贮存量较大的省份为青海 325.0 万吨，占全国工业企业工业危险废物贮存量的 40.1%；新疆 273.2 万吨，占全国工业企业的 33.7%；两省占全国工业危险废物贮存量的 73.8%。

4.2.2　工业行业危险废物产生和处理情况

2013 年，工业危险废物产生量较大的行业为化学原料和化学制品制造业 681.4 万吨，占重点调查工业企业危险废物产生量的 21.6%；非金属矿采选业 652.3 万吨，占重点调查工业企业的 20.7%；有色金属冶炼和压延加工业 564.4 万吨，占重点调查工业企业的 17.9%；造纸和纸制品业 308.0 万吨，占重点调查工业企业的 9.8%。

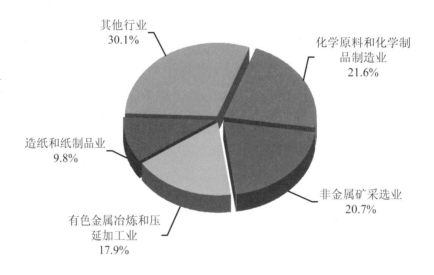

图 4-8　工业危险废物产生量行业分布情况

工业危险废物综合利用量较大的行业为化学原料和化学制品制造业 448.0 万吨，占重点调查工业企业的 26.4%，综合利用率为 65.7%；有色金属冶炼和压延加工业 363.1 万吨，占重点调查工业企业的 21.4%，综合利用率为 64.3%；造纸和纸制品业 287.6 万吨，占重点调查工业企业危险废物综合利用量的 16.9%，综合利用率为 93.4%。

工业危险废物处置量较大的行业为化学原料和化学制品制造业 231.5 万吨，占重点调查工业企业工业危险废物处置量的 33.0%，处置率为 34.0%；计算机、通信和其他电子设备制造业 68.7 万吨，占重点调查工业企业的 9.8%，处置率为 42.5%；有色金属冶炼和压延加工业 61.1 万吨，占重点调查工业企业的 8.7%，处置率为 10.8%。

工业危险废物贮存量较大的行业为非金属矿采选业 592.6 万吨，占重点调查工业企业工业危险废物贮存量的 73.1%；有色金属冶炼和压延加工业 155.8 万吨，占重点调查工业企业的 19.2%；化学原料和化学制品制造业 20.5 万吨，占重点调查工业企业的 2.5%。

化学原料和化学制品制造业产生的危险废物主要是废酸 241.7 万吨和废碱 104.0 万吨，分别占 35.5% 和 15.2%；非金属矿采选业产生的危险废物主要是石棉废物 651.0 万吨，占 99.8%；有色金属冶炼和压延加工业产生的危险废物主要是有色金属冶炼废物 285.7 万吨、无机氰化物废物 137.4 万吨和含铅废物 41.5 万吨，分别占 50.6%、24.3% 和 7.4%；造纸和纸制品业产生的危险废物主要是废碱 202.5 万吨和染料、涂料废物 103.5 万吨，分别占该行业重点调查工业企业危险废物产生量的 65.7% 和 33.6%。

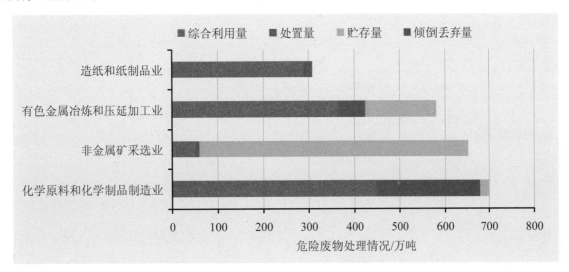

图 4-9 主要工业行业危险废物处理情况

5

环境污染治理投资

ANNUAL STATISTIC REPORT ON ENVIRONMENT IN CHINA
2013

5.1 总体情况

5.1.1 环境污染治理投资总额

环境污染治理投资包括老工业污染源治理、建设项目"三同时"、城市环境基础设施建设 3 个部分。2013 年，我国环境污染治理投资总额为 9 037.2 亿元，占国内生产总值（GDP）的 1.59%，占全社会固定资产投资总额的 2.02%，比 2012 年增加 9.5%。其中，城市环境基础设施建设投资 5 223.0 亿元，老工业污染源治理投资 849.7 亿元，建设项目"三同时"投资 2 964.5 亿元，分别占环境污染治理投资总额的 57.8%、9.4%、32.8%。

表 5-1 全国环境污染治理投资情况   单位：亿元

年份	城市环境基础设施建设投资	老工业污染源治理投资	建设项目"三同时"环保投资	投资总额
2005	1 289.7	458.2	640.1	2 388.0
2010	4 224.2	397.0	2 033.0	6 654.2
2011	3 469.4	444.4	2 112.4	6 026.2
2012	5 062.7	500.5	2 690.4	8 253.6
2013	5 223.0	849.7	2 964.5	9 037.2
变化率/%	3.2	69.8	10.2	9.5

注：①从 2012 年起，城市环境基础设施建设投资中不仅包括城市的环境基础设施建设投资，还包括县城的相关投资，下同。

②变化率表示与 2012 年相比指标的变化情况，下同。

5.1.2 污染治理设施直接投资

污染治理设施直接投资是指直接用于污染治理设施、具有直接环保效益的投资，具体包括老工业污染源、建设项目"三同时"以及城市环境基础设施投资中用于污水处理及再生利用、污泥处置和垃圾处理设施的投资。因此污染治理设施直接投资的统计口径小于污染治理投资。

2013 年，我国污染治理设施直接投资总额为 4 479.5 亿元，占污染治理投资总额的 49.6%，其中城市环境基础设施投资、老工业污染源治理投资和建设项目"三同时"环保投资分别占污染治理设施直接投资的 14.8%、19.0% 和 66.2%。建设项目"三同时"环保投资是污染治理设施直接投资的主要来源。

2013 年，环境治理设施直接投资比 2012 年增加 19.0%。其中，城市环境基础设施投资、老工业污染源治理投资和建设项目"三同时"环保投资分别比 2012 年增加 15.8%、69.8% 和 10.2%。

表 5-2 我国污染治理设施直接投资情况

年份	污染治理设施直接投资/亿元	城市环境基础设施建设投资	老工业污染源治理投资	建设项目"三同时"环保投资	占当年环境污染治理投资总额比例/%	占当年 GDP 比例/%
2005	1 346.4	248.1	458.2	640.1	56.4	0.74
2010	3 078.8	648.8	397.0	2 033.0	46.3	0.77
2011	3 076.5	519.7	444.4	2 112.4	51.1	0.65
2012	3 765.4	574.5	500.5	2 690.4	45.6	0.72
2013	4 479.5	665.3	849.7	2 964.5	49.6	0.79
变化率/%	19.0	15.8	69.8	10.2	—	—

5.1.3 各地区环境污染治理投资

2013 年，我国环境污染治理投资总额为 9 037.2 亿元，除西藏、海南、青海和宁夏 4 个地区外，其余 27 个地区环境污染治理投资总额超过 100 亿元。与 2012 年相比，除辽宁、江西、湖北、浙江、重庆、海南、河北和吉林 8 个地区外，其余 23 个地区环境污染治理投资总额均有所增长。

2013 年，13 个地区环境污染治理投资占 GDP 比重超过全国平均水平（1.59%），广东、吉林、海南、上海、河南、四川、湖南、浙江 8 个地区比重较低，污染治理投资占 GDP 的比重均低于 1%。

2013 年，全国 GDP 比 2012 年增加 7.7%，环境污染治理投资增速略低于 GDP 增速，全国环境污染治理投资弹性系数（注：环境污染治理投资弹性系数=环境污染治理投资增速/GDP 增速）为 0.99，其中，20 个地区环境污染治理投资增速超过 GDP 增速。

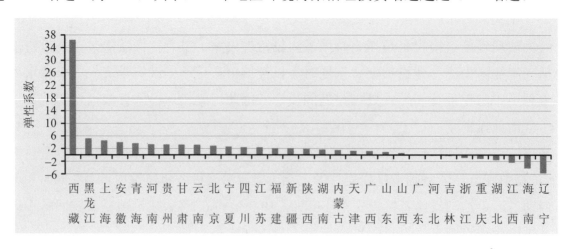

图 5-1 2013 年各地区环境污染治理弹性系数

114

5.1.4　污染治理设施运行费用情况

污染治理设施运行费用是指工业、城镇生活污染（废水、废气及固体废物）治污设施运行费用，不包括农村污染治理设施。2013 年，污染治理设施运行费用 2 665.3 亿元，比 2012 年增加 1.7%。

2013 年，工业废气治理设施 234 316 套（台），运行费用 1 497.8 亿元，占污染治理设施总运行费用的 56.2%。其中，脱硫设施和脱硝设施运行费用分别为 611.8 亿元和145.5 亿元，占废气治理设施运行费用的 40.8%和 9.7%。与 2010 年相比，工业废气治理设施运行费用增加 42.0%。

废水治理设施运行费用包括工业企业废水治理设施和污水处理厂两个部分。2013 年，废水治理设施运行费用 1 022.2 亿元。其中，工业废水治理设施费用 628.7 亿元，占废水治理设施运行费用的 61.5%，比 2012 年减少 5.8%；污水处理厂运行费用 393.6 亿元，占废水治理设施运行费用的 38.5%，比 2012 年增加 13.0%。与 2010 年相比，工业废水治理设施和污水处理厂运行费用分别增加 15.3%和 52.0%。

2013 年，生活垃圾处理场运行费用 86.6 亿元，比 2012 年减少 12.1%；危险（医疗）废物集中处理（置）场运行费用 58.7 亿元，比 2012 年增加 9.0%。

表 5-3　污染治理设施运行费用　　　　　　　　　　　　　　单位：亿元

年份	工业废水治理设施	工业废气治理设施	脱硫设施	脱硝设施	污水处理厂	生活垃圾处理场	危险（医疗）废物集中处理（置）场	合计	占污染治理设施直接投资比例/%
2005	276.7	267.1	—	—	73.7	—	7.1	624.6	46.39
2010	545.3	1 054.5	435.0	—	259.0	—	38.9	1 897.7	61.64
2011	732.1	1 579.5	587.3	44.5	307.2	59.2	48.2	2 726.1	88.61
2012	667.7	1 452.3	540.4	66.0	348.2	98.5	53.9	2 620.6	69.60
2013	628.7	1 497.8	611.8	145.5	393.6	86.6	58.7	2 665.3	59.50
变化率/%	-5.8	3.1	13.2	120.5	13.0	-12.1	9.0	1.7	—

5.2　城市环境基础设施建设

2013 年，城市环境基础设施建设投资中，燃气工程建设投资 607.9 亿元，比 2012 年增加 10.2%；集中供热工程建设投资 819.5 亿元，比 2012 年增加 2.7%；排水工程建设投资 1 055.0 亿元，比 2012 年增加 12.9%；园林绿化工程建设投资 2 234.9 亿元，比 2012

年减少 6.1%；市容环境卫生工程建设投资 505.7 亿元，比 2012 年增加 26.9%。

燃气、集中供热、排水、园林绿化和市容环境卫生投资分别占城市环境基础设施建设总投资的 11.6%、15.7%、20.2%、42.8%和 9.7%，园林绿化和排水设施投资为城市环境基础设施建设投资的重点。

表 5-4　全国近年城市环境基础设施建设投资构成　　　　单位：亿元

年份	投资总额	燃气	集中供热	排水	园林绿化	市容环境卫生
2005	1 289.7	142.4	220.2	368.0	411.3	147.8
2010	4 224.2	290.8	433.2	901.6	2 297.0	301.6
2011	3 469.4	331.4	437.6	770.1	1 546.2	384.1
2012	5 062.7	551.8	798.1	934.1	2 380.0	398.6
2013	5 223.0	607.9	819.5	1 055.0	2 234.9	505.7
变化率/%	3.2	10.2	2.7	12.9	−6.1	26.9

5.3　老工业污染源治理投资

2013 年，老工业污染源污染治理本年施工项目 5 593 个，其中，废水、废气、固体废物、噪声及其他治理项目分别 1 490 个、2 785 个、256 个、91 个和 971 个，占本年施工项目数的 26.6%、49.8%、4.6%、1.6%和 17.4%。

2013 年，老工业污染源污染治理投资中，废水、废气、固体废物、噪声及其他治理项目投资分别为 124.9 亿元、640.9 亿元、14.0 亿元、1.8 亿元和 68.1 亿元，分别占老工业污染源治理投资额的 14.7%、75.4%、1.7%、0.2%和 8.0%。与 2005 年相比，废气治理项目投资增加 200.9%，废水、固体废物、噪声治理项目及其他污染治理项目投资分别减少 6.6%、48.7%、43.1%和 16.0%。

表 5-5　老工业源污染治理投资构成　　　　单位：亿元

年份	投资总额	废水	废气	固体废物	噪声	其他
2005	458.2	133.7	213.0	27.4	3.1	81.0
2010	397.0	130.1	188.8	14.3	1.5	62.2
2011	444.4	157.7	211.7	31.4	2.2	41.4
2012	500.5	140.3	257.7	24.7	1.2	76.5
2013	849.7	124.9	640.9	14.0	1.8	68.1
变化率/%	69.8	−11.0	148.7	−43.1	46.9	−11.0

5.4 建设项目"三同时"环保投资

2013 年，建设项目"三同时"环保投资 2 964.5 亿元，比 2012 年增加 10.2%，占建设项目投资总额的 4.6%。

表 5-6 建设项目"三同时"投资情况

年份	环保投资额/亿元	占建设项目投资总额/%	占全社会固定资产投资总额/%	占环境治理投资总额/%
2001	336.4	3.6	0.9	30.4
2005	640.1	4.0	0.7	26.8
2010	2 033.0	4.1	0.7	30.6
2011	2 112.4	3.1	0.7	35.1
2012	2 690.4	2.7	0.7	31.9
2013	2 964.5	4.6	0.7	32.7
变化率/%	10.2	—	—	—

6

环境管理

ANNUAL STATISTIC REPORT ON ENVIRONMENT IN CHINA
2013

6.1 环保机构建设

2013 年，全国环保系统机构总数 14 257 个。其中，国家级机构 45 个，省级机构 400 个，地市级环保机构 2 252 个，县级环保机构 8 866 个，乡镇环保机构 2 694 个。各级环保行政机构 3 176 个，各级环境监察机构 2 923 个，各级环境监测机构 2 754 个。

全国环保系统共有 21.2 万人。其中，环保机关人员 5.3 万人，占环保系统总人数的 24.9%；环境监察人员 6.3 万人，占环保系统总人数的 29.6%；环境监测人员 5.8 万人，占环保系统总人数的 27.3%。

表 6-1 环保行政机构、监察机构、监测站年末实有人员情况

年份	年末实有人数/人	环保行政机构		环境监察机构		环境监测站	
		实有人数/人	占本级人员总数比例/%	实有人数/人	占本级人员总数比例/%	实有人数/人	占本级人员总数比例/%
2005	166 774	44 024	26.4	50 040	30.0	46 984	28.2
2006	170 290	44 141	25.9	52 845	31.2	47 689	28.2
2007	176 988	43 626	24.6	57 427	32.4	49 335	27.9
2008	183 555	44 847	24.4	59 477	32.1	51 753	28.3
2009	188 991	45 626	24.1	60 896	32.2	52 944	28.0
2010	193 911	45 938	23.7	62 468	32.2	54 698	28.2
2011	201 161	46 128	22.9	64 426	32.0	56 226	28.0
2012	205 334	53 286	26.0	61 081	29.7	56 554	27.5
2013	212 048	52 845	24.9	62 696	29.6	57 884	27.3
国家级	2 951	357	12.1	519	17.6	185	6.3
省 级	14 730	2 801	19.0	1 305	8.9	3 098	21.0
地市级	47 016	9 993	21.3	9 861	21.0	16 433	35.0
县 级	137 099	39 694	29.0	51 011	37.2	38 168	27.8

6.2 环境法制及环境信访情况

2013 年，全国各级环保系统承办的人大建议数 7 222 件，政协提案数 10 990 件。全国办理环境行政处罚案件 13.9 万件，环境行政复议案件 550 件。

全国当年颁布地方性法规 29 件，累计有效的地方性法规 379 件。当年颁布地方政府规章 41 件，累计有效的政府规章 314 件。当年备案的地方环境标准数 31 件，累计备案

的地方环境标准数 126 件。

全国各级环保系统共收到群众来信 10.4 万封，群众来访 4.6 万批次，10.7 万人次，其中，已办结来信和来访 15.2 万件。电话及网络投诉 111.0 万件，其中，已办结数为 109.7 万件。

全国发生突发环境事件 712 次，其中重大环境事件 3 次，较大环境事件 12 次，一般环境事件 697 次。

表 6-2　环境信访工作情况

年份	来信总数/封	来访批次/批	来访人次/次	来信、来访已办结数量/件	电话/网络投诉数/件	电话/网络投诉办结数/件
2005	608 245	88 237	142 360	—	—	—
2006	616 122	71 287	110 592	—	—	—
2007	123 357	43 909	77 399	—	—	—
2008	705 127	43 862	84 971	—	—	—
2009	696 134	42 170	73 798	—	—	—
2010	701 073	34 683	65 948	—	—	—
2011	201 631	53 505	107 597	251 607	852 700	834 588
2012	107 120	43 260	96 145	159 283	892 348	888 836
2013	103 776	46 162	107 165	151 635	1 110 212	1 096 595

6.3　环境监测

2013 年，全国监测用房总面积为 286.8 万米^2，监测业务经费为 37.0 亿元。环境监测仪器 26.8 万台（套），仪器设备原值为 146.3 亿元。

全国环境空气监测点位 3 001 个，酸雨监测点位 1 176 个，沙尘天气影响环境质量监测点位数 82 个，地表水水质监测断面 9 414 个，饮用水水源地监测点位数 912 个，近岸海域监测点位数 882 个，开展环境噪声监测的监测点位数 84 608 个，开展生态监测的监测点位数 163 个，开展污染源监督性监测的重点企业数 61 454 个。

6.4　自然生态保护

2013 年，全国各类自然保护区共计 2 697 个。自然保护区面积 14 631.0 万公顷，约占国土面积的 14.6%。国家级、省级自然保护区个数分别占全国自然保护区总数的 15.1%、

31.7%，其面积分别占自然保护区总面积的 64.3%、26.8%。

全国共建设国家级生态市 6 个，国家级生态县 86 个，国家级生态村镇 3 219 个。省级生态市 35 个，省级生态县 272 个。共建成国家有机食品生产基地 138 个。

表 6-3　全国自然保护区数量　　　　　　　　　　　　　单位：个

年份	自然保护区数	国家级	省级	地市级	县级
2005	2 349	243	773	421	912
2006	2 395	265	793	422	915
2007	2 531	303	780	462	986
2008	2 538	303	806	432	997
2009	2 541	319	827	416	979
2010	2 588	319	859	418	992
2011	2 640	335	870	421	1 014
2012	2 669	363	876	406	1 024
2013	2 697	407	855	—	—

表 6-4　全国自然保护区面积　　　　　　　　　　　　单位：万公顷

年份	自然保护区面积	国家级	省级	地市级	县级
2005	14 994.9	8 898.9	4 487.0	501.5	1 107.5
2006	15 153.5	9 169.7	4 441.8	522.4	1 019.6
2007	15 188.2	9 365.6	4 260.1	537.6	1 024.9
2008	14 894.3	9 120.3	4 240.2	497.1	1 036.8
2009	14 774.7	9 267.1	4 004.5	471.2	1 031.9
2010	14 944.1	9 267.6	4 174.8	468.2	1 033.4
2011	14 971.1	9 315.3	4 152.6	472.4	1 030.9
2012	14 978.7	9 414.6	4 090.9	432.5	1 040.8
2013	14 631.0	9 403.9	3 919.5	—	—

6.5　环境影响评价及环保验收

2013 年，共审批建设项目环境影响评价文件 47.6 万件，其中编制报告书的项目 3.4 万件，填报报告表的项目 19.0 万件，编制登记表的项目 25.2 万件。

2013 年，全国完成环保验收项目 15.0 万个，其中环保验收一次合格的项目 14.5 万个。其中，当年完成工业企业验收项目环保投资为 1 942.8 亿元。

6.6 污染源控制及管理

除废水、废气中主要污染物的治理投入和工程设施建设运行外（详见第 2 章、第 3 章主要污染物排放与治理情况及第 5 章环境污染治理投资），2013 年，全国完成清洁生产审核企业 9 763 家，其中进行强制性审核的企业 6 385 家。应开展监测的重金属污染防控重点企业 3 805 家，其中重金属排放达标的重点企业 3 107 家。已发放危险废物经营许可证 1 763 个，其中具有医疗废物经营范围的许可证 258 个。

6.7 污染源自动监控及排污收费

2013 年，全国已实施自动监控的国家重点监控企业 9 779 个，其中已实施自动监控的废水排放口 7 812 个，废气排放口 7 145 个。实施自动监控国家重点监控企业中，化学需氧量监控设备与环境保护部门稳定联网的企业 5 382 个，氨氮监控设备与环境保护部门稳定联网的企业 3 822 个，二氧化硫监控设备与环境保护部门稳定联网的企业 5 489 个，氮氧化物监控设备与环境保护部门稳定联网的企业 5 445 个。

全国排污费解缴入库单位共 35.2 万户，入库金额 204.8 亿元。

6.8 环境宣教

2013 年，各级环保系统共建成环境教育基地 1 902 个，组织开展社会环境宣传教育活动 17 087 次，参与社会环境宣传教育活动的人数达到 3 294.2 万人次。

7

全国辐射环境水平

ANNUAL STATISTIC REPORT ON ENVIRONMENT IN CHINA
2013

2013 年，全国辐射环境质量总体良好。

7.1 环境电离辐射

2013 年，全国环境电离辐射水平保持在天然本底涨落范围内。辐射环境自动监测站实时连续γ辐射空气吸收剂量率均在当地天然本底水平涨落范围内。气溶胶、沉降物总α和总β活度浓度、空气中氡活度浓度均为正常环境水平。长江、黄河、珠江、松花江、淮河、海河、辽河、浙闽片河流、西南诸河、西北诸河、重点湖泊（水库）人工放射性核素活度浓度与历年相比均无明显变化，天然放射性核素活度浓度与 1983—1990 年全国环境天然放射性水平调查结果处于同一水平。地下饮用水及开展监测的省会城市集中式饮用水水源地水中总α和总β活度浓度均低于《生活饮用水卫生标准》（GB 5749—2006）规定的限值。近岸海域海水中人工放射性核素锶 90 和铯 137 活度浓度均低于《海水水质标准》（GB 3097—1997）规定的限值。土壤中人工放射性核素活度浓度与历年相比无明显变化，天然放射性核素活度浓度与 1983—1990 年全国环境天然放射性水平调查结果处于同一水平。

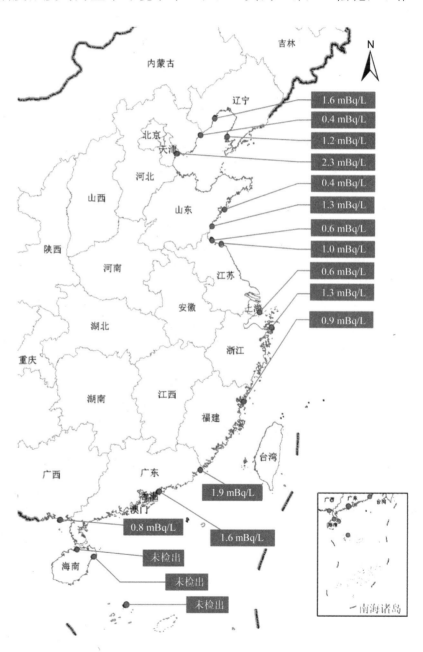

图 7-1　2013 年中国近岸海域海水中铯 137 活度浓度

图片来源：国家测绘局 1∶400 万基础地理信息底图

7.2 运行核电厂周围环境电离辐射

2013 年，辽宁红沿河核电厂和福建宁德核电厂投入运行。秦山核电基地各核电厂、大亚湾/岭澳核电厂、田湾核电厂、红沿河核电厂和宁德核电厂外围各辐射环境自动监测站实时连续γ辐射空气吸收剂量率（未扣除宇宙射线响应值）年均值分别为 100.8 纳戈/小时、123.7 纳戈/小时、99.9 纳戈/小时、76.8 纳戈/小时和 98.1 纳戈/小时，均在当地天然本底水平涨落范围内。核电厂外围气溶胶、沉降物、地表水、地下水和土壤等各种环境介质中除氚外其他放射性核素活度浓度与历年相比均无明显变化。秦山核电基地周围环境空气、降水、地表水、井水及部分生物样品中氚活度浓度，大亚湾/岭澳核电厂和田湾核电厂排放口附近海域海水中氚活度浓度与核电厂运行前本底值相比均有所升高，但对公众造成的辐射剂量均远低于国家规定的剂量限值。

7.3 民用研究堆周围环境电离辐射

中国原子能科学研究院、清华大学核能与新能源技术研究院、中国核动力研究设计院和深圳大学微堆等研究设施外围环境γ辐射空气吸收剂量率，气溶胶、沉降物、地表水、地下水和土壤中放射性核素活度浓度与历年相比均无明显变化；饮用地下水总α和总β活度浓度均低于《生活饮用水卫生标准》规定的限值。

7.4 核燃料循环设施和废物处置设施周围环境电离辐射

中核兰州铀浓缩有限公司、中核陕西铀浓缩有限公司、中核北方核燃料元件有限公司、中核建中核燃料元件公司和中核四○四有限公司等核燃料循环设施及西北低中放废物处置场、北龙低中放废物处置场和青海国营二二一厂放射性填埋坑外围环境γ辐射空气吸收剂量率均无明显变化，环境介质中未监测到由上述企业生产、加工、贮存、处理、运输等活动引起的放射性核素活度浓度升高。

7.5　铀矿冶周围环境电离辐射

铀矿冶设施周围辐射环境质量总体稳定。空气中氡活度浓度、气溶胶总α活度浓度、地表水及地下水中放射性核素铀和镭226活度浓度均无明显变化。

7.6　电磁辐射

2013年，全国环境电磁辐射质量总体良好。环境电磁综合场强均远低于《电磁辐射防护规定》（GB 8702—88）中有关公众照射参考导出限值。电磁辐射设施周围环境电磁辐射水平无明显变化。开展监测的移动通信基站天线周围环境敏感点的电磁辐射水平低于《电磁辐射防护规定》规定的公众照射导出限值，开展监测的各输电线和变电站周围环境敏感点工频电场强度和磁感应强度均低于《500 kV超高压送变电工程电磁辐射环境影响评价技术规范》（HJ/T 24—1998）规定的居民区工频电场评价标准和公众全天候辐射时的工频限值。

简 要 说 明

1．本年报资料根据全国 31 个省、自治区、直辖市及新疆生产建设兵团环境统计资料汇总整理而成，未包括香港特别行政区、澳门特别行政区以及台湾省数据。

2．本年报主要反映中国环境污染排放、治理及环境管理情况。主要内容包括废水及污染物的排放与治理情况，废气及污染物排放与治理情况，一般工业固体废物、危险废物（医疗废物）的产生、综合利用及处理处置情况，集中式污染治理设施、环境污染治理投资以及环境管理等情况。

3．调查范围

本年报数据根据"十二五"环境统计报表制度调查收集汇总，调查范围较"十一五"有所扩大。"十二五"环境统计报表制度调查范围包括工业污染源、农业污染源、城镇生活污染源、机动车、集中式污染治理设施和环境管理 6 方面内容。

（1）工业污染源调查范围：辖区内凡事实有污染物排放的所有工业企业。

（2）城镇生活污染源调查范围：城市和集镇内居民在日常生活及各种活动中产生、排放的污染物情况。

（3）机动车调查范围：辖区内载客汽车、载货汽车、三轮汽车及低速载货汽车、摩托车等类型的注册机动车，不包括非道路移动污染源类型。

（4）农业污染源调查范围：畜禽养殖业、水产养殖业和种植业。畜禽养殖业主要包括生猪、奶牛、肉牛、蛋鸡和肉鸡 5 种畜禽类型的规模化养殖场（养殖小区）和养殖专业户，不包括其他养殖类型，也不包括养殖数量较小的散户。

（5）集中式污染治理设施调查范围：辖区内所有集中式污染治理设施，包括污水处理厂、垃圾处理厂（场）、危险废物（医疗废物）处理（处置）厂。

（6）环境管理反映环保系统自身能力建设、业务工作进展及成果等情况，调查范围主要包括环保机构数/人数、环境信访与环境法制、环境保护能力建设投资、环境污染源控制与管理、环境监测、污染源自动监控、排污费征收、自然生态保护与建设、环境影响评价、建设项目竣工环境保护验收情况、突发环境事件、环境宣教 12 个方面的内容。

4．其他需要说明的问题

（1）2013 年中国环境统计年报中采用水利部水资源分区对流域进行汇总，流域分为松花江、辽河、海河、黄河、淮河、长江、珠江、东南诸河、西南诸河和西北诸河十大水系。

自 2012 年起，根据《重点流域水污染防治"十二五"规划》中的流域分区，增加了对松花江、辽河、海河、黄河中上游、淮河、长江中下游等重点规划流域的汇总，重点流域按照规划范围汇总，与水资源分区范围并不完全重合，因此数据有所不同。

（2）本年报中所指集中式污染治理设施的排放量，仅指生活垃圾处理厂（场）和危险（医疗）废物集中处置厂（场）的渗滤液和焚烧废气中的污染物。

（3）本年报的调查范围尚未完全涵盖所有产生排放污染的活动，如工业企业生产过程中或建筑工地等无组织排放，农村生活的废水、废气及固体废物排放，飞机、火车、船舶等交通移动源等，因此统计结果可能与其他相关统计或科学研究的结论存在差异。

主要环境统计指标解释附后。

8

各地区环境统计

ANNUAL STATISTIC REPORT ON ENVIRONMENT IN CHINA
2013

各地区主要污染物排放情况（一）
Discharge of Key Pollutants by Region（1）

（2013）

单位：亿吨 (100 million tons)

年 份 地 区	Year Region	废水排放总量 Total Volume of Waste Water Discharged	工业 Industrial	生活 Household	集中式 Centralized Treatment
2012		684.76	221.59	462.69	0.49
2013		695.44	209.84	485.11	0.50
北 京	Beijing	14.46	0.95	13.50	0.01
天 津	Tianjin	8.42	1.87	6.55	...
河 北	Hebei	31.09	10.99	20.10	0.01
山 西	Shanxi	13.80	4.78	9.02	...
内蒙古	Inner Mongolia	10.69	3.70	6.99	...
辽 宁	Liaoning	23.45	7.83	15.61	0.01
吉 林	Jilin	11.77	4.27	7.50	0.01
黑龙江	Heilongjiang	15.31	4.78	10.52	...
上 海	Shanghai	22.30	4.54	17.72	0.03
江 苏	Jiangsu	59.44	22.06	37.35	0.03
浙 江	Zhejiang	41.91	16.37	25.50	0.05
安 徽	Anhui	26.62	7.10	19.51	0.02
福 建	Fujian	25.91	10.47	15.43	0.02
江 西	Jiangxi	20.71	6.82	13.86	0.03
山 东	Shandong	49.46	18.12	31.31	0.03
河 南	Henan	41.26	13.08	28.17	0.01
湖 北	Hubei	29.41	8.50	20.88	0.02
湖 南	Hunan	30.72	9.23	21.45	0.04
广 东	Guangdong	86.25	17.05	69.13	0.07
广 西	Guangxi	22.53	8.95	13.56	0.02
海 南	Hainan	3.62	0.67	2.94	...
重 庆	Chongqing	14.25	3.35	10.89	0.01
四 川	Sichuan	30.76	6.49	24.26	0.02
贵 州	Guizhou	9.31	2.29	7.01	0.01
云 南	Yunnan	15.66	4.18	11.46	0.01
西 藏	Tibet	0.50	0.04	0.46	...
陕 西	Shaanxi	13.22	3.49	9.72	0.01
甘 肃	Gansu	6.50	2.02	4.48	...
青 海	Qinghai	2.20	0.84	1.36	...
宁 夏	Ningxia	3.85	1.57	2.28	...
新 疆	Xinjiang	10.07	3.47	6.59	0.01

各地区主要污染物排放情况（二）
Discharge of Key Pollutants by Region（2）
（2013）

单位：万吨 （10 000 tons）

年 份 地 区	Year Region	化学需氧量排放总量 Total Volume of COD Discharged	工业 Industrial	农业 Agricultural	生活 Household	集中式 Centralized Treatment
	2012	2 423.73	338.45	1 153.80	912.75	18.73
	2013	2 352.72	319.47	1 125.76	889.81	17.68
北 京	Beijing	17.85	0.61	7.47	8.99	0.78
天 津	Tianjin	22.15	2.62	10.99	8.49	0.05
河 北	Hebei	130.99	17.44	89.48	23.33	0.74
山 西	Shanxi	46.13	7.88	17.46	20.54	0.26
内蒙古	Inner Mongolia	86.32	9.20	60.86	16.03	0.23
辽 宁	Liaoning	125.26	8.47	84.98	31.22	0.59
吉 林	Jilin	76.12	6.53	49.15	19.24	1.20
黑龙江	Heilongjiang	144.73	9.61	103.98	30.95	0.19
上 海	Shanghai	23.56	2.55	3.11	17.32	0.58
江 苏	Jiangsu	114.89	20.92	37.61	55.87	0.49
浙 江	Zhejiang	75.51	17.41	19.77	37.63	0.71
安 徽	Anhui	90.27	8.71	37.07	43.64	0.85
福 建	Fujian	63.90	8.12	20.81	34.56	0.41
江 西	Jiangxi	73.45	9.33	23.25	39.90	0.97
山 东	Shandong	184.57	13.27	129.50	41.33	0.47
河 南	Henan	135.42	17.07	78.15	39.48	0.73
湖 北	Hubei	105.82	12.81	46.14	45.51	1.36
湖 南	Hunan	124.90	14.07	55.70	53.89	1.24
广 东	Guangdong	173.39	23.43	58.13	90.38	1.45
广 西	Guangxi	75.94	17.54	21.08	36.93	0.39
海 南	Hainan	19.44	1.25	10.15	7.93	0.10
重 庆	Chongqing	39.18	5.15	12.10	21.86	0.06
四 川	Sichuan	123.20	10.65	52.88	59.12	0.55
贵 州	Guizhou	32.82	6.32	6.15	19.81	0.53
云 南	Yunnan	54.72	16.75	7.24	29.44	1.30
西 藏	Tibet	2.58	0.06	0.41	2.09	0.02
陕 西	Shaanxi	51.93	9.54	19.25	22.62	0.52
甘 肃	Gansu	37.91	9.05	14.16	14.54	0.16
青 海	Qinghai	10.34	4.21	2.21	3.71	0.22
宁 夏	Ningxia	22.19	10.31	10.12	1.68	0.08
新 疆	Xinjiang	67.24	18.59	36.39	11.81	0.45

各地区主要污染物排放情况（三）
Discharge of Key Pollutants by Region（3）
（2013）

单位：万吨 （10 000 tons）

年 份 地 区	Year Region	氨氮排放 总量 Total Volume of Ammonia Nitrogen Discharged	工业 Industrial	农业 Agricultural	生活 Household	集中式 Centralized Treatment
	2012	253.59	26.41	80.62	144.63	1.93
	2013	245.66	24.58	77.92	141.36	1.79
北 京	Beijing	1.97	0.03	0.45	1.42	0.07
天 津	Tianjin	2.47	0.33	0.56	1.57	...
河 北	Hebei	10.71	1.43	4.31	4.92	0.06
山 西	Shanxi	5.53	0.76	1.21	3.53	0.02
内蒙古	Inner Mongolia	5.12	1.14	1.20	2.76	0.02
辽 宁	Liaoning	10.33	0.71	3.36	6.17	0.09
吉 林	Jilin	5.47	0.42	1.70	3.22	0.13
黑龙江	Heilongjiang	8.77	0.60	3.33	4.82	0.02
上 海	Shanghai	4.58	0.19	0.32	4.03	0.03
江 苏	Jiangsu	14.74	1.44	3.82	9.43	0.06
浙 江	Zhejiang	10.75	1.11	2.59	6.99	0.06
安 徽	Anhui	10.33	0.77	3.68	5.81	0.08
福 建	Fujian	9.09	0.61	3.20	5.23	0.04
江 西	Jiangxi	8.88	0.90	2.87	5.04	0.08
山 东	Shandong	16.15	1.02	7.10	7.98	0.05
河 南	Henan	14.42	1.23	6.11	7.00	0.09
湖 北	Hubei	12.49	1.36	4.52	6.46	0.15
湖 南	Hunan	15.77	2.30	6.13	7.22	0.12
广 东	Guangdong	21.64	1.46	5.54	14.49	0.15
广 西	Guangxi	8.10	0.72	2.61	4.74	0.03
海 南	Hainan	2.26	0.09	0.87	1.30	0.01
重 庆	Chongqing	5.22	0.33	1.25	3.62	0.02
四 川	Sichuan	13.70	0.50	5.59	7.54	0.07
贵 州	Guizhou	3.83	0.35	0.79	2.62	0.06
云 南	Yunnan	5.80	0.43	1.15	4.06	0.17
西 藏	Tibet	0.32	...	0.05	0.26	...
陕 西	Shaanxi	5.96	0.83	1.50	3.56	0.06
甘 肃	Gansu	3.92	1.26	0.55	2.09	0.01
青 海	Qinghai	0.97	0.21	0.09	0.66	0.01
宁 夏	Ningxia	1.71	0.85	0.21	0.64	0.01
新 疆	Xinjiang	4.65	1.18	1.26	2.18	0.03

各地区主要污染物排放情况（四）
Discharge of Key Pollutants by Region（4）
（2013）

单位：万吨 (10 000 tons)

年 份 地 区	Year Region	二氧化硫排放总量 Total Volume of Sulphur Dioxide Discharged	工业 Industrial	生活 Household	集中式 Centralized Treatment
	2012	2 117.63	1 911.71	205.66	0.26
	2013	2 043.92	1 835.19	208.54	0.19
北 京	Beijing	8.70	5.20	3.50	…
天 津	Tianjin	21.68	20.78	0.90	0.01
河 北	Hebei	128.47	117.31	11.15	…
山 西	Shanxi	125.54	114.08	11.46	…
内 蒙 古	Inner Mongolia	135.87	123.64	12.23	…
辽 宁	Liaoning	102.70	94.73	7.97	0.01
吉 林	Jilin	38.15	33.10	5.05	…
黑 龙 江	Heilongjiang	48.91	35.27	13.64	…
上 海	Shanghai	21.58	17.29	4.29	…
江 苏	Jiangsu	94.17	90.95	3.20	0.03
浙 江	Zhejiang	59.34	57.91	1.40	0.02
安 徽	Anhui	50.13	45.02	5.09	0.02
福 建	Fujian	36.10	34.20	1.90	…
江 西	Jiangxi	55.77	54.35	1.42	…
山 东	Shandong	164.50	144.53	19.94	0.02
河 南	Henan	125.40	110.27	15.13	…
湖 北	Hubei	59.94	52.40	7.53	…
湖 南	Hunan	64.13	58.87	5.26	…
广 东	Guangdong	76.19	73.20	2.95	0.04
广 西	Guangxi	47.20	43.80	3.39	…
海 南	Hainan	3.24	3.17	0.08	…
重 庆	Chongqing	54.77	49.44	5.33	…
四 川	Sichuan	81.67	74.64	7.01	0.03
贵 州	Guizhou	98.64	77.86	20.78	…
云 南	Yunnan	66.31	61.30	5.01	…
西 藏	Tibet	0.42	0.13	0.29	…
陕 西	Shaanxi	80.62	70.71	9.90	…
甘 肃	Gansu	56.20	47.28	8.92	…
青 海	Qinghai	15.67	13.08	2.59	…
宁 夏	Ningxia	38.97	36.82	2.15	…
新 疆	Xinjiang	82.94	73.86	9.09	…

各地区主要污染物排放情况（五）
Discharge of Key Pollutants by Region（5）

（2013）

单位：万吨 (10 000 tons)

年 份 地 区	Year Region	氮氧化物 排放总量 Total Volume of Nitrogen Oxide Discharged	工业 Industrial	生活 Household	机动车 Vehicle	集中式 Centralized Treatment
	2012	2 337.76	1 658.05	39.31	640.03	0.37
	2013	2 227.36	1 545.61	40.75	640.55	0.44
北 京	Beijing	16.63	7.59	1.36	7.65	0.03
天 津	Tianjin	31.17	25.06	0.52	5.57	0.02
河 北	Hebei	165.25	110.56	2.33	52.35	...
山 西	Shanxi	115.78	86.36	3.04	26.37	0.01
内蒙古	Inner Mongolia	137.76	110.45	2.30	25.01	...
辽 宁	Liaoning	95.54	67.33	1.74	26.46	...
吉 林	Jilin	56.05	37.05	1.19	17.81	...
黑龙江	Heilongjiang	75.16	44.19	5.49	25.47	...
上 海	Shanghai	38.04	26.23	2.35	9.41	0.04
江 苏	Jiangsu	133.80	98.53	0.61	34.62	0.04
浙 江	Zhejiang	75.30	57.35	0.30	17.63	0.02
安 徽	Anhui	86.37	62.64	1.07	22.63	0.03
福 建	Fujian	43.83	33.06	0.24	10.53	...
江 西	Jiangxi	57.04	34.60	0.31	22.13	...
山 东	Shandong	165.13	117.16	2.64	45.31	0.02
河 南	Henan	156.56	102.88	2.43	51.25	0.01
湖 北	Hubei	61.24	40.45	1.30	19.49	...
湖 南	Hunan	58.82	39.78	0.80	18.23	...
广 东	Guangdong	120.42	72.26	0.69	47.31	0.16
广 西	Guangxi	50.43	34.89	0.36	15.17	0.01
海 南	Hainan	10.02	6.68	0.10	3.24	...
重 庆	Chongqing	36.20	24.79	0.45	10.96	...
四 川	Sichuan	62.43	40.88	0.99	20.53	0.03
贵 州	Guizhou	55.73	44.69	0.79	10.25	...
云 南	Yunnan	52.37	31.67	0.67	20.02	...
西 藏	Tibet	4.43	0.25	0.03	4.15	...
陕 西	Shaanxi	75.89	55.21	2.66	18.02	...
甘 肃	Gansu	44.29	30.64	1.44	12.21	...
青 海	Qinghai	13.23	9.32	0.62	3.29	...
宁 夏	Ningxia	43.74	35.72	0.27	7.75	...
新 疆	Xinjiang	88.69	57.30	1.66	29.73	...

各地区主要污染物排放情况（六）
Discharge of Key Pollutants by Region（6）
（2013）

单位：万吨 (10 000 tons)

年　份 地　区	Year Region	烟（粉）尘排放 总量 Total Volume of Soot and Dust Discharged	工业 Industrial	生活 Household	机动车 Vehicle	集中式 Centralized Treatment
2012		1 234. 31	1 029. 31	142. 67	62. 14	0. 19
2013		1 278. 14	1 094. 62	123. 90	59. 42	0. 19
北　京	Beijing	5. 93	2. 72	2. 83	0. 38	...
天　津	Tianjin	8. 75	6. 28	1. 84	0. 63	...
河　北	Hebei	131. 33	118. 72	7. 70	4. 91	...
山　西	Shanxi	102. 67	89. 79	10. 55	2. 33	...
内蒙古	Inner Mongolia	82. 21	68. 41	10. 90	2. 90	...
辽　宁	Liaoning	67. 06	57. 28	7. 06	2. 71	0. 01
吉　林	Jilin	32. 02	25. 09	5. 11	1. 81	
黑龙江	Heilongjiang	72. 25	52. 01	17. 63	2. 60	
上　海	Shanghai	8. 09	6. 72	0. 65	0. 72	0. 01
江　苏	Jiangsu	50. 00	45. 56	1. 71	2. 70	0. 03
浙　江	Zhejiang	31. 97	29. 66	0. 69	1. 61	0. 02
安　徽	Anhui	41. 86	35. 18	4. 45	2. 22	0. 02
福　建	Fujian	25. 94	24. 06	0. 98	0. 90	...
江　西	Jiangxi	35. 63	32. 47	0. 62	2. 53	...
山　东	Shandong	69. 67	54. 24	10. 81	4. 62	0. 01
河　南	Henan	64. 13	54. 72	4. 27	5. 13	...
湖　北	Hubei	35. 95	29. 48	4. 82	1. 66	...
湖　南	Hunan	35. 87	31. 68	2. 62	1. 57	...
广　东	Guangdong	35. 40	29. 67	1. 21	4. 47	0. 05
广　西	Guangxi	28. 95	26. 00	1. 34	1. 60	...
海　南	Hainan	1. 80	1. 40	0. 04	0. 36	...
重　庆	Chongqing	19. 12	17. 98	0. 44	0. 70	...
四　川	Sichuan	29. 60	26. 89	1. 14	1. 57	0. 01
贵　州	Guizhou	30. 13	26. 21	2. 91	1. 01	...
云　南	Yunnan	38. 69	35. 48	1. 52	1. 69	...
西　藏	Tibet	0. 68	0. 10	0. 10	0. 47	...
陕　西	Shaanxi	53. 77	46. 85	5. 50	1. 42	...
甘　肃	Gansu	22. 66	17. 46	4. 36	0. 84	...
青　海	Qinghai	17. 38	14. 92	2. 16	0. 29	...
宁　夏	Ningxia	23. 06	21. 29	0. 91	0. 86	...
新　疆	Xinjiang	75. 59	66. 33	7. 04	2. 22	0. 01

各地区工业废水排放及处理情况（一）

Discharge and Treatment of Industrial Waste Water by Region（1）

（2013）

单位：万吨 （10 000 tons）

年 份 地 区	Year Region	工业废水排放量 Total Volume of Industrial Waste Water Discharged	直接排入环境的 Direct Discharged to Environment	排入污水处理厂的 Discharged to Centralized Waste Water Treatment	工业废水处理量 Industrial Waste Water Treated
	2012	2 215 857	1 732 817	483 039	5 274 705
	2013	2 098 398	1 585 283	513 115	4 924 811
北　京	Beijing	9 486	1 959	7 527	10 498
天　津	Tianjin	18 692	7 015	11 677	29 161
河　北	Hebei	109 876	74 448	35 428	766 322
山　西	Shanxi	47 795	42 204	5 591	126 012
内蒙古	Inner Mongolia	36 986	20 340	16 646	107 223
辽　宁	Liaoning	78 286	61 962	16 323	213 534
吉　林	Jilin	42 656	35 915	6 741	61 806
黑龙江	Heilongjiang	47 796	41 219	6 578	94 946
上　海	Shanghai	45 426	18 071	27 355	65 019
江　苏	Jiangsu	220 559	142 872	77 687	395 899
浙　江	Zhejiang	163 674	63 019	100 655	256 857
安　徽	Anhui	70 972	61 797	9 175	203 173
福　建	Fujian	104 658	90 584	14 074	168 167
江　西	Jiangxi	68 230	66 223	2 007	167 422
山　东	Shandong	181 179	115 921	65 258	346 256
河　南	Henan	130 789	112 129	18 660	173 052
湖　北	Hubei	84 993	71 869	13 124	181 540
湖　南	Hunan	92 311	89 340	2 970	263 399
广　东	Guangdong	170 463	133 568	36 895	293 037
广　西	Guangxi	89 508	85 877	3 631	304 911
海　南	Hainan	6 744	5 775	969	6 533
重　庆	Chongqing	33 451	32 206	1 245	34 451
四　川	Sichuan	64 864	56 567	8 297	171 902
贵　州	Guizhou	22 898	22 789	109	111 295
云　南	Yunnan	41 844	40 989	855	174 187
西　藏	Tibet	400	400	0	722
陕　西	Shaanxi	34 871	27 147	7 724	59 928
甘　肃	Gansu	20 171	17 428	2 743	30 816
青　海	Qinghai	8 395	8 395	0	25 064
宁　夏	Ningxia	15 708	14 274	1 434	23 902
新　疆	Xinjiang	34 718	22 980	11 738	57 778

各地区工业废水排放及处理情况（二）
Discharge and Treatment of Industrial Waste Water by Region（2）
（2013）

年　份 地　区	Year Region	废水治理设施数/ 套 Number of Facilities for Treatment of Waste Water （set）	废水治理设施 治理能力/ （万吨/日） Capacity of Facilities for Treatment of Waste Water （10 000 tons/day）	废水治理设施 运行费用/ 万元 Annual Expenditure for Operation （10 000 yuan）
	2012	85 673	26 620	6 677 024.6
	2013	80 298	25 642	6 286 649.2
北　京	Beijing	518	60	36 310.9
天　津	Tianjin	1 088	151	171 400.4
河　北	Hebei	4 573	4 029	477 030.5
山　西	Shanxi	3 037	804	207 983.3
内蒙古	Inner Mongolia	1 128	549	109 181.1
辽　宁	Liaoning	2 290	1 181	228 872.6
吉　林	Jilin	660	266	68 476.9
黑龙江	Heilongjiang	1 187	691	289 257.8
上　海	Shanghai	1 743	319	170 673.6
江　苏	Jiangsu	7 452	1 927	733 168.3
浙　江	Zhejiang	8 283	1 425	572 983.5
安　徽	Anhui	2 444	952	217 062.0
福　建	Fujian	3 503	682	152 311.7
江　西	Jiangxi	2 543	1 072	187 451.5
山　东	Shandong	5 189	1 858	541 410.3
河　南	Henan	3 306	1 028	231 883.3
湖　北	Hubei	2 084	1 029	168 277.0
湖　南	Hunan	3 036	1 154	188 548.8
广　东	Guangdong	9 918	1 506	562 850.4
广　西	Guangxi	2 340	1 069	194 088.2
海　南	Hainan	309	53	33 453.5
重　庆	Chongqing	1 669	243	61 117.8
四　川	Sichuan	3 866	1 053	199 680.5
贵　州	Guizhou	1 589	616	60 099.4
云　南	Yunnan	2 602	799	129 646.8
西　藏	Tibet	33	7	712.1
陕　西	Shaanxi	1 846	414	106 477.2
甘　肃	Gansu	596	172	38 324.1
青　海	Qinghai	182	77	13 852.7
宁　夏	Ningxia	367	175	35 309.2
新　疆	Xinjiang	917	283	98 753.8

各地区工业废水排放及处理情况（三）

Discharge and Treatment of Industrial Waste Water by Region（3）

（2013）

单位：吨 （ton）

年　份 地　区	Year Region	工业废水中污染物产生量 Amount of Pollutants Produced in the Industrial Waste Water				
		化学需氧量 COD	氨氮 Ammonia Nitrogen	石油类 Petroleum	挥发酚 Volatile Phenol	氰化物 Cyanide
	2012	24 305 917. 4	1 460 908. 1	296 813. 7	65 801. 3	5 180. 2
	2013	22 148 450. 8	1 391 197. 7	261 789. 7	69 996. 5	5 772. 0
北　京	Beijing	60 648. 4	2 451. 0	3 259. 8	900. 3	5. 1
天　津	Tianjin	153 419. 7	8 553. 4	969. 0	2. 8	1. 5
河　北	Hebei	1 491 954. 6	89 201. 4	22 644. 6	10 362. 5	682. 7
山　西	Shanxi	531 953. 3	39 295. 1	10 153. 5	10 144. 8	441. 4
内蒙古	Inner Mongolia	552 867. 5	72 775. 8	4 740. 1	3 152. 7	332. 2
辽　宁	Liaoning	410 535. 7	42 905. 2	17 657. 4	1 372. 5	167. 5
吉　林	Jilin	389 195. 7	9 800. 0	2 601. 0	882. 9	19. 0
黑龙江	Heilongjiang	1 015 001. 3	65 576. 5	42 065. 1	4 982. 2	51. 0
上　海	Shanghai	231 856. 0	10 092. 4	5 197. 6	777. 0	176. 7
江　苏	Jiangsu	1 802 232. 3	71 658. 2	11 986. 3	3 660. 7	351. 6
浙　江	Zhejiang	2 141 998. 8	78 235. 4	24 175. 4	398. 8	753. 9
安　徽	Anhui	713 334. 9	46 325. 8	7 966. 6	1 722. 9	291. 9
福　建	Fujian	776 119. 5	35 105. 8	4 229. 0	386. 1	110. 3
江　西	Jiangxi	493 312. 9	34 243. 6	6 007. 8	2 230. 3	31. 1
山　东	Shandong	2 272 342. 2	98 197. 2	26 723. 4	6 391. 3	449. 6
河　南	Henan	1 595 698. 5	57 888. 1	15 173. 9	2 760. 6	159. 7
湖　北	Hubei	527 492. 4	31 890. 4	3 799. 9	1 059. 8	51. 2
湖　南	Hunan	611 337. 3	58 849. 7	5 675. 4	989. 7	213. 4
广　东	Guangdong	1 343 356. 8	50 014. 9	4 711. 9	185. 8	666. 4
广　西	Guangxi	1 222 898. 4	52 719. 6	4 135. 4	1 987. 2	33. 7
海　南	Hainan	94 608. 4	2 352. 5	108. 3	…	…
重　庆	Chongqing	327 201. 4	24 216. 3	2 922. 9	962. 0	91. 8
四　川	Sichuan	647 004. 0	72 218. 1	5 469. 9	1 791. 7	125. 9
贵　州	Guizhou	198 691. 6	47 801. 1	1 995. 1	1 514. 0	37. 5
云　南	Yunnan	663 611. 1	36 516. 0	2 436. 1	2 512. 5	105. 9
西　藏	Tibet	4 493. 9	144. 7	4. 0	…	…
陕　西	Shaanxi	438 454. 0	37 860. 8	7 175. 1	6 091. 2	189. 6
甘　肃	Gansu	198 368. 9	24 276. 6	2 444. 0	262. 9	3. 9
青　海	Qinghai	86 933. 9	3 075. 2	1 361. 7	6. 1	9. 0
宁　夏	Ningxia	401 115. 2	62 372. 5	2 873. 1	615. 0	26. 8
新　疆	Xinjiang	750 412. 1	124 584. 5	11 126. 3	1 890. 3	191. 6

各地区工业废水排放及处理情况（四）
Discharge and Treatment of Industrial Waste Water by Region（4）
（2013）

单位：吨 （ton）

年　份　Year 地　区　Region	工业废水中污染物产生量 Amount of Pollutants Produced in the Industrial Waste Water					
	汞 Mercury	镉 Cadmium	六价铬 Hexavalent Chromium	总铬 Total Chrome	铅 Lead	砷 Arsenic
2012	19.961	2 318.538	3 273.441	7 322.903	3 260.838	11 300.867
2013	20.936	2 243.096	3 250.031	6 990.343	2 962.439	11 228.723
北　京　Beijing	…	0.001	11.643	12.328	0.216	3.291
天　津　Tianjin	0.077	0.065	27.576	37.818	2.483	0.125
河　北　Hebei	0.040	1.289	309.374	375.532	9.337	42.062
山　西　Shanxi	0.503	13.264	12.734	13.294	9.376	54.845
内蒙古　Inner Mongolia	2.414	93.276	20.261	20.429	115.150	242.557
辽　宁　Liaoning	0.184	40.791	20.523	30.044	40.812	38.335
吉　林　Jilin	0.207	0.032	4.199	4.213	0.161	2.251
黑龙江　Heilongjiang	0.567	0.130	7.555	7.602	1.352	9.433
上　海　Shanghai	0.010	0.070	69.586	89.702	0.740	0.202
江　苏　Jiangsu	0.026	2.367	235.310	397.376	34.391	5.925
浙　江　Zhejiang	0.006	4.010	1 580.121	2 791.663	8.330	32.532
安　徽　Anhui	0.013	16.210	15.524	20.912	60.696	656.725
福　建　Fujian	0.076	3.909	392.184	755.642	26.542	34.512
江　西　Jiangxi	0.932	462.614	44.903	53.640	112.243	5 707.455
山　东　Shandong	1.126	286.845	45.870	108.244	283.438	918.931
河　南　Henan	1.536	269.988	14.667	184.587	366.092	119.711
湖　北　Hubei	0.259	7.563	40.241	51.941	18.627	115.430
湖　南　Hunan	2.791	163.241	33.634	50.716	390.125	340.151
广　东　Guangdong	0.105	1.645	146.994	1 558.396	15.981	10.053
广　西　Guangxi	2.370	73.052	21.050	52.971	86.856	129.346
海　南　Hainan	…	0.014	…	15.119	0.004	0.005
重　庆　Chongqing	…	0.028	107.302	146.155	7.410	1.646
四　川　Sichuan	0.569	40.018	59.450	128.031	46.231	378.892
贵　州　Guizhou	0.093	0.431	1.295	3.028	0.778	21.606
云　南　Yunnan	0.908	338.280	3.105	3.779	513.682	1 055.773
西　藏　Tibet	0.059	0.029	…	6.362	0.744	15.762
陕　西　Shaanxi	0.362	68.438	14.254	16.452	88.145	54.267
甘　肃　Gansu	0.265	333.731	2.297	36.125	693.908	1 078.721
青　海　Qinghai	0.510	19.677	0.044	0.054	22.045	64.446
宁　夏　Ningxia	0.711	0.086	0.799	0.958	1.296	73.280
新　疆　Xinjiang	4.214	2.001	7.538	17.232	5.248	20.449

144

各地区工业废水排放及处理情况（五）
Discharge and Treatment of Industrial Waste Water by Region（5）
（2013）

单位：吨 (ton)

| 年　份
地　区 | Year
Region | 工业废水中污染物排放量
Amount of Pollutants Discharged in the Industrial Waste Water | | | | |
		化学需氧量 COD	氨氮 Ammonia Nitrogen	石油类 Petroleum	挥发酚 Volatile Phenol	氰化物 Cyanide
	2012	3 384 539.3	264 142.3	17 327.2	1 481.4	171.8
	2013	3 194 669.9	245 845.4	17 389.2	1 259.1	162.0
北　京	Beijing	6 055.0	330.4	48.3	0.1	0.1
天　津	Tianjin	26 214.7	3 338.7	118.2	1.8	1.0
河　北	Hebei	174 439.1	14 270.9	889.5	26.2	8.6
山　西	Shanxi	78 806.9	7 618.5	980.8	620.8	29.3
内蒙古	Inner Mongolia	91 977.3	11 383.1	939.6	240.4	1.7
辽　宁	Liaoning	84 692.0	7 133.1	951.5	7.2	3.2
吉　林	Jilin	65 288.2	4 222.6	329.6	4.2	2.4
黑龙江	Heilongjiang	96 123.3	5 977.1	289.6	6.4	1.5
上　海	Shanghai	25 503.2	1 934.0	619.6	2.0	3.0
江　苏	Jiangsu	209 175.0	14 393.4	1 314.6	40.3	14.6
浙　江	Zhejiang	174 083.2	11 051.9	741.6	6.6	6.6
安　徽	Anhui	87 135.1	7 651.6	750.1	6.1	7.1
福　建	Fujian	81 169.0	6 150.0	500.5	1.8	4.3
江　西	Jiangxi	93 266.0	8 987.5	728.3	14.1	7.4
山　东	Shandong	132 726.9	10 223.9	525.2	38.6	3.7
河　南	Henan	170 682.6	12 298.7	1 225.8	139.3	17.5
湖　北	Hubei	128 087.6	13 608.9	1 002.6	17.8	6.8
湖　南	Hunan	140 688.7	23 046.5	575.4	17.9	10.2
广　东	Guangdong	234 281.2	14 599.3	643.2	12.5	9.0
广　西	Guangxi	175 417.1	7 221.0	261.9	12.8	4.9
海　南	Hainan	12 526.0	913.2	8.5	…	…
重　庆	Chongqing	51 533.7	3 265.7	335.8	9.5	1.5
四　川	Sichuan	106 492.5	4 975.0	634.3	3.3	2.3
贵　州	Guizhou	63 214.2	3 508.8	435.3	0.2	0.6
云　南	Yunnan	167 521.8	4 308.5	365.5	1.3	1.1
西　藏	Tibet	620.3	46.0	0.5	…	…
陕　西	Shaanxi	95 399.9	8 346.5	657.7	2.5	4.0
甘　肃	Gansu	90 480.5	12 636.0	294.3	2.1	0.1
青　海	Qinghai	42 071.0	2 075.5	340.9	1.2	…
宁　夏	Ningxia	103 083.3	8 540.8	153.0	6.8	2.7
新　疆	Xinjiang	185 914.7	11 788.5	727.5	15.1	7.1

各地区工业废水排放及处理情况（六）
Discharge and Treatment of Industrial Waste Water by Region（6）
（2013）

单位：吨 （ton）

年份 地区	Year Region	工业废水中污染物排放量 Amount of Pollutants Discharged in the Industrial Waste Water					
		汞 Mercury	镉 Cadmium	六价铬 Hexavalent Chromium	总铬 Total Chrome	铅 Lead	砷 Arsenic
	2012	1.082	26.662	70.375	188.636	97.137	127.692
	2013	0.783	17.875	58.139	161.863	74.084	111.607
北 京	Beijing	…	0.001	0.320	0.422	0.054	0.012
天 津	Tianjin	0.006	0.003	0.133	0.356	0.086	0.012
河 北	Hebei	…	0.003	2.588	5.352	0.230	0.054
山 西	Shanxi	0.010	0.780	0.141	0.233	0.172	0.192
内蒙古	Inner Mongolia	0.091	0.518	0.037	0.123	3.675	6.046
辽 宁	Liaoning	0.004	0.020	0.235	0.643	0.166	0.228
吉 林	Jilin	0.002	0.004	0.058	0.084	0.045	0.567
黑龙江	Heilongjiang	…	0.001	0.097	0.108	0.025	0.024
上 海	Shanghai	0.008	0.001	1.067	2.381	0.105	0.043
江 苏	Jiangsu	0.003	0.014	4.003	8.829	1.052	0.214
浙 江	Zhejiang	0.001	0.195	8.788	18.595	0.393	0.264
安 徽	Anhui	0.002	0.105	0.254	0.637	1.443	4.797
福 建	Fujian	0.013	0.233	3.089	11.395	3.069	1.418
江 西	Jiangxi	0.072	1.952	17.186	17.453	5.568	9.303
山 东	Shandong	0.015	1.034	0.509	7.817	0.840	2.383
河 南	Henan	0.013	0.941	0.886	30.043	3.943	1.175
湖 北	Hubei	0.021	0.508	9.043	9.845	2.525	6.063
湖 南	Hunan	0.225	6.701	1.005	11.263	24.203	42.519
广 东	Guangdong	0.012	0.352	5.170	22.098	2.507	1.022
广 西	Guangxi	0.105	1.067	0.146	1.755	6.257	8.283
海 南	Hainan	…	0.001	…	0.122	0.003	0.001
重 庆	Chongqing	…	0.003	0.218	0.423	0.081	0.033
四 川	Sichuan	0.010	0.075	0.729	1.794	1.079	2.067
贵 州	Guizhou	0.016	0.021	0.029	0.045	0.238	0.477
云 南	Yunnan	0.006	1.191	0.006	0.064	6.154	8.135
西 藏	Tibet	…	…	…	…	…	8.942
陕 西	Shaanxi	0.025	0.556	0.222	1.636	1.412	0.698
甘 肃	Gansu	0.094	1.286	0.415	6.127	8.050	4.280
青 海	Qinghai	0.008	0.298	0.007	0.011	0.626	1.479
宁 夏	Ningxia	0.005	…	0.126	0.237	0.010	0.050
新 疆	Xinjiang	0.016	0.011	1.631	1.974	0.072	0.825

各地区工业废气排放及处理情况（一）

Discharge and Treatment of Industrial Waste Gas by Region（1）

（2013）

年 份 地 区	Year Region	工业废气排放总量（标态）/ 亿米³ Total Volume of Industrial Waste Gas Emission （100 millon cu.m）	工业废气中污染物排放量/吨 Volume of Pollutants Emission in the Industrial Waste Gas （ton）		
			二氧化硫 Sulphur Dioxide	氮氧化物 Nitrogen Oxide	烟（粉）尘 Soot and Dust
	2012	635 519	19 117 055	16 580 515	10 293 082
	2013	669 361	18 351 904	15 456 148	10 946 235
北 京	Beijing	3 692	52 041	75 927	27 182
天 津	Tianjin	8 080	207 793	250 646	62 766
河 北	Hebei	79 121	1 173 147	1 105 634	1 187 198
山 西	Shanxi	41 276	1 140 835	863 611	897 937
内蒙古	Inner Mongolia	31 128	1 236 409	1 104 482	684 054
辽 宁	Liaoning	29 443	947 330	673 335	572 774
吉 林	Jilin	9 804	330 956	370 505	250 928
黑龙江	Heilongjiang	10 622	352 732	441 898	520 144
上 海	Shanghai	13 344	172 867	262 346	67 174
江 苏	Jiangsu	49 797	909 478	985 313	455 568
浙 江	Zhejiang	24 565	579 104	573 498	296 586
安 徽	Anhui	28 335	450 223	626 449	351 757
福 建	Fujian	16 183	342 040	330 628	240 583
江 西	Jiangxi	15 574	543 497	345 992	324 694
山 东	Shandong	47 160	1 445 348	1 171 600	542 371
河 南	Henan	37 665	1 102 662	1 028 822	547 210
湖 北	Hubei	19 987	524 005	404 489	294 788
湖 南	Hunan	17 276	588 696	397 849	316 770
广 东	Guangdong	28 434	731 995	722 639	296 664
广 西	Guangxi	21 369	437 996	348 938	260 028
海 南	Hainan	4 721	31 652	66 831	14 029
重 庆	Chongqing	9 532	494 415	247 905	179 842
四 川	Sichuan	19 761	746 363	408 796	268 866
贵 州	Guizhou	24 467	778 581	446 899	262 072
云 南	Yunnan	15 958	612 954	316 711	354 770
西 藏	Tibet	115	1 256	2 491	1 005
陕 西	Shaanxi	16 279	707 146	552 078	468 507
甘 肃	Gansu	12 677	472 798	306 404	174 556
青 海	Qinghai	5 621	130 818	93 180	149 239
宁 夏	Ningxia	8 909	368 201	357 233	212 914
新 疆	Xinjiang	18 464	738 566	573 017	663 257

各地区工业废气排放及处理情况（二）
Discharge and Treatment of Industrial Waste Gas by Region（2）

（2013）

单位：吨 (ton)

| 年　份
地　区 | Year
Region | 工业废气中污染物产生量
Volume of Pollutants Emission in the Industrial Waste Gas | | |
		二氧化硫 Sulphur Dioxide	氮氧化物 Nitrogen Oxide	烟（粉）尘 Soot and Dust
2012		59 328 890	17 932 874	777 994 342
2013		63 203 149	19 040 550	759 408 732
北　京	Beijing	156 984	95 394	4 229 328
天　津	Tianjin	1 425 856	349 931	6 703 176
河　北	Hebei	3 368 736	1 305 414	52 508 510
山　西	Shanxi	3 781 766	1 126 855	35 466 745
内蒙古	Inner Mongolia	4 357 987	1 199 330	47 034 874
辽　宁	Liaoning	2 200 800	757 738	29 086 308
吉　林	Jilin	581 214	399 551	13 848 050
黑龙江	Heilongjiang	549 350	463 517	17 441 934
上　海	Shanghai	538 840	328 829	6 608 157
江　苏	Jiangsu	3 198 660	1 362 931	40 764 970
浙　江	Zhejiang	1 701 631	755 404	25 826 955
安　徽	Anhui	3 016 006	804 589	40 758 228
福　建	Fujian	961 931	491 625	19 458 436
江　西	Jiangxi	2 591 327	385 691	17 867 212
山　东	Shandong	5 714 807	1 372 935	63 837 220
河　南	Henan	3 214 999	1 152 166	57 570 010
湖　北	Hubei	2 812 878	486 058	25 097 688
湖　南	Hunan	1 986 370	573 543	26 080 125
广　东	Guangdong	2 775 251	955 790	23 029 751
广　西	Guangxi	1 691 921	404 415	20 658 217
海　南	Hainan	133 266	76 811	1 725 152
重　庆	Chongqing	1 404 981	302 028	20 346 826
四　川	Sichuan	2 074 960	475 063	27 025 696
贵　州	Guizhou	2 625 824	563 703	30 701 470
云　南	Yunnan	2 398 990	340 903	22 658 001
西　藏	Tibet	1 304	2 494	353 738
陕　西	Shaanxi	2 098 764	905 565	24 173 238
甘　肃	Gansu	2 610 169	373 634	14 772 457
青　海	Qinghai	244 830	95 324	4 571 598
宁　夏	Ningxia	1 475 251	539 116	19 107 341
新　疆	Xinjiang	1 507 496	594 202	20 097 320

各地区工业废气排放及处理情况（三）
Discharge and Treatment of Industrial Waste Gas by Region（3）
（2013）

单位：套 （set）

年 份 地 区	Year Region	废气治理 设施数 Facilities for Treatment of Waste Gas	脱硫设施数 Desulfurization Facilities	脱硝设施数 Denitrification Facilities	除尘设施数 Dedusting Facilities	废气治理设施处理 能力（标态）/ （万米³/小时） Capacity of Facilities for Treatment of Waste Gas （10 000 cu.m/hour）
	2012	225 913	20 968	832	177 985	1 649 353
	2013	234 316	21 677	2 152	180 154	1 435 110
北 京	Beijing	3 316	436	58	2 309	10 729
天 津	Tianjin	4 262	1 027	24	2 452	21 118
河 北	Hebei	18 172	1 327	120	15 879	179 690
山 西	Shanxi	14 353	3 446	95	10 364	97 187
内蒙古	Inner Mongolia	7 590	502	41	6 922	73 372
辽 宁	Liaoning	11 601	969	52	10 229	78 778
吉 林	Jilin	3 594	96	25	3 307	24 027
黑龙江	Heilongjiang	4 882	189	23	4 490	28 800
上 海	Shanghai	4 896	238	48	2 651	29 883
江 苏	Jiangsu	17 964	1 676	179	11 297	104 344
浙 江	Zhejiang	17 512	1 440	191	11 154	57 354
安 徽	Anhui	6 100	346	77	5 134	37 496
福 建	Fujian	8 112	174	66	6 922	35 192
江 西	Jiangxi	5 673	385	36	4 727	27 464
山 东	Shandong	16 436	2 752	113	12 588	120 289
河 南	Henan	10 524	1 035	80	9 147	67 321
湖 北	Hubei	6 640	435	59	5 538	36 296
湖 南	Hunan	5 711	462	67	4 949	30 196
广 东	Guangdong	19 392	1 808	199	9 920	80 528
广 西	Guangxi	6 301	481	53	5 446	37 486
海 南	Hainan	823	17	8	757	3 743
重 庆	Chongqing	4 439	227	38	3 275	23 084
四 川	Sichuan	9 246	658	208	7 637	48 879
贵 州	Guizhou	3 464	441	85	2 818	14 965
云 南	Yunnan	7 346	363	47	5 875	36 227
西 藏	Tibet	255	0	0	166	352
陕 西	Shaanxi	3 835	266	75	3 281	34 907
甘 肃	Gansu	3 424	114	38	3 225	23 726
青 海	Qinghai	1 425	23	4	1 321	7 414
宁 夏	Ningxia	1 700	87	26	1 568	18 600
新 疆	Xinjiang	5 328	257	17	4 806	45 664

各地区工业废气排放及处理情况（四）
Discharge and Treatment of Industrial Waste Gas by Region（4）
（2013）

单位：万元 （10 000 yuan）

年 份 Year 地 区 Region	废气治理设施 运行费用 Annul Expenditure for Operation	脱硫设施运行费用 Annul Expenditure for Desulfurization Facilities	脱硝设施运行费用 Annul Expenditure for Denitrification Facilities	除尘设施运行费用 Annul Expenditure for Dedusting Facilities
2012	14 522 519. 8	5 404 357. 7	660 273. 6	6 474 310. 3
2013	14 977 779. 0	6 118 246. 4	1 455 422. 3	6 179 677. 1
北 京 Beijing	90 066. 3	27 586. 6	22 487. 5	26 160. 4
天 津 Tianjin	256 505. 5	124 620. 2	23 760. 2	79 565. 3
河 北 Hebei	1 331 602. 6	388 177. 7	102 087. 0	758 041. 1
山 西 Shanxi	861 522. 3	356 124. 7	115 462. 3	363 471. 0
内蒙古 Inner Mongolia	655 995. 2	324 440. 2	30 173. 3	264 156. 8
辽 宁 Liaoning	569 595. 5	185 599. 3	22 547. 4	332 518. 4
吉 林 Jilin	193 931. 1	61 430. 0	9 385. 8	93 541. 4
黑龙江 Heilongjiang	148 971. 3	45 977. 9	4 188. 0	86 454. 4
上 海 Shanghai	471 487. 1	136 694. 3	47 862. 3	144 172. 5
江 苏 Jiangsu	1 269 571. 6	511 141. 4	182 648. 5	413 039. 4
浙 江 Zhejiang	945 787. 3	446 203. 3	132 125. 3	269 417. 9
安 徽 Anhui	558 278. 0	208 002. 8	61 875. 9	274 930. 1
福 建 Fujian	410 417. 5	141 281. 5	56 621. 0	171 688. 3
江 西 Jiangxi	377 429. 6	171 484. 3	19 402. 3	169 403. 8
山 东 Shandong	1 308 651. 9	587 992. 5	82 427. 6	543 949. 2
河 南 Henan	616 656. 9	270 165. 0	56 400. 3	268 444. 7
湖 北 Hubei	458 055. 8	172 746. 9	49 014. 5	218 369. 8
湖 南 Hunan	355 282. 1	136 306. 3	41 787. 7	159 578. 9
广 东 Guangdong	1 060 764. 1	515 384. 9	183 820. 0	248 343. 9
广 西 Guangxi	303 377. 0	108 193. 0	13 560. 8	162 054. 9
海 南 Hainan	36 377. 4	13 640. 6	1 370. 0	15 776. 1
重 庆 Chongqing	231 695. 8	95 064. 8	27 362. 8	93 946. 9
四 川 Sichuan	473 889. 5	184 774. 8	23 805. 4	212 083. 5
贵 州 Guizhou	417 579. 9	241 244. 7	19 375. 7	127 266. 5
云 南 Yunnan	396 752. 2	145 222. 4	18 368. 9	191 003. 1
西 藏 Tibet	1 964. 4	…	…	1 376. 4
陕 西 Shaanxi	319 056. 6	135 668. 7	41 160. 1	112 246. 0
甘 肃 Gansu	230 589. 3	112 945. 0	22 590. 1	89 231. 9
青 海 Qinghai	97 044. 8	12 423. 4	705. 4	77 651. 2
宁 夏 Ningxia	243 668. 2	124 425. 4	41 216. 8	75 249. 3
新 疆 Xinjiang	285 212. 2	133 283. 8	1 829. 4	136 544. 0

各地区一般工业固体废物产生及处置利用情况
Generation and Utilization of Industrial Solid Wastes by Region
（2013）

单位：万吨 （10 000 tons）

年 份 地 区 Year Region	一般工业固体废物产生量 Industrial Solid Wastes Generated	一般工业固体废物综合利用量 Industrial Solid Wastes Utilized	一般工业固体废物处置量 Industrial Solid Wastes Disposed	一般工业固体废物贮存量 Stock of Industrial Solid Wastes	一般工业固体废物倾倒丢弃量 Industrial Solid Wastes Discharged	一般工业固体废物综合利用率/% Industrial Solid Wastes Utilization Rate（%）
2012	329 044	202 462	70 745	59 786	144	61.0
2013	327 702	205 916	82 969	42 634	129	62.2
北 京 Beijing	1 044	904	140	0	0	86.6
天 津 Tianjin	1 592	1 582	10	0	0	99.4
河 北 Hebei	43 289	18 356	23 429	1 847	0	42.1
山 西 Shanxi	30 520	19 815	8 187	2 749	0	64.6
内蒙古 Inner Mongolia	20 081	9 984	8 296	2 233	1	48.7
辽 宁 Liaoning	26 759	11 742	11 289	3 883	9	43.8
吉 林 Jilin	4 591	3 712	522	522	0	78.1
黑龙江 Heilongjiang	6 094	4 145	419	1 557	0	67.7
上 海 Shanghai	2 054	1 995	58	5	0	96.9
江 苏 Jiangsu	10 856	10 502	287	197	0	95.7
浙 江 Zhejiang	4 300	4 091	177	49	0	94.8
安 徽 Anhui	11 937	10 462	1 374	933	0	84.0
福 建 Fujian	8 535	7 544	962	51	0	88.2
江 西 Jiangxi	11 518	6 431	397	4 738	2	55.7
山 东 Shandong	18 172	17 134	788	436	0	93.4
河 南 Henan	16 270	12 466	3 470	450	0	76.1
湖 北 Hubei	8 181	6 196	1 646	412	1	75.1
湖 南 Hunan	7 806	5 011	1 964	888	1	63.7
广 东 Guangdong	5 912	5 024	732	169	2	84.8
广 西 Guangxi	7 676	5 425	1 609	1 198	0	66.4
海 南 Hainan	415	271	46	98	0	65.4
重 庆 Chongqing	3 162	2 695	415	79	11	84.2
四 川 Sichuan	14 007	5 780	5 301	3 107	7	41.0
贵 州 Guizhou	8 194	4 160	1 607	2 470	19	50.4
云 南 Yunnan	16 040	8 414	4 834	2 863	49	52.2
西 藏 Tibet	362	5	26	346	0	1.5
陕 西 Shaanxi	7 491	4 758	1 622	1 131	0	63.4
甘 肃 Gansu	5 907	3 300	1 859	768	0	55.7
青 海 Qinghai	12 377	6 798	7	5 602	0	54.8
宁 夏 Ningxia	3 277	2 398	613	292	0	73.1
新 疆 Xinjiang	9 283	4 814	886	3 562	27	51.8

各地区工业危险废物产生及处置利用情况
Generation and Utilization of Hazardous Wastes by Region
（2013）

单位：吨 (ton)

年　份 地　区 Year Region		危险废物产生量 Hazardous Wastes Generated	危险废物综合利用量 Hazardous Wastes Utilized	危险废物处置量 Hazardous Wastes Disposed	危险废物贮存量 Stock of Hazardous Wastes	危险废物倾倒丢弃量 Hazardous Wastes Discharged	危险废物处置利用率/% Hazardous Wastes Disposition and Utilization Rate（%）
	2012	34 652 427	20 046 450	6 982 060	8 469 139	16	76.1
	2013	31 568 910	17 000 926	7 012 005	8 108 791	0	74.8
北　京	Beijing	132 177	57 878	68 126	6 270	0	95.3
天　津	Tianjin	118 136	35 285	82 853	37	0	100
河　北	Hebei	644 962	387 521	254 503	3 083	0	99.5
山　西	Shanxi	194 895	141 985	51 354	2 033	0	99.0
内蒙古	Inner Mongolia	1 177 050	593 513	266 381	319 024	0	72.9
辽　宁	Liaoning	1 046 400	766 342	397 540	53 879	0	95.6
吉　林	Jilin	740 346	393 856	346 523	49	0	100
黑龙江	Heilongjiang	225 809	43 565	176 765	5 679	0	97.5
上　海	Shanghai	543 108	287 633	256 423	3 882	0	99.3
江　苏	Jiangsu	2 180 856	1 073 407	1 091 659	37 605	0	98.3
浙　江	Zhejiang	1 046 829	319 657	709 886	43 266	0	96.0
安　徽	Anhui	558 402	489 103	113 280	2 806	0	99.5
福　建	Fujian	212 350	71 321	92 380	49 214	0	76.9
江　西	Jiangxi	441 671	364 325	74 015	11 959	0	97.3
山　东	Shandong	5 090 742	4 424 234	618 176	87 873	0	98.3
河　南	Henan	592 673	421 533	170 971	504	0	99.9
湖　北	Hubei	596 342	305 024	198 963	99 365	0	83.5
湖　南	Hunan	2 842 859	2 428 499	207 904	290 683	0	90.1
广　东	Guangdong	1 331 151	742 188	586 364	5 871	0	99.6
广　西	Guangxi	962 271	796 994	67 194	149 538	0	85.2
海　南	Hainan	23 816	10	22 441	2 360	0	90.5
重　庆	Chongqing	466 801	327 787	132 333	10 279	0	97.8
四　川	Sichuan	417 167	159 072	253 971	6 935	0	98.3
贵　州	Guizhou	357 827	268 956	52 364	37 952	0	89.4
云　南	Yunnan	1 937 582	981 601	286 972	692 392	0	64.7
西　藏	Tibet	0	0	0	0	0	…
陕　西	Shaanxi	303 388	95 703	119 042	92 099	0	70.0
甘　肃	Gansu	307 184	84 873	137 770	102 794	0	68.4
青　海	Qinghai	3 998 501	696 646	80 989	3 250 394	0	19.3
宁　夏	Ningxia	50 880	39 330	2 714	8 882	0	82.6
新　疆	Xinjiang	3 026 736	203 088	92 146	2 732 084	0	9.8

各地区汇总工业企业概况（一）

Summarization of Industrial Enterprises Investigeted by Region（1）

（2013）

年 份 地 区	Year Region	汇总工业 企业数/个 Number of Industrial Enterprises Investigated （unit）	工业总产值 （现价）/ 亿元 Gross Industrial Output Value （current rate） （100 million yuan）	工业 炉窑数/ 台 Number of Industrial Furnaces （unit）	工业 锅炉数/ 台 Number of Industrial Boilers （unit）	35 蒸吨 及以上的 Beyond 35 tons of steam	20（含）～ 35 蒸 吨的 Between 20 and 35 tons of steam	10（含）～ 20 蒸 吨的 Between 10 and 35 tons of steam	10 蒸吨 以下的 Below 10 tons of steam
	2012	147 996	368 365.3	99 205	103 469	13 683	7 400	13 726	68 660
	2013	147 657	375 719.7	94 149	102 466	14 044	7 447	13 816	67 159
北 京	Beijing	932	6 896.7	242	1 962	329	239	402	992
天 津	Tianjin	2 657	13 003.0	866	2 484	424	302	369	1 389
河 北	Hebei	10 488	20 823.9	6 696	6 481	774	367	863	4 477
山 西	Shanxi	6 258	11 721.7	4 914	6 057	722	450	638	4 247
内蒙古	Inner Mongolia	3 110	8 227.2	3 119	4 342	889	444	827	2 182
辽 宁	Liaoning	6 305	18 689.8	6 449	6 933	1 219	955	1 243	3 516
吉 林	Jilin	1 483	5 601.1	583	2 843	612	237	420	1 574
黑龙江	Heilongjiang	1 884	6 398.4	3 518	4 885	764	510	802	2 809
上 海	Shanghai	2 089	18 255.7	1 179	2 019	153	65	153	1 648
江 苏	Jiangsu	10 743	40 809.9	4 416	6 774	979	257	688	4 850
浙 江	Zhejiang	13 211	26 306.2	4 316	7 710	660	268	918	5 864
安 徽	Anhui	8 365	11 646.8	4 614	2 839	269	94	300	2 176
福 建	Fujian	5 755	10 180.4	2 597	3 359	170	118	411	2 660
江 西	Jiangxi	5 115	6 438.1	3 376	2 076	131	59	216	1 670
山 东	Shandong	7 708	39 058.7	4 965	6 512	1 829	484	851	3 348
河 南	Henan	6 494	13 292.3	5 364	4 025	555	250	484	2 736
湖 北	Hubei	3 590	11 727.7	2 298	2 413	243	130	285	1 755
湖 南	Hunan	4 668	9 704.0	4 519	2 151	182	76	218	1 675
广 东	Guangdong	14 959	35 685.6	5 071	7 610	529	297	1 134	5 650
广 西	Guangxi	3 532	7 023.2	2 110	1 785	373	183	187	1 042
海 南	Hainan	458	1 365.6	206	293	36	21	25	211
重 庆	Chongqing	3 145	7 667.5	1 768	1 455	116	62	120	1 157
四 川	Sichuan	7 510	11 459.3	6 886	3 476	257	181	361	2 677
贵 州	Guizhou	3 452	4 196.1	3 029	1 358	130	60	112	1 056
云 南	Yunnan	4 143	6 647.1	2 036	1 592	212	149	188	1 043
西 藏	Tibet	95	57.9	22	35	0	0	2	33
陕 西	Shaanxi	3 650	7 984.1	2 433	2 383	326	197	420	1 440
甘 肃	Gansu	2 341	3 859.4	2 574	2 237	230	260	435	1 312
青 海	Qinghai	603	1 435.9	744	490	51	57	74	308
宁 夏	Ningxia	819	2 236.3	1 443	1 138	256	146	207	529
新 疆	Xinjiang	2 095	7 320.1	1 796	2 749	624	529	463	1 133

各地区汇总工业企业概况（二）

Summarization of Industrial Enterprises Investigeted by Region（2）

（2013）

年 份 地 区	Year Region	工业用水总量/万吨 Quantity of Water Used by Industry （10 000 tons）	取水量 Fresh Water	重复用水量 Recycled Water	工业煤炭消耗量/万吨 Total Amount of Industrial Coal Consumed （10 000 tons）	燃料煤 Coal Used as Fuel
	2012	38 493 488	5 020 781	33 472 707	382 786	290 056
	2013	38 667 789	4 987 373	33 680 415	401 728	302 513
北 京	Beijing	307 114	19 460	287 654	1 535	1 462
天 津	Tianjin	932 262	40 010	892 253	4 879	4 588
河 北	Hebei	4 222 295	253 072	3 969 224	29 622	19 015
山 西	Shanxi	1 663 930	125 265	1 538 665	32 329	19 306
内蒙古	Inner Mongolia	1 329 603	90 296	1 239 307	33 471	24 291
辽 宁	Liaoning	2 292 411	285 828	2 006 583	17 042	13 518
吉 林	Jilin	1 030 371	97 300	933 071	10 408	7 617
黑龙江	Heilongjiang	852 267	182 851	669 416	12 480	8 592
上 海	Shanghai	798 569	71 480	727 088	5 362	4 111
江 苏	Jiangsu	3 308 733	511 146	2 797 587	27 494	23 851
浙 江	Zhejiang	2 243 209	775 845	1 467 364	14 493	13 543
安 徽	Anhui	1 435 194	126 589	1 308 606	14 038	11 672
福 建	Fujian	680 075	90 663	589 413	7 950	6 788
江 西	Jiangxi	636 735	98 381	538 354	6 610	5 149
山 东	Shandong	3 447 685	316 902	3 130 784	32 003	24 412
河 南	Henan	2 427 541	221 204	2 206 337	29 443	18 666
湖 北	Hubei	1 355 173	224 746	1 130 427	9 603	7 367
湖 南	Hunan	926 884	198 605	728 279	8 562	6 730
广 东	Guangdong	1 380 212	290 796	1 089 416	17 829	17 054
广 西	Guangxi	1 502 946	316 204	1 186 742	6 789	5 489
海 南	Hainan	239 926	11 519	228 408	1 034	1 033
重 庆	Chongqing	626 175	130 733	495 442	4 505	3 700
四 川	Sichuan	964 131	120 655	843 476	8 078	6 204
贵 州	Guizhou	1 133 855	56 265	1 077 590	8 851	7 813
云 南	Yunnan	744 047	107 901	636 146	7 696	5 819
西 藏	Tibet	2 415	831	1 584	43	40
陕 西	Shaanxi	585 847	59 581	526 266	20 872	10 795
甘 肃	Gansu	281 285	45 103	236 183	6 150	5 460
青 海	Qinghai	127 088	14 446	112 643	1 753	1 358
宁 夏	Ningxia	469 791	36 551	433 241	8 710	7 076
新 疆	Xinjiang	720 017	67 149	652 868	12 096	9 993

各地区汇总工业企业概况（三）

Summarization of Industrial Enterprises Investigeted by Region（3）

（2013）

年　份 地　区	Year Region	燃料油消耗量/ 万吨 Total Amount of Fuel Oil Consumed （10 000 tons）	焦炭消耗量/ 万吨 Total Amount of Coke Consumed （10 000 tons）	天然气消耗量/ 亿米^3 Total Amount of Natural Gas Consumed （100 million cu.m）	其他燃料消耗量/ 万吨标煤 Total Amount of other Fuel Consumed （10 000 tons of standard coal）
	2012	1 912	28 442	2 126	12 479
	2013	1 553	29 986	1 044	10 765
北　京	Beijing	2	0	43	113
天　津	Tianjin	9	577	19	184
河　北	Hebei	14	6 989	49	1 141
山　西	Shanxi	4	1 690	16	568
内蒙古	Inner Mongolia	5	1 060	24	168
辽　宁	Liaoning	129	2 360	31	759
吉　林	Jilin	18	449	17	128
黑龙江	Heilongjiang	28	320	28	204
上　海	Shanghai	47	634	50	193
江　苏	Jiangsu	589	2 336	94	666
浙　江	Zhejiang	99	401	59	264
安　徽	Anhui	11	1 007	11	737
福　建	Fujian	48	634	42	180
江　西	Jiangxi	14	945	4	220
山　东	Shandong	102	2 644	59	1 027
河　南	Henan	30	1 004	42	444
湖　北	Hubei	30	418	24	118
湖　南	Hunan	7	652	9	227
广　东	Guangdong	282	467	92	1 078
广　西	Guangxi	11	892	10	953
海　南	Hainan	3	0	45	191
重　庆	Chongqing	1	298	38	29
四　川	Sichuan	2	1 063	58	244
贵　州	Guizhou	3	301	6	39
云　南	Yunnan	11	1 033	29	196
西　藏	Tibet	14	0	0	0
陕　西	Shaanxi	12	274	21	113
甘　肃	Gansu	12	532	7	122
青　海	Qinghai	1	169	26	15
宁　夏	Ningxia	2	184	4	18
新　疆	Xinjiang	15	653	88	427

各地区工业污染防治投资情况（一）

Treatment Investment for Industrial Pollution by Region（1）

（2013）

单位：个 (unit)

年 份 地 区	Year Region	汇总工业 企业数 Number of Industrial Enterprises Collected	本年施工 项目总数 Numer of Projects under Construction	工业废水 治理项目 Treatment of Waste Water	工业废气 治理项目 Treatment of Waste Gas	脱硫治理 项目 Treatment of Desulfurization	脱硝治理 项目 Treatment of Denitration	工业固体 废物治理 项目 Treatment of Solid Wastes	噪声治理 项目 Treatment of Noise Pollution	其他治理 项目 Treatment of Other Pollution
	2012	6 203	5 390	1 801	2 144	501	293	256	105	1 084
	2013	5 839	5 593	1 490	2 785	629	823	256	91	971
北 京	Beijing	38	45	7	24	1	8	0	1	13
天 津	Tianjin	81	66	14	34	16	3	0	0	18
河 北	Hebei	329	350	41	233	64	36	52	2	22
山 西	Shanxi	339	323	53	161	53	36	32	6	71
内蒙古	Inner Mongolia	136	167	26	115	37	43	11	2	13
辽 宁	Liaoning	93	98	3	60	12	22	8	3	24
吉 林	Jilin	61	44	13	25	11	7	0	1	5
黑龙江	Heilongjiang	85	80	8	63	17	26	1	0	8
上 海	Shanghai	116	112	30	39	2	0	2	6	35
江 苏	Jiangsu	443	383	169	168	29	46	1	7	38
浙 江	Zhejiang	751	893	332	297	34	42	9	6	249
安 徽	Anhui	142	132	28	85	16	52	2	3	14
福 建	Fujian	275	239	83	74	14	17	50	4	28
江 西	Jiangxi	82	97	17	60	15	16	5	1	14
山 东	Shandong	506	571	86	374	114	97	6	11	94
河 南	Henan	185	138	37	80	18	37	5	1	15
湖 北	Hubei	147	113	28	59	17	25	2	1	23
湖 南	Hunan	156	115	35	64	13	22	3	1	12
广 东	Guangdong	544	510	152	200	52	30	17	12	129
广 西	Guangxi	159	143	52	55	10	26	6	3	27
海 南	Hainan	14	13	2	10	2	7	0	0	1
重 庆	Chongqing	108	51	19	25	3	5	0	1	6
四 川	Sichuan	238	201	73	89	15	45	5	4	30
贵 州	Guizhou	135	82	20	52	4	41	4	1	5
云 南	Yunnan	282	299	72	156	21	32	27	5	39
西 藏	Tibet	30	12	7	1	0	0	1	1	2
陕 西	Shaanxi	131	111	25	64	9	42	1	5	16
甘 肃	Gansu	66	54	25	25	2	19	1	0	3
青 海	Qinghai	26	20	1	14	1	5	0	0	5
宁 夏	Ningxia	71	69	15	43	13	20	5	1	5
新 疆	Xinjiang	70	62	17	36	14	16	0	2	7

各地区工业污染防治投资情况（二）

Treatment Investment for Industrial Pollution by Region（2）

（2013）

单位：个 　　　（unit）

年 份 地 区 Year Region	本年竣工项目总数 Number of Projects Completed	工业废水治理项目 Treatment of Waste Water	工业废气治理项目 Treatment of Waste Gas	脱硫治理项目 Treatment of Desulfurization	脱硝治理项目 Treatment of Denitration	工业固体废物治理项目 Treatment of Solid Wastes	噪声治理项目 Treatment of Noise Pollution	其他治理项目 Treatment of Other Pollution
2012	5 565	2 017	2 153	513	209	229	113	1 053
2013	5 958	1 649	3 004	716	840	255	93	957
北 京 Beijing	45	9	25	1	7	0	1	10
天 津 Tianjin	66	13	39	18	5	0	0	14
河 北 Hebei	390	50	262	68	39	52	2	24
山 西 Shanxi	360	67	184	66	48	30	5	74
内蒙古 Inner Mongolia	158	26	101	33	35	15	2	14
辽 宁 Liaoning	137	13	83	19	26	10	3	28
吉 林 Jilin	60	12	33	14	11	1	1	13
黑龙江 Heilongjiang	80	11	59	16	22	2	0	8
上 海 Shanghai	115	28	40	5	1	3	6	38
江 苏 Jiangsu	430	174	204	32	51	3	7	42
浙 江 Zhejiang	775	286	269	38	44	8	7	205
安 徽 Anhui	158	42	96	16	59	2	3	15
福 建 Fujian	256	94	79	18	30	48	5	30
江 西 Jiangxi	97	22	57	12	16	4	1	13
山 东 Shandong	552	106	338	108	68	5	10	93
河 南 Henan	187	62	95	24	40	9	3	18
湖 北 Hubei	148	42	83	34	23	2	0	21
湖 南 Hunan	138	37	79	17	26	4	2	16
广 东 Guangdong	596	184	257	61	39	18	12	125
广 西 Guangxi	157	53	71	11	29	5	3	25
海 南 Hainan	8	2	4	2	2	0	0	2
重 庆 Chongqing	70	27	35	2	9	0	1	7
四 川 Sichuan	216	85	93	19	42	3	4	31
贵 州 Guizhou	117	34	68	7	51	3	2	10
云 南 Yunnan	290	65	161	24	29	23	6	35
西 藏 Tibet	27	16	7	3	0	1	1	2
陕 西 Shaanxi	114	39	51	9	31	1	4	19
甘 肃 Gansu	59	23	27	2	20	0	0	9
青 海 Qinghai	25	1	19	2	7	0	0	5
宁 夏 Ningxia	65	8	51	16	22	2	0	4
新 疆 Xinjiang	62	18	34	19	8	1	2	7

各地区工业污染防治投资情况（三）
Treatment Investment for Industrial Pollution by Region（3）
（2013）

单位：万元 (10 000 yuan)

年 份 Year 地 区 Region	施工项目本年完成投资 Investment Completed in the Treatment of Industrial Pollution This Year	工业废水治理项目 Treatment of Waste Water	工业废气治理项目 Treatment of Waste Gas	脱硫治理项目 Treatment of Desulfurization	脱硝治理项目 Treatment of Denitration	工业固体废物治理项目 Treatment of Solid Wastes	噪声治理项目 Treatment of Noise Pollution	其他治理项目 Treatment of Other Pollution
2012	5 004 572. 7	1 403 447. 5	2 577 138. 7	801 975. 2	909 416. 3	247 499. 3	11 626. 8	764 860. 4
2013	8 496 646. 6	1 248 822. 3	6 409 108. 7	1 628 515. 3	3 856 113. 5	140 480. 1	17 627. 9	680 607. 6
北 京 Beijing	42 767. 7	8 427. 9	31 732. 1	50. 0	14 173. 3	…	35. 0	2 572. 7
天 津 Tianjin	148 365. 8	6 435. 8	74 491. 5	32 222. 2	28 973. 8	…	24. 7	67 413. 8
河 北 Hebei	511 769. 2	59 636. 9	440 113. 2	139 600. 1	207 325. 1	513. 2	130. 0	11 375. 9
山 西 Shanxi	555 609. 5	43 014. 5	416 619. 0	85 059. 6	310 237. 0	22 519. 4	593. 3	72 863. 4
内蒙古 Inner Mongolia	626 746. 3	53 676. 6	477 778. 8	144 257. 1	313 303. 7	27 483. 9	101. 0	67 706. 0
辽 宁 Liaoning	276 907. 5	25 596. 3	237 364. 3	54 824. 1	142 315. 5	1 426. 5	549. 2	11 971. 3
吉 林 Jilin	93 730. 6	9 484. 6	81 643. 3	28 350. 6	38 321. 8	988. 0	8. 4	1 606. 3
黑龙江 Heilongjiang	206 987. 6	17 508. 7	184 618. 5	99 593. 7	66 860. 8	1 192. 7	…	3 667. 6
上 海 Shanghai	52 076. 6	7 809. 1	19 056. 0	1 300. 0	15 000. 0	216. 3	275. 3	24 720. 0
江 苏 Jiangsu	593 775. 7	102 500. 6	456 159. 9	103 322. 5	251 654. 7	16 534. 2	460. 6	18 120. 4
浙 江 Zhejiang	576 644. 6	150 634. 2	312 508. 2	40 645. 6	190 968. 4	547. 0	1 034. 6	111 920. 6
安 徽 Anhui	233 195. 2	19 127. 3	204 806. 8	32 190. 0	165 588. 5	102. 0	146. 0	9 013. 1
福 建 Fujian	383 963. 6	139 470. 6	209 778. 0	24 186. 4	106 726. 1	7 699. 9	571. 9	26 443. 2
江 西 Jiangxi	155 192. 2	32 334. 1	104 202. 1	30 309. 6	61 618. 5	461. 5	25. 1	18 169. 4
山 东 Shandong	843 493. 5	101 280. 7	701 239. 8	209 801. 7	341 106. 8	2 775. 0	5 767. 8	32 430. 2
河 南 Henan	439 719. 7	48 111. 8	349 728. 8	66 584. 5	261 822. 0	22 011. 0	185. 0	19 683. 1
湖 北 Hubei	251 745. 0	15 872. 6	216 672. 9	32 989. 9	168 321. 8	1 700. 0	52. 0	17 447. 5
湖 南 Hunan	233 654. 5	53 022. 3	143 636. 3	39 949. 6	67 694. 7	3 763. 8	56. 0	33 176. 1
广 东 Guangdong	324 634. 4	43 911. 7	262 342. 5	39 364. 3	163 083. 9	2 725. 5	508. 8	15 146. 0
广 西 Guangxi	183 218. 2	66 235. 2	110 217. 0	20 976. 2	76 120. 8	540. 0	276. 0	5 950. 0
海 南 Hainan	35 094. 2	572. 0	28 497. 2	10 143. 0	17 097. 6	…	…	6 025. 0
重 庆 Chongqing	78 880. 2	6 399. 1	71 855. 8	8 667. 4	56 579. 0	…	44. 0	581. 4
四 川 Sichuan	188 391. 6	29 791. 4	148 926. 4	58 164. 5	75 366. 1	1 406. 6	2 215. 5	6 051. 7
贵 州 Guizhou	195 561. 9	22 867. 0	170 471. 9	1 739. 4	158 604. 8	197. 6	542. 7	1 482. 8
云 南 Yunnan	238 930. 4	35 224. 1	163 633. 1	40 942. 2	98 122. 7	8 768. 3	3 115. 4	28 189. 5
西 藏 Tibet	9 889. 0	8 450. 0	466. 0	145. 0	…	845. 0	15. 0	113. 0
陕 西 Shaanxi	417 561. 8	60 626. 1	321 029. 3	112 657. 7	204 719. 7	2 300. 0	850. 9	32 755. 5
甘 肃 Gansu	182 143. 8	21 782. 7	138 314. 5	29 515. 0	101 462. 9	400. 0	…	21 646. 6
青 海 Qinghai	30 456. 0	2 985. 0	27 174. 0	1 129. 1	11 135. 1	…	…	297. 0
宁 夏 Ningxia	165 486. 4	18 947. 0	138 888. 9	22 351. 6	102 874. 3	5 175. 7	5. 5	2 469. 2
新 疆 Xinjiang	220 054. 2	37 086. 5	165 142. 8	117 482. 9	38 934. 0	8 187. 0	38. 2	9 599. 7

158

各地区工业污染防治投资情况（四）
Treatment Investment for Industrial Pollution by Region（4）
（2013）

年　份　　Year 地　区　　Region	本年竣工项目新增设计处理能力 Capacity for Treatment of Industrial Pollution New-added		
	治理废水/ （万吨/日） Treatment of Waste Water （10 000 tons/day）	治理废气（标态）/ （万米³/小时） Treatment of Waste Gas （10 000 cu.m/hour）	治理固体废物/ （万吨/日） Treatment of Solid Wastes （10 000 tons/day）
2012	485. 8	50 060. 7	71. 8
2013	345. 9	107 944. 8	15. 1
北　京　Beijing	3. 2	219. 9	…
天　津　Tianjin	3. 4	1 221. 1	…
河　北　Hebei	7. 3	9 944. 2	2. 5
山　西　Shanxi	18. 7	7 210. 4	1. 0
内蒙古　Inner Mongolia	2. 7	9 445. 8	0. 6
辽　宁　Liaoning	7. 0	3 768. 4	0. 1
吉　林　Jilin	3. 6	93. 4	…
黑龙江　Heilongjiang	2. 4	1 961. 3	…
上　海　Shanghai	2. 8	459. 6	…
江　苏　Jiangsu	23. 5	9 024. 5	0. 1
浙　江　Zhejiang	36. 7	4 416. 1	…
安　徽　Anhui	16. 7	6 154. 3	0. 2
福　建　Fujian	28. 4	3 031. 4	0. 6
江　西　Jiangxi	2. 6	2 140. 5	0. 2
山　东　Shandong	36. 5	12 468. 7	0. 2
河　南　Henan	10. 9	6 168. 2	0. 6
湖　北　Hubei	5. 6	3 299. 4	0. 1
湖　南　Hunan	22. 1	1 991. 5	0. 5
广　东　Guangdong	20. 0	4 339. 7	0. 7
广　西　Guangxi	25. 4	2 723. 5	0. 1
海　南　Hainan	…	491. 3	…
重　庆　Chongqing	0. 5	2 333. 5	…
四　川　Sichuan	22. 5	3 041. 9	0. 2
贵　州　Guizhou	5. 6	1 471. 2	…
云　南　Yunnan	17. 1	1 192. 7	7. 2
西　藏　Tibet	0. 7	…	…
陕　西　Shaanxi	7. 3	5 894. 0	…
甘　肃　Gansu	2. 6	965. 3	…
青　海　Qinghai	…	400. 5	…
宁　夏　Ningxia	5. 1	1 721. 4	…
新　疆　Xinjiang	4. 7	351. 0	0. 1

各地区农业污染排放情况（一）
Discharge of Agricultural Pollution by Region

（2013）

单位：万吨 (10 000 tons)

| 年 份　Year | 农业污染物排放（流失）总量 Total Amount of Discharge of Agricultural Pollutant | | | |
地 区　Region	化学需氧量 COD	氨氮 Ammonial Nitrogen	总氮 Total Nitrogen	总磷 Total Phosphorus
2012	1 153.7	80.6	437.5	51.2
2013	1 125.8	77.9	425.0	50.0
北 京　Beijing	7.5	0.5	2.6	0.3
天 津　Tianjin	11.0	0.6	2.8	0.3
河 北　Hebei	89.5	4.3	37.2	4.4
山 西　Shanxi	17.5	1.2	8.2	0.9
内蒙古　Inner Mongolia	60.9	1.2	14.4	1.3
辽 宁　Liaoning	85.0	3.4	18.6	2.6
吉 林　Jilin	49.2	1.7	11.5	1.5
黑龙江　Heilongjiang	104.0	3.3	24.4	2.4
上 海　Shanghai	3.1	0.3	1.5	0.2
江 苏　Jiangsu	37.6	3.8	17.9	2.1
浙 江　Zhejiang	19.8	2.6	9.2	1.1
安 徽　Anhui	37.1	3.7	17.2	1.9
福 建　Fujian	20.8	3.2	9.6	1.4
江 西　Jiangxi	23.2	2.9	9.4	1.2
山 东　Shandong	129.5	7.1	56.6	6.8
河 南　Henan	78.2	6.1	39.3	4.9
湖 北　Hubei	46.1	4.5	19.2	2.5
湖 南　Hunan	55.7	6.1	19.1	2.2
广 东　Guangdong	58.1	5.5	18.4	2.8
广 西　Guangxi	21.1	2.6	11.1	1.3
海 南　Hainan	10.1	0.9	2.9	0.4
重 庆　Chongqing	12.1	1.3	5.2	0.7
四 川　Sichuan	52.9	5.6	21.7	2.4
贵 州　Guizhou	6.2	0.8	4.5	0.5
云 南　Yunnan	7.2	1.1	7.7	0.8
西 藏　Tibet	0.4	…	0.6	…
陕 西　Shaanxi	19.2	1.5	9.2	0.9
甘 肃　Gansu	14.2	0.5	4.7	0.4
青 海　Qinghai	2.2	0.1	0.7	0.1
宁 夏　Ningxia	10.1	0.2	3.0	0.3
新 疆　Xinjiang	36.4	1.3	16.5	1.3

160

各地区农业污染排放情况（二）

Discharge of Agricultural Pollution by Region（2）

（2013）

单位：吨 (ton)

年 份 地 区	Year Region	化学需氧量排放总量 Total Amount of COD Discharged	水产养殖业 Aquiculture	畜禽养殖业 Livestock and Poultry	规模化畜禽养殖场/小区 Large-Scale Farms	养殖专业户 Small-Scale Farms
	2012	11 537 999	548 418	10 989 581	3 794 478	7 195 103
	2013	11 257 575	540 078	10 717 497	3 469 983	7 247 514
北 京	Beijing	74 717	3 833	70 884	39 224	31 660
天 津	Tianjin	109 934	15 188	94 746	29 733	65 013
河 北	Hebei	894 825	9 052	885 773	435 443	450 330
山 西	Shanxi	174 570	244	174 326	73 382	100 944
内蒙古	Inner Mongolia	608 634	770	607 864	135 341	472 523
辽 宁	Liaoning	849 828	13 113	836 715	149 290	687 425
吉 林	Jilin	491 501	1 022	490 479	94 779	395 701
黑龙江	Heilongjiang	1 039 828	4 525	1 035 303	236 969	798 334
上 海	Shanghai	31 112	5 561	25 551	12 619	12 932
江 苏	Jiangsu	376 111	53 303	322 808	103 372	219 436
浙 江	Zhejiang	197 659	23 689	173 970	74 820	99 150
安 徽	Anhui	370 704	13 706	356 998	177 362	179 637
福 建	Fujian	208 054	32 827	175 227	76 109	99 118
江 西	Jiangxi	232 479	12 979	219 501	108 158	111 343
山 东	Shandong	1 294 985	8 688	1 286 298	336 870	949 428
河 南	Henan	781 510	8 523	772 988	421 533	351 454
湖 北	Hubei	461 417	138 330	323 087	141 465	181 622
湖 南	Hunan	557 025	32 209	524 816	179 590	345 226
广 东	Guangdong	581 323	114 508	466 815	150 915	315 900
广 西	Guangxi	210 770	12 672	198 098	50 070	148 027
海 南	Hainan	101 489	14 035	87 454	22 095	65 358
重 庆	Chongqing	121 034	2 821	118 213	37 589	80 624
四 川	Sichuan	528 814	8 065	520 749	154 493	366 256
贵 州	Guizhou	61 542	1 013	60 528	15 715	44 813
云 南	Yunnan	72 356	3 168	69 188	17 085	52 104
西 藏	Tibet	4 051	…	4 051	632	3 419
陕 西	Shaanxi	192 494	1 457	191 037	69 837	121 200
甘 肃	Gansu	141 636	377	141 259	15 460	125 799
青 海	Qinghai	22 066	…	22 066	7 685	14 381
宁 夏	Ningxia	101 232	2 501	98 731	23 622	75 109
新 疆	Xinjiang	363 875	1 900	361 975	78 726	283 249

各地区农业污染排放情况（三）
Discharge of Agricultural Pollution by Region（3）

（2013）

单位：吨 (ton)

年 份 地 区 Year Region	氨氮排放总量 Total Amount of Ammonia Nitrogen Discharged	种植业 Crops	水产养殖业 Aquiculture	畜禽养殖业 Livestock and Poultry	规模化畜禽养殖场/小区 Large-Scale Farms	养殖专业户 Small-Scale Farms
2012	806 216	151 566	23 400	631 250	347 878	283 372
2013	779 199	152 071	23 079	604 050	321 899	282 151
北 京 Beijing	4 504	372	188	3 944	2 772	1 171
天 津 Tianjin	5 633	488	703	4 442	1 749	2 693
河 北 Hebei	43 060	5 762	769	36 529	21 192	15 337
山 西 Shanxi	12 145	2 820	24	9 301	5 545	3 756
内蒙古 Inner Mongolia	12 024	3 352	39	8 633	3 804	4 828
辽 宁 Liaoning	33 559	2 487	907	30 165	10 316	19 849
吉 林 Jilin	16 993	2 713	65	14 215	5 242	8 973
黑龙江 Heilongjiang	33 320	7 198	590	25 532	9 112	16 420
上 海 Shanghai	3 204	908	53	2 243	1 368	875
江 苏 Jiangsu	38 204	13 723	1 732	22 749	8 934	13 815
浙 江 Zhejiang	25 906	5 720	1 479	18 707	11 611	7 096
安 徽 Anhui	36 828	9 107	472	27 249	18 021	9 228
福 建 Fujian	32 035	3 876	6 982	21 176	12 821	8 355
江 西 Jiangxi	28 698	5 590	164	22 944	16 840	6 104
山 东 Shandong	71 038	8 939	719	61 380	26 991	34 388
河 南 Henan	61 065	4 900	418	55 746	41 085	14 661
湖 北 Hubei	45 167	13 742	1 222	30 204	18 628	11 576
湖 南 Hunan	61 306	8 855	363	52 088	29 288	22 800
广 东 Guangdong	55 381	9 165	3 631	42 585	24 039	18 546
广 西 Guangxi	26 079	9 333	778	15 969	7 245	8 724
海 南 Hainan	8 656	2 414	795	5 447	2 235	3 212
重 庆 Chongqing	12 517	2 425	60	10 031	5 438	4 594
四 川 Sichuan	55 949	9 758	281	45 910	21 817	24 093
贵 州 Guizhou	7 876	3 753	239	3 883	1 512	2 371
云 南 Yunnan	11 454	7 515	109	3 830	1 952	1 877
西 藏 Tibet	491	407	…	85	56	28
陕 西 Shaanxi	15 027	3 263	68	11 696	5 579	6 117
甘 肃 Gansu	5 496	1 638	21	3 837	850	2 987
青 海 Qinghai	867	138	…	729	462	267
宁 夏 Ningxia	2 105	412	120	1 573	565	1 008
新 疆 Xinjiang	12 612	1 297	87	11 228	4 828	6 400

各地区农业污染排放情况（四）
Discharge of Agricultural Pollution by Region（4）

（2013）

单位：吨 （ton）

年　份 地　区	Year Region	总氮排放总量 Total Amount of Total Nitrogen Discharged	种植业 Crops	水产养殖业 Aquiculture	畜禽养殖业 Livestock and Poultry	规模化畜禽养 殖场/小区 Large-Scale Farms	养殖专业户 Small-Scale Farms
	2012	4 374 552	1 578 163	81 668	2 714 721	1 246 629	1 468 092
	2013	4 249 607	1 565 832	78 658	2 605 118	1 239 279	1 365 838
北　京	Beijing	25 809	8 262	629	16 917	11 681	5 236
天　津	Tianjin	28 419	9 098	2 263	17 058	5 826	11 232
河　北	Hebei	371 841	132 310	2 343	237 188	167 489	69 699
山　西	Shanxi	81 799	36 196	17	45 586	28 800	16 786
内蒙古	Inner Mongolia	143 933	26 882	94	116 956	33 127	83 829
辽　宁	Liaoning	186 036	17 201	2 260	166 575	46 449	120 126
吉　林	Jilin	114 599	20 126	177	94 296	36 058	58 238
黑龙江	Heilongjiang	243 951	25 157	356	218 439	51 538	166 900
上　海	Shanghai	15 403	5 885	383	9 135	6 671	2 464
江　苏	Jiangsu	179 089	88 580	6 788	83 721	43 215	40 506
浙　江	Zhejiang	91 909	39 750	3 331	48 829	28 255	20 574
安　徽	Anhui	171 775	94 064	3 040	74 671	43 039	31 631
福　建	Fujian	95 628	28 144	11 245	56 240	33 957	22 282
江　西	Jiangxi	94 326	36 867	1 329	56 131	35 588	20 543
山　东	Shandong	566 226	256 931	2 074	307 221	150 798	156 423
河　南	Henan	392 779	149 893	1 509	241 377	184 098	57 279
湖　北	Hubei	191 705	76 669	14 858	100 178	64 929	35 249
湖　南	Hunan	191 151	67 967	4 705	118 479	46 876	71 603
广　东	Guangdong	183 597	66 403	13 512	103 682	46 200	57 482
广　西	Guangxi	111 405	66 648	2 432	42 325	14 649	27 676
海　南	Hainan	29 485	11 871	407	17 207	5 494	11 713
重　庆	Chongqing	52 116	22 089	355	29 672	14 526	15 146
四　川	Sichuan	216 901	78 594	1 705	136 602	61 545	75 058
贵　州	Guizhou	45 454	29 042	812	15 599	6 821	8 779
云　南	Yunnan	77 264	58 620	1 089	17 555	8 791	8 764
西　藏	Tibet	5 845	4 897	0	947	268	680
陕　西	Shaanxi	92 287	49 540	214	42 533	19 322	23 211
甘　肃	Gansu	47 374	20 406	65	26 903	6 135	20 768
青　海	Qinghai	6 617	1 445	0	5 172	2 075	3 097
宁　夏	Ningxia	30 017	5 431	384	24 202	11 450	12 752
新　疆	Xinjiang	164 866	30 867	280	133 719	23 608	110 111

各地区农业污染排放情况（五）
Discharge of Agricultural Pollution by Region（5）
（2013）

单位：吨   （ton）

年 份 地 区 Year Region	总磷排放总量 Total Amount of Total Phosphorus Discharged	种植业 Crops	水产养殖业 Aquiculture	畜禽养殖业 Livestock and Poultry	规模化畜禽养殖场/小区 Large-Scale Farms	养殖专业户 Small-Scale Farms
2012	511 675	108 716	16 513	386 445	221 309	165 136
2013	499 824	106 034	15 138	378 652	221 166	157 486
北 京 Beijing	3 496	388	124	2 984	2 248	736
天 津 Tianjin	3 255	498	445	2 311	1 057	1 255
河 北 Hebei	44 094	7 091	474	36 529	27 604	8 925
山 西 Shanxi	9 296	1 976	6	7 314	5 472	1 843
内蒙古 Inner Mongolia	12 692	1 410	63	11 220	4 703	6 517
辽 宁 Liaoning	25 982	1 105	268	24 610	9 358	15 252
吉 林 Jilin	14 624	1 054	27	13 543	6 361	7 182
黑龙江 Heilongjiang	24 416	1 827	211	22 378	7 515	14 863
上 海 Shanghai	1 976	427	76	1 473	1 100	373
江 苏 Jiangsu	20 734	4 353	1 510	14 871	9 115	5 756
浙 江 Zhejiang	11 338	3 387	552	7 400	4 825	2 575
安 徽 Anhui	18 760	5 020	563	13 177	8 518	4 659
福 建 Fujian	14 478	2 975	1 921	9 583	7 301	2 282
江 西 Jiangxi	12 201	3 637	245	8 319	5 653	2 666
山 东 Shandong	68 053	13 520	450	54 084	31 018	23 065
河 南 Henan	48 640	7 529	283	40 828	33 078	7 749
湖 北 Hubei	24 790	5 473	2 896	16 421	12 144	4 277
湖 南 Hunan	21 859	6 295	874	14 690	7 287	7 403
广 东 Guangdong	27 652	7 383	2 851	17 418	8 351	9 066
广 西 Guangxi	13 476	6 492	428	6 557	2 486	4 071
海 南 Hainan	3 906	1 314	62	2 531	1 028	1 503
重 庆 Chongqing	6 553	2 178	68	4 308	2 500	1 807
四 川 Sichuan	24 267	6 554	318	17 395	9 517	7 878
贵 州 Guizhou	4 960	2 579	173	2 208	1 256	952
云 南 Yunnan	7 915	5 133	64	2 719	1 644	1 075
西 藏 Tibet	430	329	0	100	49	52
陕 西 Shaanxi	8 612	3 204	43	5 365	3 073	2 292
甘 肃 Gansu	4 162	1 383	14	2 765	1 021	1 744
青 海 Qinghai	635	78	0	557	321	236
宁 夏 Ningxia	3 072	300	76	2 696	1 719	977
新 疆 Xinjiang	13 498	1 142	56	12 300	3 845	8 455

各地区城镇生活污染排放及处理情况（一）

Discharge and Treatment of Household Pollution by Region（1）

（2013）

年　份 地　区	Year Region	城镇人口/ 万人 Urban Population （10 000 people）	生活煤炭消费量/ 万吨 Total Amount of Household Coal Consumed （10 000 tons）	生活用水总量/ 万吨 Quantity of Water Consumed by Household （10 000 tons）	居民家庭用水 总量 Water Consumed in Household	公共服务用水 总量 Water Consumed by Public Services
	2012	71 707. 8	16 164. 5	5 455 716. 8	4 552 151. 6	903 565. 2
	2013	73 726. 2	17 343. 4	5 725 662. 5	4 745 015. 7	980 646. 8
北　京	Beijing	1 825	412	158 674	90 700	67 974
天　津	Tianjin	1 214	204	73 554	59 554	14 000
河　北	Hebei	3 516	922	224 095	176 188	47 907
山　西	Shanxi	1 914	1 390	116 247	93 894	22 352
内蒙古	Inner Mongolia	1 466	1 029	85 097	68 209	16 888
辽　宁	Liaoning	2 917	735	187 027	151 131	35 896
吉　林	Jilin	1 491	574	88 463	61 960	26 503
黑龙江	Heilongjiang	2 188	2 540	129 816	116 606	13 211
上　海	Shanghai	2 156	230	196 900	181 148	15 752
江　苏	Jiangsu	5 095	201	432 205	346 496	85 709
浙　江	Zhejiang	3 520	119	301 715	249 260	52 455
安　徽	Anhui	2 912	459	222 863	193 513	29 349
福　建	Fujian	2 294	116	179 035	163 049	15 987
江　西	Jiangxi	2 202	66	163 708	138 600	25 108
山　东	Shandong	5 233	1 859	362 452	321 725	40 727
河　南	Henan	4 183	1 209	324 522	274 073	50 449
湖　北	Hubei	3 161	788	244 543	218 517	26 026
湖　南	Hunan	3 211	390	250 210	219 675	30 534
广　东	Guangdong	7 277	194	833 353	690 815	142 538
广　西	Guangxi	2 115	187	164 188	140 358	23 831
海　南	Hainan	472	6	38 275	28 641	9 634
重　庆	Chongqing	1 754	146	127 455	114 709	12 745
四　川	Sichuan	3 685	463	291 703	223 575	68 128
贵　州	Guizhou	1 353	511	88 388	68 808	19 580
云　南	Yunnan	1 859	254	137 956	113 160	24 796
西　藏	Tibet	100	11	5 425	4 722	703
陕　西	Shaanxi	1 931	709	119 822	96 079	23 743
甘　肃	Gansu	1 047	536	55 446	46 867	8 579
青　海	Qinghai	281	311	16 614	11 701	4 913
宁　夏	Ningxia	340	101	26 601	21 172	5 429
新　疆	Xinjiang	1 013	672	79 311	60 111	19 200

各地区城镇生活污染排放及处理情况（二）

Discharge and Treatment of Household Pollution by Region（2）

（2013）

年　份　Year 地　区　Region	城镇生活污水排放量/万吨 Amount of Household Waste Water Discharged（10 000 tons）	城镇生活化学需氧量产生量/吨 Amount of Household COD Generated（ton）	城镇生活氨氮产生量/吨 Amount of Household Ammonia Nitrogen Generated（ton）	城镇生活化学需氧量排放量/吨 Amount of Household COD Discharged（ton）	城镇生活氨氮排放量/吨 Amount of Household Ammonia Nitrogen Discharged（ton）
2012	4 626 897.1	17 325 170.1	2 220 738.5	9 127 509.5	1 446 275.8
2013	4 851 057.8	17 927 150.9	2 277 511.4	8 898 127.7	1 413 566.9
北　京　Beijing	134 991.0	527 173.0	64 729.0	89 868.0	14 189.0
天　津　Tianjin	65 468.6	304 702.3	41 599.1	84 885.5	15 670.8
河　北　Hebei	200 952.5	760 039.1	102 232.2	233 275.5	49 176.5
山　西　Shanxi	90 203.2	437 470.7	59 256.9	205 356.8	35 320.7
内蒙古　Inner Mongolia	69 900.4	339 498.3	42 462.4	160 290.1	27 601.0
辽　宁　Liaoning	156 106.5	676 907.0	91 594.6	312 225.1	61 719.9
吉　林　Jilin	74 980.2	346 376.3	44 746.0	192 438.8	32 214.2
黑龙江　Heilongjiang	105 243.5	481 048.2	64 340.0	309 467.3	48 229.0
上　海　Shanghai	177 210.0	597 935.7	76 315.5	173 181.1	40 321.7
江　苏　Jiangsu	373 526.5	1 313 951.6	168 712.0	558 697.7	94 253.4
浙　江　Zhejiang	254 971.8	970 451.7	125 839.8	376 253.0	69 923.7
安　徽　Anhui	195 091.5	693 071.7	81 888.7	436 387.9	58 059.7
福　建　Fujian	154 257.9	560 922.0	72 911.3	345 647.4	52 345.1
江　西　Jiangxi	138 617.3	566 287.9	62 599.1	399 003.3	50 357.7
山　东　Shandong	313 124.4	1 234 796.0	165 116.4	413 285.7	79 757.4
河　南　Henan	281 650.1	914 263.5	116 069.0	394 774.9	69 980.6
湖　北　Hubei	208 836.3	782 730.5	93 133.6	455 111.8	64 560.0
湖　南　Hunan	214 509.7	766 479.0	93 043.5	538 897.4	72 186.8
广　东　Guangdong	691 295.5	1 801 900.4	234 872.6	903 799.0	144 901.0
广　西　Guangxi	135 640.7	512 129.8	61 240.3	369 299.2	47 388.2
海　南　Hainan	29 374.0	126 574.1	15 083.4	79 315.7	12 960.5
重　庆　Chongqing	108 936.7	435 221.2	55 682.7	218 600.5	36 210.8
四　川　Sichuan	242 574.4	939 425.0	114 622.4	591 206.1	75 449.2
贵　州　Guizhou	70 111.7	292 889.7	37 522.3	198 068.8	26 248.7
云　南　Yunnan	114 635.4	482 884.8	55 970.9	294 368.9	40 636.0
西　藏　Tibet	4 603.6	25 756.8	3 032.6	20 948.0	2 645.8
陕　西　Shaanxi	97 166.1	449 117.3	57 691.2	226 165.9	35 567.1
甘　肃　Gansu	44 769.0	214 974.8	28 808.7	145 378.2	20 925.4
青　海　Qinghai	13 550.2	63 040.1	8 213.8	37 056.4	6 597.8
宁　夏　Ningxia	22 809.8	76 981.7	9 684.8	16 782.8	6 373.0
新　疆　Xinjiang	65 949.3	232 150.7	28 496.5	118 090.9	21 796.2

各地区城镇生活污染排放及处理情况（三）

Discharge and Treatment of Household Pollution by Region（3）

（2013）

单位：吨 (ton)

年 份 地 区	Year Region	城镇生活二氧化硫排放量 Amount of Household Sulphur Dioxide Emission	城镇生活氮氧化物排放量 Amount of Household Nitrogen Oxide Emission	城镇生活烟尘排放量 Amount of Household Soot Emission
	2012	2 056 646.1	393 122.9	1 426 733.5
	2013	2 085 373.0	407 493.1	1 239 000.4
北 京	Beijing	34 966.6	13 637.8	28 258.0
天 津	Tianjin	8 959.0	5 220.8	18 400.0
河 北	Hebei	111 523.9	23 318.9	77 015.2
山 西	Shanxi	114 564.9	30 427.6	105 474.6
内蒙古	Inner Mongolia	122 281.3	22 980.1	109 018.3
辽 宁	Liaoning	79 663.0	17 423.9	70 623.7
吉 林	Jilin	50 495.4	11 873.7	51 123.6
黑龙江	Heilongjiang	136 359.2	54 948.0	176 326.1
上 海	Shanghai	42 947.0	23 474.0	6 451.0
江 苏	Jiangsu	31 950.1	6 111.7	17 090.8
浙 江	Zhejiang	14 031.8	2 981.6	6 902.1
安 徽	Anhui	50 927.6	10 651.0	44 489.6
福 建	Fujian	18 950.4	2 355.7	9 766.6
江 西	Jiangxi	14 203.1	3 106.4	6 213.0
山 东	Shandong	199 410.7	26 412.5	108 087.1
河 南	Henan	151 285.1	24 270.2	42 737.2
湖 北	Hubei	75 322.8	13 003.3	48 153.8
湖 南	Hunan	52 623.8	7 985.0	26 225.7
广 东	Guangdong	29 517.4	6 888.3	12 076.6
广 西	Guangxi	33 948.9	3 630.8	13 378.2
海 南	Hainan	760.9	1 004.4	407.2
重 庆	Chongqing	53 261.4	4 487.3	4 401.0
四 川	Sichuan	70 082.9	9 913.4	11 386.6
贵 州	Guizhou	207 839.6	7 933.3	29 078.4
云 南	Yunnan	50 130.6	6 734.6	15 214.8
西 藏	Tibet	2 935.5	311.9	1 039.5
陕 西	Shaanxi	98 999.2	26 595.1	54 973.3
甘 肃	Gansu	89 182.1	14 401.2	43 609.3
青 海	Qinghai	25 875.4	6 161.2	21 599.7
宁 夏	Ningxia	21 510.5	2 679.3	9 074.0
新 疆	Xinjiang	90 863.0	16 569.9	70 405.5

各地区机动车污染排放情况
Discharge of Motor Vehicle Pollution by Region
（2013）

单位：万吨 (10 000 tons)

年　份　Year 地　区　Region	机动车污染物排放总量 Total Amount of Discharge of Motor Vehicle Pollution			
	总颗粒物 Total Particulate	氮氧化物 Nitrogen Oxide	一氧化碳 Carbon Monoxide	碳氢化合物 Hydrocarbon
2012	62. 139	640. 029	3 471. 669	438. 158
2013	59. 425	640. 553	3 439. 730	431. 205
北　京　Beijing	0. 381	7. 647	69. 009	7. 709
天　津　Tianjin	0. 627	5. 567	45. 237	5. 131
河　北　Hebei	4. 906	52. 348	251. 125	31. 851
山　西　Shanxi	2. 328	26. 368	137. 225	16. 852
内蒙古　Inner Mongolia	2. 904	25. 010	129. 351	17. 870
辽　宁　Liaoning	2. 710	26. 458	120. 326	15. 927
吉　林　Jilin	1. 813	17. 813	110. 504	14. 039
黑龙江　Heilongjiang	2. 598	25. 473	152. 263	19. 217
上　海　Shanghai	0. 722	9. 412	42. 491	6. 188
江　苏　Jiangsu	2. 704	34. 617	181. 492	21. 576
浙　江　Zhejiang	1. 606	17. 630	123. 569	14. 232
安　徽　Anhui	2. 215	22. 630	79. 646	10. 770
福　建　Fujian	0. 900	10. 532	58. 295	6. 906
江　西　Jiangxi	2. 534	22. 129	78. 474	11. 244
山　东　Shandong	4. 621	45. 312	226. 958	27. 945
河　南　Henan	5. 131	51. 249	211. 041	27. 206
湖　北　Hubei	1. 655	19. 487	105. 354	12. 545
湖　南　Hunan	1. 567	18. 232	89. 040	10. 887
广　东　Guangdong	4. 470	47. 308	314. 880	35. 797
广　西　Guangxi	1. 603	15. 166	86. 970	10. 848
海　南　Hainan	0. 357	3. 241	15. 956	2. 008
重　庆　Chongqing	0. 695	10. 963	62. 012	7. 976
四　川　Sichuan	1. 565	20. 531	114. 553	13. 643
贵　州　Guizhou	1. 015	10. 245	57. 239	7. 691
云　南　Yunnan	1. 690	20. 021	136. 572	19. 391
西　藏　Tibet	0. 471	4. 152	27. 288	3. 392
陕　西　Shaanxi	1. 421	18. 020	102. 994	13. 386
甘　肃　Gansu	0. 840	12. 212	107. 651	13. 054
青　海　Qinghai	0. 292	3. 291	27. 556	3. 422
宁　夏　Ningxia	0. 863	7. 752	32. 861	4. 699
新　疆　Xinjiang	2. 220	29. 733	141. 797	17. 806

各地区城镇污水处理情况（一）
Urban Waste Water Treatment by Region（1）
（2013）

年　份 地　区	Year Region	污水处理厂数/ 座 Number of Urban Waste Water Treatment Plants （unit）	污水处理厂 设计处理能力/ （万吨/日） Treatment Capacity （10 000 tons/day）	本年运行费用/ 万元 Annul Expenditure for Operation （10 000 yuan）	污水处理厂累计 完成投资/万元 Total Investment of Urban Waste Water Treatment （10 000 yuan）	新增固定资产/ 万元 Newly-Added Fixed Assets （10 000 yuan）
	2012	4 628	15 314	3 482 302. 1	34 077 389. 5	1 758 523. 4
	2013	5 364	16 574	3 935 524. 9	38 681 674. 9	2 158 148. 5
北　京	Beijing	138	434	116 032. 3	971 057. 4	11 139. 1
天　津	Tianjin	60	272	86 783. 0	702 224. 1	1 749. 4
河　北	Hebei	255	849	191 195. 8	1 988 608. 4	99 797. 3
山　西	Shanxi	185	330	77 237. 9	919 998. 0	90 280. 2
内蒙古	Inner Mongolia	134	334	90 896. 4	1 223 040. 0	125 233. 1
辽　宁	Liaoning	158	771	141 281. 4	1 296 457. 7	108 669. 8
吉　林	Jilin	59	288	48 629. 9	706 995. 6	27 911. 6
黑龙江	Heilongjiang	107	348	77 774. 4	874 960. 4	38 979. 1
上　海	Shanghai	56	771	161 761. 9	1 410 181. 0	25 405. 3
江　苏	Jiangsu	626	1 498	443 594. 2	4 307 381. 3	266 293. 7
浙　江	Zhejiang	361	1 098	425 591. 3	2 890 271. 0	169 078. 4
安　徽	Anhui	204	543	91 424. 3	1 198 274. 4	132 208. 3
福　建	Fujian	230	466	75 841. 2	829 451. 2	62 564. 0
江　西	Jiangxi	137	339	66 253. 2	683 905. 7	35 016. 7
山　东	Shandong	406	1 376	348 746. 7	2 704 713. 2	218 888. 8
河　南	Henan	231	853	149 892. 5	1 758 703. 9	120 811. 6
湖　北	Hubei	172	666	127 011. 0	1 356 272. 1	65 460. 0
湖　南	Hunan	161	543	109 742. 5	1 336 532. 9	52 045. 5
广　东	Guangdong	471	2 075	463 818. 0	4 421 692. 8	144 666. 9
广　西	Guangxi	117	383	92 417. 0	942 937. 8	20 420. 7
海　南	Hainan	56	108	16 867. 6	297 922. 0	952. 0
重　庆	Chongqing	149	286	107 786. 6	750 738. 0	20 250. 9
四　川	Sichuan	340	574	155 586. 8	1 333 909. 6	69 673. 2
贵　州	Guizhou	118	176	29 758. 3	498 626. 8	17 605. 9
云　南	Yunnan	114	291	66 460. 3	962 601. 3	45 157. 8
西　藏	Tibet	2	14	2 181. 0	18 210. 0	…
陕　西	Shaanxi	145	346	70 179. 0	1 022 201. 1	72 967. 7
甘　肃	Gansu	66	184	36 725. 5	436 881. 1	29 700. 8
青　海	Qinghai	19	40	10 731. 2	122 514. 6	2 450. 0
宁　夏	Ningxia	20	75	15 174. 2	165 315. 6	218. 3
新　疆	Xinjiang	67	243	38 149. 5	549 096. 3	82 552. 6

各地区城镇污水处理情况（二）
Urban Waste Water Treatment by Region（2）

（2013）

单位：万吨 (10 000 tons)

年 份 地 区	Year Region	污水实际 处理量 Quantity of Waste Water Treated	污水再生 利用量 Waste Water Recycled	工业用水量 Waste Water Recycled for Industry	市政用水量 Waste Water Recycled for Municipal Services	景观用水量 Waste Water Recycled for Landscape
	2012	4 161 551	120 597	74 015	12 544	34 037
	2013	4 561 081	154 281	99 778	13 202	41 301
北 京	Beijing	130 929	10 410	5 034	1 529	3 847
天 津	Tianjin	72 118	2 327	1 538	400	390
河 北	Hebei	231 896	21 920	14 654	2 684	4 582
山 西	Shanxi	76 032	9 684	8 878	100	706
内蒙古	Inner Mongolia	70 955	13 894	11 506	1 326	1 063
辽 宁	Liaoning	182 895	9 888	8 927	1	961
吉 林	Jilin	67 822	998	706	0	292
黑龙江	Heilongjiang	76 745	203	197	5	1
上 海	Shanghai	208 183	165	17	148	0
江 苏	Jiangsu	408 432	11 214	7 959	411	2 843
浙 江	Zhejiang	313 193	16 442	12 986	1 833	1 622
安 徽	Anhui	167 234	3 163	1 141	918	1 104
福 建	Fujian	124 800	2 316	38	95	2 183
江 西	Jiangxi	96 010	554	342	11	201
山 东	Shandong	401 300	12 863	8 757	166	3 940
河 南	Henan	249 499	11 456	8 934	361	2 161
湖 北	Hubei	186 963	554	29	403	123
湖 南	Hunan	158 552	146	29	101	16
广 东	Guangdong	624 962	14 646	1 794	333	12 519
广 西	Guangxi	102 321	337	235	13	89
海 南	Hainan	28 181	363	0	255	109
重 庆	Chongqing	92 973	658	449	135	73
四 川	Sichuan	171 371	195	97	44	54
贵 州	Guizhou	53 608	593	0	0	593
云 南	Yunnan	79 688	144	144	0	0
西 藏	Tibet	2 038	0	0	0	0
陕 西	Shaanxi	78 724	2 002	1 047	107	848
甘 肃	Gansu	31 103	4 220	3 044	516	660
青 海	Qinghai	10 158	704	704	0	0
宁 夏	Ningxia	19 596	172	0	172	0
新 疆	Xinjiang	42 800	2 048	592	1 134	321

各地区城镇污水处理情况（三）
Urban Waste Water Treatment by Region（3）
（2013）

单位：万吨 （10 000 tons）

年　份 地　区	Year Region	污泥产生量 Quantity of Sludge Generated	污泥处置量 Quantity of Sludge Disposed	土地利用量 Landuse	填埋处置量 Landfill	建筑材料利 用量 As Building Material	焚烧处置量 Incineration	污泥倾倒丢 弃量 Quantity of Sludge Discharged
	2012	2 418.55	2 418.54	453.57	1 266.51	235.38	463.08	0.01
	2013	2 635.84	2 635.84	466.31	1 263.58	305.91	600.03	…
北　京	Beijing	123.96	123.96	60.01	30.21	14.94	18.81	…
天　津	Tianjin	39.06	39.06	2.64	34.15	1.63	0.64	…
河　北	Hebei	157.89	157.89	32.74	120.21	0.44	4.50	…
山　西	Shanxi	42.59	42.59	20.26	20.00	0.12	2.21	…
内蒙古	Inner Mongolia	27.97	27.97	0.69	24.69	0.16	2.43	…
辽　宁	Liaoning	99.53	99.53	6.58	84.86	6.14	1.94	…
吉　林	Jilin	32.67	32.67	0.27	27.72	…	4.68	…
黑龙江	Heilongjiang	31.89	31.89	3.30	27.67	0.02	0.91	…
上　海	Shanghai	126.13	126.13	0.40	102.68	11.04	12.01	…
江　苏	Jiangsu	287.06	287.06	22.33	45.29	35.74	183.71	…
浙　江	Zhejiang	394.52	394.52	36.76	33.07	93.00	231.69	…
安　徽	Anhui	81.04	81.04	24.51	30.67	3.38	22.48	…
福　建	Fujian	61.74	61.74	11.88	33.15	8.32	8.39	…
江　西	Jiangxi	32.69	32.69	2.01	18.14	7.23	5.31	…
山　东	Shandong	232.48	232.48	69.03	80.14	23.21	60.09	…
河　南	Henan	158.04	158.04	33.65	111.57	8.44	4.38	…
湖　北	Hubei	61.76	61.76	8.21	47.82	3.01	2.71	…
湖　南	Hunan	52.48	52.48	0.26	44.46	6.89	0.86	…
广　东	Guangdong	254.14	254.14	71.44	119.26	39.62	23.83	…
广　西	Guangxi	24.97	24.97	11.53	8.95	4.31	0.18	…
海　南	Hainan	6.20	6.20	6.07	0.13	…	…	…
重　庆	Chongqing	40.25	40.25	2.84	21.87	13.92	1.61	…
四　川	Sichuan	69.71	69.71	17.60	31.15	15.70	5.26	…
贵　州	Guizhou	21.47	21.47	0.09	21.29	0.09	…	…
云　南	Yunnan	35.89	35.89	4.30	31.52	0.02	0.06	…
西　藏	Tibet	36.14	36.14	0.03	36.11	…	…	…
陕　西	Shaanxi	49.90	49.90	5.42	35.96	8.45	0.06	…
甘　肃	Gansu	23.92	23.92	2.65	21.20	…	0.08	…
青　海	Qinghai	5.86	5.86	…	5.77	0.09	…	…
宁　夏	Ningxia	11.13	11.13	7.21	2.92	…	1.01	…
新　疆	Xinjiang	12.76	12.76	1.63	10.94	…	0.19	…

各地区城镇污水处理情况（四）
Urban Waste Water Treatment by Region（4）
（2013）

单位：吨 (ton)

年　份 Year 地　区 Region	污染物去除量 Quantity of Pollutants Removed by Urban Waste Water Treatment						
	化学需氧量 COD	氨氮 Ammonia Nitrogen	油类 Oil	总氮 Total Nitrogen	总磷 Total Phosphorus	挥发酚 Volatile Phenols	氰化物 Cyanide
2012	10 196 874.7	889 960.7	46 071.5	739 926.7	126 332.4	997.9	696.5
2013	11 142 762.3	1 005 640.0	46 981.0	827 240.5	116 256.2	997.7	1 266.7
北　京 Beijing	427 243.9	58 232.9	7 718.7	57 816.4	5 231.0	5.4	0.9
天　津 Tianjin	258 587.0	25 909.8	1 723.8	26 591.0	2 298.8	6.7	0.1
河　北 Hebei	657 457.7	66 327.4	437.6	33 583.0	5 181.2	…	…
山　西 Shanxi	225 047.2	25 088.0	1 543.6	24 343.5	2 738.9	4.1	0.3
内蒙古 Inner Mongolia	234 145.7	19 726.7	698.8	16 469.2	1 926.4	2.0	11.9
辽　宁 Liaoning	405 826.5	33 730.4	1 957.5	23 941.5	2 850.7	17.5	…
吉　林 Jilin	172 235.4	12 849.1	1 505.6	11 718.4	1 820.4	36.6	2.3
黑龙江 Heilongjiang	188 332.4	18 900.6	465.9	16 900.6	2 361.8	0.8	1.2
上　海 Shanghai	603 536.7	44 347.7	4 607.1	40 528.0	7 242.6	209.8	8.1
江　苏 Jiangsu	1 031 476.6	86 199.7	3 415.1	59 171.6	9 324.6	128.9	23.4
浙　江 Zhejiang	1 218 898.2	73 453.7	3 665.7	42 523.5	9 071.2	0.1	594.6
安　徽 Anhui	269 483.9	26 257.1	801.9	22 353.9	2 917.8	43.3	0.3
福　建 Fujian	333 496.7	24 714.4	1 289.7	21 186.3	2 410.8	13.0	22.5
江　西 Jiangxi	142 796.8	12 678.8	44.2	8 970.7	1 438.2	5.5	…
山　东 Shandong	1 114 564.2	101 948.0	966.1	74 484.5	10 813.2	69.4	41.9
河　南 Henan	621 085.3	60 406.5	2 617.5	73 080.7	9 425.0	88.5	32.1
湖　北 Hubei	293 506.9	27 194.1	363.6	18 272.7	3 239.0	173.4	0.4
湖　南 Hunan	251 534.7	23 298.4	1 390.8	18 930.3	2 225.8	30.2	8.9
广　东 Guangdong	1 086 689.2	104 698.9	3 692.1	86 431.0	16 307.2	11.8	505.8
广　西 Guangxi	146 139.0	17 084.2	460.1	11 689.4	2 000.1	…	…
海　南 Hainan	49 347.6	2 113.3	1.2	2 973.4	835.5	…	…
重　庆 Chongqing	231 104.3	19 712.7	962.1	26 197.6	2 868.7	0.4	…
四　川 Sichuan	348 663.4	39 159.3	2 265.0	40 398.8	3 010.3	25.2	11.5
贵　州 Guizhou	104 873.8	12 571.0	82.0	5 333.0	512.2	0.5	…
云　南 Yunnan	174 636.1	16 426.9	1 398.0	16 023.5	2 681.2	7.5	…
西　藏 Tibet	1 336.0	72.0	198.0	81.0	12.6	…	…
陕　西 Shaanxi	218 703.7	23 718.9	1 686.2	26 816.7	2 100.0	…	…
甘　肃 Gansu	94 211.4	9 394.9	667.5	9 134.8	1 074.3	84.5	0.4
青　海 Qinghai	25 983.7	1 616.0	…	51.7	11.6	…	…
宁　夏 Ningxia	56 982.0	4 790.4	58.9	2 126.9	1 196.8	31.8	…
新　疆 Xinjiang	154 836.3	13 018.3	296.7	9 117.1	1 128.4	0.9	…

各地区生活垃圾处理情况（一）
Centralized Treatment of Garbage by Region（1）

（2013）

年　份　Year 地　区　Region		生活垃圾处理厂数/ 座 Number of Garbage Treatment Plants（unit）	本年运行费用/ 万元 Annul Expenditure for Operation （10 000 yuan）	生活垃圾处理厂累计 完成投资/万元 Total Investment of Garbage Treatment Plants （10 000 yuan）	新增固定资产/ 万元 Newly-Added Fixed Assets （10 000 yuan）
	2012	2 125	985 337. 9	9 472 672. 2	465 385. 1
	2013	2 135	865 648. 3	10 145 780. 1	415 574. 5
北　京	Beijing	19	81 752. 7	355 471. 0	18 718. 1
天　津	Tianjin	10	16 386. 3	232 900. 4	2. 4
河　北	Hebei	131	26 917. 4	415 483. 7	10 921. 4
山　西	Shanxi	78	29 560. 8	227 880. 7	5 791. 3
内蒙古	Inner Mongolia	63	19 920. 9	290 202. 7	7 837. 7
辽　宁	Liaoning	35	12 208. 1	161 915. 0	488. 1
吉　林	Jilin	47	22 776. 7	201 171. 0	11 920. 6
黑龙江	Heilongjiang	27	11 170. 8	158 310. 1	4 414. 9
上　海	Shanghai	25	68 257. 5	486 570. 6	3 845. 5
江　苏	Jiangsu	63	57 713. 5	661 732. 9	7 965. 7
浙　江	Zhejiang	96	89 732. 4	825 449. 6	61 369. 4
安　徽	Anhui	127	22 707. 6	290 018. 0	69 763. 2
福　建	Fujian	89	15 529. 4	339 695. 4	12 027. 0
江　西	Jiangxi	83	11 968. 9	189 692. 9	4 500. 7
山　东	Shandong	84	36 547. 2	490 260. 1	15 646. 9
河　南	Henan	120	29 388. 5	420 128. 6	7 752. 1
湖　北	Hubei	96	15 174. 2	340 175. 6	6 315. 7
湖　南	Hunan	94	44 343. 0	497 376. 5	21 975. 0
广　东	Guangdong	149	120 662. 7	1 310 542. 5	63 748. 9
广　西	Guangxi	53	14 731. 2	313 711. 0	14 859. 2
海　南	Hainan	21	4 034. 9	104 112. 0	663. 0
重　庆	Chongqing	45	25 258. 7	386 642. 2	2 756. 7
四　川	Sichuan	107	30 294. 6	486 993. 0	14 760. 7
贵　州	Guizhou	37	8 522. 5	142 480. 4	13 282. 8
云　南	Yunnan	90	14 624. 8	197 430. 8	7 152. 0
西　藏	Tibet	11	1 297. 3	26 067. 4	1 130. 0
陕　西	Shaanxi	70	9 848. 7	219 841. 7	6 418. 7
甘　肃	Gansu	66	6 202. 0	87 636. 0	1 484. 5
青　海	Qinghai	54	2 251. 1	57 640. 7	1 487. 4
宁　夏	Ningxia	20	2 583. 8	48 089. 6	2 525. 3
新　疆	Xinjiang	125	13 280. 2	180 158. 0	14 049. 6

各地区生活垃圾处理情况（二）
Centralized Treatment of Garbage by Region（2）

（2013）

单位：吨 （ton）

年 份 Year 地 区 Region	渗滤液中污染物排放量 Amount of Pollutants Discharged in Landfill Leachate					
	化学需氧量 COD	氨氮 Ammonia Nitrogen	油类 Oil	总磷 Total Phosphorus	挥发酚 Volatile Phenol	氰化物 Cyanide
2012	185 815.765	19 222.846	153.986	254.067	19.932	0.859
2013	175 875.542	17 782.202	158.992	264.396	18.205	0.795
北 京 Beijing	7 835.101	680.879	2.056	7.301	0.331	0.014
天 津 Tianjin	473.200	38.470	0.030	0.090	…	…
河 北 Hebei	7 406.895	559.015	8.007	12.288	1.247	0.032
山 西 Shanxi	2 577.870	247.200	4.216	4.675	0.332	0.007
内蒙古 Inner Mongolia	2 310.150	189.970	8.131	7.137	1.065	0.009
辽 宁 Liaoning	5 575.609	921.236	7.010	9.219	0.900	0.024
吉 林 Jilin	11 977.774	1 286.091	11.777	31.377	1.508	0.042
黑龙江 Heilongjiang	1 898.934	219.979	0.718	1.110	0.141	0.004
上 海 Shanghai	5 803.993	293.547	2.750	5.458	1.022	0.020
江 苏 Jiangsu	4 877.503	571.907	4.692	5.378	0.271	0.018
浙 江 Zhejiang	6 983.288	603.487	5.659	12.677	0.645	0.048
安 徽 Anhui	8 455.871	788.465	8.556	13.074	0.269	0.054
福 建 Fujian	4 117.321	398.648	14.746	6.259	0.240	0.035
江 西 Jiangxi	9 720.524	783.454	8.829	15.365	1.651	0.092
山 东 Shandong	4 703.043	497.448	5.281	6.340	0.642	0.029
河 南 Henan	7 263.963	885.095	6.085	12.615	1.348	0.031
湖 北 Hubei	13 573.783	1 517.186	6.704	15.888	1.249	0.057
湖 南 Hunan	12 398.372	1 164.078	9.587	20.975	1.636	0.097
广 东 Guangdong	14 075.274	1 382.455	5.590	13.124	0.075	0.045
广 西 Guangxi	3 897.317	306.996	3.217	6.327	0.112	0.018
海 南 Hainan	1 049.245	97.179	0.145	1.183	…	0.002
重 庆 Chongqing	626.994	159.276	0.870	0.874	…	0.005
四 川 Sichuan	5 451.924	662.246	7.133	5.700	0.197	0.014
贵 州 Guizhou	5 329.483	625.177	2.068	4.263	0.037	0.011
云 南 Yunnan	12 979.550	1 651.020	8.138	21.114	1.425	0.031
西 藏 Tibet	154.224	10.282	0.103	0.193	0.019	0.001
陕 西 Shaanxi	5 200.926	618.564	4.925	5.885	0.145	0.020
甘 肃 Gansu	1 631.395	113.510	0.791	1.322	0.675	0.003
青 海 Qinghai	2 196.936	134.186	1.519	1.803	0.034	0.002
宁 夏 Ningxia	826.730	57.510	0.428	0.878	0.099	0.002
新 疆 Xinjiang	4 502.350	317.646	9.231	14.504	0.891	0.031

174

各地区生活垃圾处理情况（三）
Centralized Treatment of Garbage by Region（3）
（2013）

单位：千克 (kg)

年 份 Year 地 区 Region	渗滤液中污染物排放量 Amount of Pollutants Discharged in Landfill Leachate					
	汞 Mercury	镉 Cadmium	六价铬 Hexavalent Chromium	总铬 Total Chromium	铅 Plumbum	砷 Arsenic
2012	135.359	581.102	151.097	1 405.849	2 188.515	787.178
2013	131.359	553.350	142.700	1 204.445	1 984.675	606.280
北 京 Beijing	0.640	16.342	1.785	15.877	147.385	3.336
天 津 Tianjin	0.010	0.030	…	0.090	0.350	0.170
河 北 Hebei	3.609	25.786	0.944	49.150	116.208	35.347
山 西 Shanxi	0.942	4.551	0.140	9.704	19.258	5.070
内蒙古 Inner Mongolia	3.672	19.228	0.440	29.168	71.673	25.989
辽 宁 Liaoning	8.747	21.570	0.785	39.077	81.271	18.943
吉 林 Jilin	3.790	24.995	1.802	50.011	116.249	23.211
黑龙江 Heilongjiang	1.646	3.142	1.500	7.313	14.280	4.631
上 海 Shanghai	2.412	11.526	0.140	44.637	67.882	25.445
江 苏 Jiangsu	13.109	14.293	6.425	38.997	76.371	22.666
浙 江 Zhejiang	6.182	25.546	6.946	87.692	155.321	27.095
安 徽 Anhui	5.213	27.074	18.504	77.352	106.591	30.258
福 建 Fujian	1.774	84.650	7.531	30.352	48.369	12.245
江 西 Jiangxi	7.909	28.396	9.761	106.788	141.793	54.861
山 东 Shandong	2.872	19.580	5.240	32.526	73.455	18.216
河 南 Henan	5.984	23.402	4.314	47.867	111.382	32.497
湖 北 Hubei	10.604	32.416	10.207	79.484	98.953	46.604
湖 南 Hunan	10.126	44.937	19.900	103.116	114.455	50.834
广 东 Guangdong	17.570	14.037	7.947	79.200	57.089	26.547
广 西 Guangxi	2.836	9.275	9.077	24.949	44.781	8.024
海 南 Hainan	0.220	0.680	0.006	1.577	2.677	1.038
重 庆 Chongqing	0.624	2.262	0.738	4.754	14.288	4.092
四 川 Sichuan	2.823	9.346	5.278	41.319	57.785	17.143
贵 州 Guizhou	4.093	5.760	1.313	19.334	27.117	9.142
云 南 Yunnan	4.920	19.418	12.939	53.542	94.148	26.235
西 藏 Tibet	0.064	0.514	…	1.028	2.570	0.643
陕 西 Shaanxi	1.741	10.333	1.968	29.417	37.584	9.516
甘 肃 Gansu	0.925	2.736	…	7.353	11.995	3.696
青 海 Qinghai	0.536	1.116	0.019	2.105	5.302	1.368
宁 夏 Ningxia	0.336	2.412	0.060	4.777	9.543	2.722
新 疆 Xinjiang	5.430	47.997	6.991	85.889	58.550	58.696

各地区生活垃圾处理情况（四）
Centralized Treatment of Garbage by Region（4）

（2013）

单位：吨 （ton）

年 份 Year	焚烧废气中污染物排放量 Volume of Pollutants Emission in the Incineration Gas		
地 区 Region	二氧化硫 Sulfur Dioxide	氮氧化物 Nitrogen Oxide	烟尘 Soot
2012	1 585. 30	2 402. 22	1 046. 61
2013	1 176. 38	2 874. 88	833. 85
北 京 Beijing	33. 85	283. 98	39. 27
天 津 Tianjin	18. 50	148. 00	10. 00
河 北 Hebei	2. 67	…	1. 11
山 西 Shanxi	24. 66	64. 08	14. 06
内 蒙 古 Inner Mongolia	…	…	…
辽 宁 Liaoning	37. 96	7. 72	32. 84
吉 林 Jilin	…	…	…
黑 龙 江 Heilongjiang	…	…	…
上 海 Shanghai	…	0. 01	0. 01
江 苏 Jiangsu	163. 12	193. 44	41. 96
浙 江 Zhejiang	108. 65	96. 38	136. 96
安 徽 Anhui	178. 32	231. 53	180. 14
福 建 Fujian	3. 91	3. 71	1. 63
江 西 Jiangxi	…	…	…
山 东 Shandong	42. 72	99. 89	28. 36
河 南 Henan	21. 50	38. 00	11. 10
湖 北 Hubei	…	…	…
湖 南 Hunan	…	…	…
广 东 Guangdong	276. 24	1 412. 99	266. 68
广 西 Guangxi	33. 72	62. 10	7. 91
海 南 Hainan	…	…	…
重 庆 Chongqing	…	…	…
四 川 Sichuan	224. 60	233. 02	59. 19
贵 州 Guizhou	…	…	…
云 南 Yunnan	5. 96	0. 03	2. 63
西 藏 Tibet	…	…	…
陕 西 Shaanxi	…	…	…
甘 肃 Gansu	…	…	…
青 海 Qinghai	…	…	…
宁 夏 Ningxia	…	…	…
新 疆 Xinjiang	…	…	…

各地区危险（医疗）废物集中处置情况（一）

Centralized Treatment of Hazardous（Medical）Wastes by Region（1）

（2013）

年 份 地 区	Year Region	危险废物集中 处置厂数/个 Number of Centralized Hazardous Wastes Treatment Plants （unit）	医疗废物集中 处置厂数/个 Number of Centralized Medical Wastes Treatment Plants（unit）	本年运行费用/ 万元 Annul Expenditure for Operation （10 000 yuan）	累计完成投资/ 万元 Total Investment of Hazardous/Medical Wastes Treatment Plants （10 000 yuan）	新增固定资产/ 万元 Newly-Added Fixed Assets （10 000 yuan）
	2012	722	236	539 438.5	2 563 959.5	191 780.3
	2013	767	243	587 441.8	3 128 806.5	205 524.1
北 京	Beijing	2	2	27 413.9	99 577.7	2 837.3
天 津	Tianjin	11	0	14 574.6	35 943.3	364.5
河 北	Hebei	31	12	9 246.8	44 384.7	3 729.1
山 西	Shanxi	5	8	2 055.1	13 396.2	1 759.7
内蒙古	Inner Mongolia	2	16	3 524.5	31 039.8	386.8
辽 宁	Liaoning	37	7	39 860.8	118 487.7	2 475.0
吉 林	Jilin	2	5	4 353.0	13 459.5	50.5
黑龙江	Heilongjiang	6	8	3 988.3	46 741.0	524.0
上 海	Shanghai	35	2	71 461.5	196 890.2	9 663.4
江 苏	Jiangsu	156	3	80 432.0	396 767.0	17 582.3
浙 江	Zhejiang	71	7	56 176.3	234 026.4	21 594.1
安 徽	Anhui	13	10	9 367.7	82 294.8	4 797.4
福 建	Fujian	14	7	8 529.0	46 206.2	2 264.3
江 西	Jiangxi	38	9	16 455.0	306 956.7	16 788.7
山 东	Shandong	52	14	27 407.0	132 061.8	10 630.4
河 南	Henan	0	18	9 390.4	27 705.4	1 063.2
湖 北	Hubei	33	10	18 227.9	146 862.5	16 942.0
湖 南	Hunan	25	9	12 763.6	67 290.0	238.7
广 东	Guangdong	103	17	74 628.3	436 561.3	14 204.0
广 西	Guangxi	17	7	6 745.3	48 240.1	1 752.9
海 南	Hainan	1	2	2 161.8	17 525.0	143.0
重 庆	Chongqing	9	4	11 991.9	59 644.1	1 349.3
四 川	Sichuan	26	16	33 756.1	151 753.3	16 382.4
贵 州	Guizhou	7	5	4 816.0	34 202.5	1 165.0
云 南	Yunnan	1	6	2 281.0	19 449.2	122.0
西 藏	Tibet	0	0	…	…	…
陕 西	Shaanxi	36	9	24 005.0	133 129.3	43 507.5
甘 肃	Gansu	8	12	4 838.0	58 536.5	2 859.4
青 海	Qinghai	6	1	963.3	46 930.6	…
宁 夏	Ningxia	2	1	1 684.9	40 776.2	3 418.3
新 疆	Xinjiang	18	16	4 343.1	41 967.5	6 929.0

各地区危险（医疗）废物集中处置情况（二）

Centralized Treatment of Hazardous（Medical）Wastes by Region（2）

（2013）

年 份 地 区	Year Region	危险废物设计 处置能力/ （吨/日） Treatment Capacity （ton/day）	危险废物实际 处置量/吨 Volume of Hazardous Wastes Disposed （ton）	工业危险废物 处置量 Industrial Hazardous Wastes	医疗废物 处置量 Medical Hazardous Wastes	其他危险废物 处置量 Other Hazardous Wastes	危险废物综合 利用量/吨 Volume of Hazardous Wastes Utilized （ton）
	2012	59 805	3 406 914	2 231 828	502 732	673 254	3 751 721
	2013	85 886	2 823 081	1 978 089	596 854	248 138	4 578 281
北 京	Beijing	855	94 685	79 631	15 054	0	58 880
天 津	Tianjin	1 120	34 582	14 591	19 991	0	165 176
河 北	Hebei	1 175	53 259	38 250	14 877	132	16 455
山 西	Shanxi	132	13 409	86	13 323	0	888
内蒙古	Inner Mongolia	226	9 675	0	9 675	0	0
辽 宁	Liaoning	5 162	153 875	116 657	16 021	21 197	250 768
吉 林	Jilin	232	32 068	25 174	6 879	14	627
黑龙江	Heilongjiang	243	12 906	35	12 870	0	694
上 海	Shanghai	2 187	199 778	135 132	30 182	34 464	173 160
江 苏	Jiangsu	12 720	296 436	238 128	32 351	25 957	1 125 151
浙 江	Zhejiang	12 376	393 009	342 371	48 941	1 697	475 785
安 徽	Anhui	466	35 363	19 807	15 556	0	17 687
福 建	Fujian	1 405	55 714	36 320	19 394	0	28 827
江 西	Jiangxi	3 907	18 008	6 955	11 053	0	288 603
山 东	Shandong	3 631	145 587	94 897	43 998	6 693	208 004
河 南	Henan	133	36 323	0	36 323	0	0
湖 北	Hubei	5 273	74 404	51 717	22 680	8	46 002
湖 南	Hunan	1 033	48 745	31 840	16 905	0	66 710
广 东	Guangdong	9 545	425 996	348 161	73 585	4 250	704 512
广 西	Guangxi	414	38 234	11 468	17 648	9 117	15 177
海 南	Hainan	70	11 428	7 750	3 678	0	0
重 庆	Chongqing	2 343	48 797	35 741	10 647	2 409	331 825
四 川	Sichuan	3 393	47 050	15 476	31 574	0	128 880
贵 州	Guizhou	364	16 029	5 378	10 334	317	1 435
云 南	Yunnan	49	12 880	0	12 880	0	0
西 藏	Tibet	0	0	0	0	0	0
陕 西	Shaanxi	12 654	294 095	161 674	30 077	102 344	306 720
甘 肃	Gansu	676	114 608	106 980	7 605	23	2 425
青 海	Qinghai	1 295	36 927	0	2 547	34 380	24 349
宁 夏	Ningxia	1 166	12 280	6 328	1 994	3 958	12 087
新 疆	Xinjiang	1 640	56 933	47 541	8 214	1 178	127 455

各地区危险（医疗）废物集中处置情况（三）
Centralized Treatment of Hazardous（Medical）Wastes by Region（3）
（2013）

年　份 地　区	Year Region	渗滤液中污染物排放量 Amount of Pollutants Discharged in Landfill Leachate					
		化学需氧量/ 吨 COD（ton）	氨氮/吨 Ammonia Nitrogen（ton）	油类/吨 Oil（ton）	总磷/吨 Total Phosphorus （ton）	挥发酚/千克 Volatile Phenol （kg）	氰化物/千克 Cyanide（kg）
	2012	1 472. 135	36. 052	12. 657	0. 285	1. 289	25. 564
	2013	953. 446	159. 463	2. 007	0. 324	13. 215	29. 307
北　京	Beijing	…	…	…	…	…	…
天　津	Tianjin	7. 535	0. 100	0. 007	…	…	0. 025
河　北	Hebei	0. 179	0. 005	…	0. 001	…	…
山　西	Shanxi	…	…	…	…	…	…
内蒙古	Inner Mongolia	0. 530	0. 350	0. 010	…	…	…
辽　宁	Liaoning	306. 500	10. 791	0. 105	0. 001	0. 001	0. 001
吉　林	Jilin	0. 130	0. 040	0. 001	…	0. 027	0. 043
黑龙江	Heilongjiang	0. 695	0. 101	0. 003	…	…	…
上　海	Shanghai	23. 987	2. 009	0. 076	0. 144	0. 900	4. 560
江　苏	Jiangsu	26. 278	7. 012	0. 136	0. 025	0. 010	0. 250
浙　江	Zhejiang	134. 267	3. 877	0. 152	0. 065	…	…
安　徽	Anhui	1. 115	0. 218	0. 001	0. 001	0. 007	…
福　建	Fujian	7. 730	0. 031	0. 001	…	0. 001	4. 711
江　西	Jiangxi	0. 532	0. 343	0. 143	…	10. 000	0. 020
山　东	Shandong	4. 699	0. 387	0. 443	…	…	1. 050
河　南	Henan	…	…	…	…	…	…
湖　北	Hubei	24. 820	2. 334	0. 026	…	…	…
湖　南	Hunan	2. 173	0. 664	0. 041	…	0. 027	0. 405
广　东	Guangdong	391. 904	122. 689	0. 199	0. 044	0. 714	12. 519
广　西	Guangxi	0. 134	…	0. 001	…	…	…
海　南	Hainan	…	…	…	…	…	…
重　庆	Chongqing	17. 813	7. 719	0. 548	0. 002	…	…
四　川	Sichuan	0. 740	0. 089	0. 024	0. 040	0. 138	0. 500
贵　州	Guizhou	0. 680	0. 320	0. 004	…	…	…
云　南	Yunnan	…	…	…	…	…	…
西　藏	Tibet	…	…	…	…	…	…
陕　西	Shaanxi	0. 199	…	0. 025	…	0. 004	0. 013
甘　肃	Gansu	0. 051	0. 002	0. 038	0. 001	0. 840	3. 160
青　海	Qinghai	…	…	…	…	…	…
宁　夏	Ningxia	…	…	…	…	…	…
新　疆	Xinjiang	0. 752	0. 383	0. 025	…	0. 546	2. 050

各地区危险（医疗）废物集中处置情况（四）

Centralized Treatment of Hazardous（Medical）Wastes by Region（4）

（2013）

单位：千克 (kg)

年　份 地　区	Year Region	汞 Mercury	镉 Cadmium	渗滤液中污染物排放量 Amount of Pollutants Discharged in Landfill Leachate 六价铬 Hexavalent Chromium	总铬 Total Chromium	铅 Plumbum	砷 Arsenic
	2012	5.987	6.448	7.889	37.012	33.620	14.837
	2013	1.797	6.999	10.040	49.981	43.637	16.490
北　京	Beijing	…	…	…	…	…	…
天　津	Tianjin	…	0.053	…	0.032	14.400	0.007
河　北	Hebei	…	…	0.003	0.003	…	…
山　西	Shanxi	…	…	…	…	…	…
内蒙古	Inner Mongolia	…	…	…	…	…	…
辽　宁	Liaoning	0.003	0.750	0.650	0.650	0.015	0.100
吉　林	Jilin	…	0.027	0.038	0.437	0.006	0.164
黑龙江	Heilongjiang	…	…	…	…	…	…
上　海	Shanghai	1.390	…	6.000	18.214	0.400	…
江　苏	Jiangsu	0.016	0.066	0.047	2.056	0.411	0.423
浙　江	Zhejiang	…	…	…	2.080	7.300	1.650
安　徽	Anhui	0.001	0.002	…	0.013	0.369	…
福　建	Fujian	…	0.001	0.006	0.016	0.009	0.009
江　西	Jiangxi	…	3.477	1.000	4.000	7.169	6.671
山　东	Shandong	…	0.010	…	0.160	0.100	0.030
河　南	Henan	…	…	…	…	…	…
湖　北	Hubei	…	…	…	…	…	…
湖　南	Hunan	0.037	0.591	0.054	0.374	1.029	2.967
广　东	Guangdong	0.002	0.483	0.462	3.819	0.706	0.018
广　西	Guangxi	…	…	…	…	…	…
海　南	Hainan	…	…	…	…	…	…
重　庆	Chongqing	…	0.154	0.319	0.677	0.216	0.055
四　川	Sichuan	0.002	0.011	0.040	0.075	0.565	0.500
贵　州	Guizhou	…	0.335	…	…	0.521	0.419
云　南	Yunnan	…	…	…	…	…	…
西　藏	Tibet	…	…	…	…	…	…
陕　西	Shaanxi	…	…	…	…	…	…
甘　肃	Gansu	0.210	0.630	…	10.540	6.320	2.110
青　海	Qinghai	…	…	…	…	…	…
宁　夏	Ningxia	…	…	…	…	…	…
新　疆	Xinjiang	0.136	0.409	1.421	6.835	4.101	1.367

各地区危险（医疗）废物集中处置情况（五）

Centralized Treatment of Hazardous（Medical）Wastes by Region（5）

（2013）

单位：吨 (ton)

年　份 地　区	Year Region	焚烧废气中污染物排放量 Volume of Pollutants Emission in the Incineration Gas		
		二氧化硫 Sulfur Dioxide	氮氧化物 Nitrogen Oxide	烟尘 Soot
2012		1 033.306	1 287.103	845.676
2013		764.311	1 543.084	1 095.972
北　京	Beijing	0.205	8.364	0.813
天　津	Tianjin	61.800	35.751	13.370
河　北	Hebei	23.529	38.874	41.514
山　西	Shanxi	2.024	16.167	12.882
内蒙古	Inner Mongolia	1.226	6.774	11.902
辽　宁	Liaoning	12.872	36.233	55.951
吉　林	Jilin	0.866	5.206	2.977
黑龙江	Heilongjiang	2.076	10.065	3.856
上　海	Shanghai	33.722	416.149	81.685
江　苏	Jiangsu	88.160	247.336	219.110
浙　江	Zhejiang	119.617	102.991	61.762
安　徽	Anhui	20.232	35.563	35.838
福　建	Fujian	9.486	39.949	7.219
江　西	Jiangxi	3.823	17.813	27.681
山　东	Shandong	166.162	92.997	24.466
河　南	Henan	14.943	22.520	6.078
湖　北	Hubei	25.530	27.322	30.427
湖　南	Hunan	1.775	6.928	7.436
广　东	Guangdong	107.925	218.951	263.209
广　西	Guangxi	7.965	20.642	30.155
海　南	Hainan	0.930	1.085	1.440
重　庆	Chongqing	9.476	19.414	9.118
四　川	Sichuan	35.518	58.015	39.392
贵　州	Guizhou	2.275	8.901	1.742
云　南	Yunnan	0.715	12.186	4.050
西　藏	Tibet	0	0	0
陕　西	Shaanxi	7.216	22.615	42.888
甘　肃	Gansu	0.718	3.870	4.641
青　海	Qinghai	0.830	2.430	0.263
宁　夏	Ningxia	0.590	2.350	1.920
新　疆	Xinjiang	2.105	5.623	52.187

各地区环境污染治理投资情况
Investment in the Treatment of Environmental Pollution by Region
（2013）

单位：亿元 (100 million yuan)

年 份 地 区	Year Region	环境污染治理投资总额 Total Investment in Treatment of Environmental Pollution	城市环境基础设施 建设投资 Investment in Urban Environment Infrastructure Facilities	工业污染源 治理投资 Investment in Treatment of Industrial Pollution Sources	建设项目"三同时" 环保投资 Investments in Environment Components for New Construction Projects
2012		8 253.6	5 062.7	500.5	2 690.4
2013		9 037.2	5 223.0	849.7	2 964.5
国家级	State Level	598.7	—	—	598.7
北 京	Beijing	430.3	404.5	4.3	21.5
天 津	Tianjin	180.2	90.9	14.8	74.4
河 北	Hebei	477.6	316.1	51.2	110.3
山 西	Shanxi	336.0	204.7	55.6	75.7
内蒙古	Inner Mongolia	484.3	324.5	62.7	97.1
辽 宁	Liaoning	331.3	206.6	27.7	97.1
吉 林	Jilin	100.4	55.7	9.4	35.3
黑龙江	Heilongjiang	274.9	200.0	20.7	54.2
上 海	Shanghai	176.5	77.0	5.2	94.3
江 苏	Jiangsu	800.4	509.9	59.4	231.1
浙 江	Zhejiang	346.4	156.5	57.7	132.2
安 徽	Anhui	467.9	319.8	23.3	124.8
福 建	Fujian	269.1	137.0	38.4	93.8
江 西	Jiangxi	234.7	182.3	15.5	36.8
山 东	Shandong	800.1	528.4	84.3	187.4
河 南	Henan	268.8	157.4	44.0	67.4
湖 北	Hubei	234.6	154.5	25.2	54.9
湖 南	Hunan	223.7	167.2	23.4	33.1
广 东	Guangdong	263.5	60.9	32.5	170.1
广 西	Guangxi	213.5	135.3	18.3	59.9
海 南	Hainan	25.6	12.5	3.5	9.5
重 庆	Chongqing	162.2	111.4	7.9	43.0
四 川	Sichuan	220.0	119.4	18.8	81.8
贵 州	Guizhou	106.3	66.8	19.6	20.0
云 南	Yunnan	189.0	35.7	23.9	129.4
西 藏	Tibet	25.5	1.5	1.0	22.9
陕 西	Shaanxi	216.8	141.1	41.8	34.0
甘 肃	Gansu	163.6	82.1	18.2	63.3
青 海	Qinghai	33.6	20.5	3.0	10.1
宁 夏	Ningxia	69.1	27.8	16.5	24.7
新 疆	Xinjiang	312.5	214.8	22.0	75.7

各地区城市环境基础设施建设投资情况
Investment in Urban Environment Infrastructure by Region
（2013）

单位：亿元 （100 million yuan）

年 份 地 区	Year Region	投资总额 Total Investment	燃气 Gas Supply	集中供热 Gerneral Heating	排水 Sewerage Projects	园林绿化 Gardening and Greening	市容环境 卫生 Sanitation
	2012	5 062.7	551.8	798.1	934.1	2 380.0	398.6
	2013	5 223.0	607.9	819.5	1 055.0	2 234.9	505.7
北 京	Beijing	404.5	68.4	60.6	52.2	95.0	128.4
天 津	Tianjin	90.9	37.2	6.5	9.5	37.2	0.4
河 北	Hebei	316.1	43.4	72.6	47.4	129.7	23.0
山 西	Shanxi	204.7	30.4	95.1	14.0	52.2	12.9
内蒙古	Inner Mongolia	324.5	17.4	88.2	58.0	137.5	23.5
辽 宁	Liaoning	206.6	20.2	84.9	31.1	66.9	3.5
吉 林	Jilin	55.7	4.6	20.8	8.0	20.6	1.8
黑龙江	Heilongjiang	200.0	11.7	109.4	27.7	33.8	17.4
上 海	Shanghai	77.0	23.6	—	15.0	32.2	6.1
江 苏	Jiangsu	509.9	35.1	—	126.4	311.1	37.4
浙 江	Zhejiang	156.5	21.0	0.9	43.0	82.2	9.3
安 徽	Anhui	319.8	37.8	2.8	80.0	180.6	18.8
福 建	Fujian	137.0	3.8	—	31.9	87.4	13.9
江 西	Jiangxi	182.3	36.0	—	32.2	104.0	10.1
山 东	Shandong	528.4	45.3	120.7	100.9	238.5	23.0
河 南	Henan	157.4	18.8	20.0	38.5	69.1	10.9
湖 北	Hubei	154.5	12.3	—	52.9	74.9	14.3
湖 南	Hunan	167.2	11.1	—	50.7	53.0	52.4
广 东	Guangdong	60.9	10.7	—	24.6	11.5	14.1
广 西	Guangxi	135.3	14.7	—	40.8	70.0	9.8
海 南	Hainan	12.5	0.5	—	9.2	1.6	1.1
重 庆	Chongqing	111.4	5.1	—	14.0	82.0	10.2
四 川	Sichuan	119.4	12.2	0.2	41.1	60.3	5.5
贵 州	Guizhou	66.8	10.1	—	22.5	28.7	5.5
云 南	Yunnan	35.7	5.9	—	10.6	15.0	4.2
西 藏	Tibet	1.5	—	—	0.7	—	0.8
陕 西	Shaanxi	141.1	16.1	24.8	28.4	64.7	7.1
甘 肃	Gansu	82.1	18.5	18.7	7.6	34.1	3.3
青 海	Qinghai	20.5	2.7	6.7	4.9	4.4	1.8
宁 夏	Ningxia	27.8	2.6	8.6	5.6	10.2	0.7
新 疆	Xinjiang	214.8	30.8	77.9	25.3	46.4	34.4

9

重点城市环境统计

ANNUAL STATISTIC REPORT ON ENVIRONMENT IN CHINA
2013

重点城市工业废水排放及处理情况（一）

（2013）

城 市 名 称	工业废水排放量/万吨	直接排入环境的	排入污水处理厂的	工业废水处理量/万吨	废水治理设施数/套	废水治理设施治理能力/（万吨/日）	废水治理设施运行费用/万元
总　计	1 125 581	752 588	372 993	2 734 777	41 600	13 520	3 697 404.6
北　京	9 486	1 959	7 527	10 498	518	60	36 310.9
天　津	18 692	7 015	11 677	29 161	1 088	151	171 400.4
石 家 庄	25 753	13 897	11 855	49 633	623	220	46 931.8
唐　山	12 589	8 955	3 633	324 208	780	1 988	194 385.2
秦 皇 岛	6 156	2 483	3 673	17 207	267	91	13 523.8
邯　郸	7 125	6 611	513	61 805	382	266	51 441.0
保　定	14 271	8 729	5 543	19 436	616	167	24 550.0
太　原	4 085	1 431	2 654	31 328	245	139	56 107.7
大　同	5 386	4 658	729	4 350	275	26	6 336.7
阳　泉	583	282	301	3 472	57	17	6 598.4
长　治	8 065	7 739	326	21 349	418	93	37 150.0
临　汾	6 444	6 409	34	11 252	325	84	26 531.9
呼和浩特	2 082	1 601	481	2 414	77	13	13 527.0
包　头	7 426	1 678	5 748	67 135	106	298	34 051.3
赤　峰	2 677	1 830	847	4 076	148	33	6 624.6
沈　阳	8 533	3 046	5 487	8 387	341	65	18 830.2
大　连	26 154	23 550	2 604	7 720	505	50	20 986.1
鞍　山	6 322	5 092	1 230	58 430	193	199	36 958.6
抚　顺	2 090	799	1 291	8 024	114	69	15 508.2
本　溪	2 953	2 839	114	69 676	196	344	32 978.6
锦　州	4 940	3 948	993	4 233	82	25	23 974.6
长　春	5 482	4 127	1 355	6 814	134	32	16 037.6
吉　林	10 366	6 474	3 892	13 113	95	48	15 006.9
哈 尔 滨	4 487	1 602	2 885	4 783	171	28	27 521.5
齐齐哈尔	8 937	8 647	290	5 955	106	54	8 921.5
牡 丹 江	1 917	1 876	41	1 300	62	18	2 544.6
上　海	45 426	18 071	27 355	65 019	1 743	319	170 673.6
南　京	25 291	19 429	5 862	121 869	678	450	141 470.5
无　锡	23 093	6 492	16 601	37 262	976	205	73 381.2
徐　州	15 376	13 092	2 285	19 294	320	85	27 865.3
常　州	12 017	6 438	5 579	21 461	514	75	28 104.6
苏　州	66 916	41 596	25 319	72 445	1 759	405	208 945.9
南　通	14 584	7 029	7 555	13 357	868	87	63 381.7
连 云 港	5 618	2 764	2 854	10 197	293	103	24 120.0
扬　州	9 731	6 360	3 371	6 425	313	46	29 094.2
镇　江	9 665	8 251	1 414	8 478	431	53	23 711.8
杭　州	39 186	8 589	30 597	121 953	1 059	560	102 686.1

重点城市工业废水排放及处理情况（一）（续表）

（2013）

城 市 名 称	工业废水 排放量/ 万吨	直接排入 环境的	排入污水处理 厂的	工业废水 处理量/ 万吨	废水治理 设施数/套	废水治理设施 治理能力/ （万吨/日）	废水治理设施 运行费用/ 万元
宁 波	19 666	12 219	7 447	30 005	928	151	105 258.6
温 州	7 433	4 425	3 009	7 071	925	69	29 427.8
湖 州	10 789	5 144	5 645	9 141	500	59	21 861.2
绍 兴	27 245	3 198	24 047	26 223	1 041	192	111 625.5
合 肥	6 018	4 091	1 927	25 563	288	163	19 595.4
芜 湖	3 779	1 945	1 834	4 100	227	32	11 465.9
马 鞍 山	6 745	6 138	607	102 644	195	314	72 605.2
福 州	4 682	3 389	1 293	17 944	401	138	22 873.4
厦 门	27 261	23 652	3 609	28 669	326	36	17 355.3
泉 州	20 080	12 474	7 606	30 237	827	124	34 085.2
南 昌	10 602	9 806	796	14 097	247	67	14 183.3
九 江	11 050	10 869	181	9 147	200	74	27 319.5
济 南	8 596	7 243	1 353	31 919	280	124	44 204.6
青 岛	10 661	2 790	7 871	47 099	478	159	34 788.8
淄 博	15 460	8 524	6 936	25 556	672	175	64 136.0
枣 庄	10 188	8 462	1 726	11 954	146	57	13 914.8
烟 台	9 530	5 753	3 777	16 075	498	116	28 976.8
潍 坊	28 103	7 787	20 317	29 365	605	148	63 023.9
济 宁	15 680	13 546	2 133	21 127	278	125	37 945.3
泰 安	8 224	6 988	1 236	11 954	204	128	13 904.2
日 照	7 911	6 332	1 579	8 540	138	46	18 831.3
郑 州	11 835	6 840	4 995	12 170	315	100	12 094.1
开 封	7 520	7 390	130	3 308	82	21	5 659.0
洛 阳	7 741	5 970	1 771	10 893	451	114	26 790.0
平 顶 山	6 352	6 352	0	10 232	136	76	10 575.3
安 阳	5 204	5 133	71	33 124	292	145	26 908.9
焦 作	10 868	9 246	1 621	10 114	181	57	13 056.1
三 门 峡	6 861	6 861	0	7 558	263	46	16 457.8
武 汉	18 814	15 535	3 278	101 100	293	376	56 707.5
宜 昌	18 419	16 445	1 974	14 643	336	85	27 306.8
荆 州	10 387	9 857	530	4 794	106	25	10 962.5
长 沙	4 049	3 675	373	3 542	264	28	7 105.6
株 洲	7 231	6 649	582	11 553	411	69	16 714.4
湘 潭	5 526	5 232	294	6 460	187	18	8 458.2
岳 阳	12 686	12 281	404	10 830	205	45	35 049.3
常 德	11 601	11 412	189	8 727	158	55	9 277.3
张 家 界	552	552	0	93	18	1	96.0
广 州	21 391	10 614	10 777	22 939	888	125	65 993.6

重点城市工业废水排放及处理情况（一）（续表）

（2013）

城 市 名 称	工业废水 排放量/ 万吨	直接排入 环境的	排入污水处理 厂的	工业废水 处理量/ 万吨	废水治理 设施数/套	废水治理设施 治理能力/ （万吨/日）	废水治理设施 运行费用/ 万元
韶 关	7 978	7 690	288	108 898	312	210	32 227.5
深 圳	15 288	7 725	7 562	14 838	1 436	51	89 264.2
珠 海	5 538	3 499	2 039	4 843	365	28	22 287.7
汕 头	5 658	4 806	852	5 301	378	57	13 266.6
湛 江	6 205	6 102	102	5 998	196	43	12 044.0
南 宁	9 752	9 208	544	9 523	279	97	18 042.1
柳 州	8 495	7 161	1 334	61 276	283	306	74 379.9
桂 林	4 005	3 382	623	5 058	228	23	4 238.8
北 海	1 830	1 558	271	25 292	51	59	7 876.5
海 口	825	140	684	816	62	7	2 165.2
重 庆	33 451	32 206	1 245	34 451	1 669	243	61 117.8
成 都	10 524	6 241	4 283	22 722	952	113	33 151.6
自 贡	1 828	1 681	147	1 360	110	7	1 946.1
攀 枝 花	4 137	3 407	731	53 198	213	308	59 778.1
泸 州	3 541	3 535	6	2 862	234	21	10 270.1
德 阳	6 638	6 231	408	8 474	278	47	13 583.0
绵 阳	5 788	5 191	597	8 904	282	52	8 294.4
南 充	3 228	1 511	1 717	1 401	50	5	1 963.8
宜 宾	5 744	5 618	125	4 267	152	35	9 766.2
贵 阳	2 262	2 262	0	21 168	198	65	16 157.4
遵 义	2 563	2 559	4	2 595	151	19	6 997.5
昆 明	4 808	4 532	276	22 863	508	126	26 373.9
曲 靖	2 473	2 270	203	45 959	385	74	16 008.8
玉 溪	5 045	5 024	22	23 644	269	155	20 773.8
拉 萨	378	378	0	373	25	5	389.7
西 安	7 771	5 236	2 535	5 302	296	23	12 745.9
铜 川	424	424	0	99	8	1	326.8
宝 鸡	4 579	3 823	757	8 189	199	73	13 224.0
咸 阳	5 238	3 025	2 213	5 701	212	86	14 148.1
渭 南	3 861	2 717	1 144	8 683	149	44	10 789.2
延 安	2 360	2 253	107	3 051	134	16	15 937.2
兰 州	4 909	2 323	2 586	13 627	96	46	6 226.3
金 昌	1 720	1 629	91	2 543	69	10	5 937.6
西 宁	2 798	2 798	0	21 137	85	62	9 887.5
银 川	6 194	4 894	1 300	7 028	75	44	15 251.4
石 嘴 山	1 399	1 265	134	7 372	170	44	4 489.9
乌鲁木齐	4 889	3 731	1 158	19 170	131	30	12 977.5
克拉玛依	4 816	4 346	470	12 764	42	34	18 330.2

重点城市工业废水排放及处理情况（二）

（2013）

单位：吨

城 市 名 称	工业废水中污染物产生量				
	化学需氧量	氨氮	石油类	挥发酚	氰化物
总 计	11 317 515.9	816 085.2	141 610.1	43 074.9	3 495.2
北 京	60 648.4	2 451.0	3 259.8	900.3	5.1
天 津	153 419.7	8 553.4	969.0	2.8	1.5
石 家 庄	231 222.2	11 652.9	1 348.8	309.8	23.2
唐 山	256 497.9	11 743.4	11 544.9	5 032.9	406.1
秦 皇 岛	59 764.9	2 168.8	60.5	...	...
邯 郸	60 460.8	7 163.6	3 506.1	3 449.3	178.0
保 定	176 725.0	4 372.0	271.7	4.0	3.2
太 原	24 514.7	1 568.8	446.7	720.6	35.2
大 同	34 528.1	3 100.7	83.5	40.1	0.4
阳 泉	1 305.8	193.2	6.5	0.4	...
长 治	113 204.1	7 600.2	2 796.8	3 799.4	109.5
临 汾	58 584.9	5 208.8	2 696.8	2 802.2	115.1
呼 和 浩 特	45 485.2	22 574.4	130.0	18.9	2.5
包 头	44 377.5	20 563.1	2 338.1	971.8	30.6
赤 峰	44 362.3	626.3	206.7	227.1	4.8
沈 阳	38 683.2	1 786.1	195.3	3.0	2.6
大 连	70 875.6	5 422.5	1 261.0	15.8	7.0
鞍 山	30 760.2	943.5	327.4	0.5	14.5
抚 顺	6 862.9	171.9	516.7	12.6	2.0
本 溪	74 116.9	1 880.4	3 322.4	1 055.3	122.9
锦 州	20 611.7	873.6	3 058.7	9.1	0.3
长 春	69 211.5	3 746.2	301.1	1.7	0.2
吉 林	56 706.8	1 203.8	606.0	180.8	10.3
哈 尔 滨	150 170.4	41 502.2	319.8	3 117.4	0.1
齐 齐 哈 尔	60 868.3	4 551.8	112.1	128.6	2.4
牡 丹 江	20 403.4	4 040.6	33.1	17.5	0.7
上 海	231 856.0	10 092.4	5 197.6	777.0	176.7
南 京	179 761.2	14 045.9	3 800.0	2 764.2	147.7
无 锡	156 032.0	4 375.7	1 486.1	80.9	3.0
徐 州	99 360.8	5 469.0	392.1	194.0	17.4
常 州	81 536.1	2 244.3	1 747.4	18.5	37.8
苏 州	384 508.8	13 122.1	934.7	5.9	49.5
南 通	167 572.9	4 108.1	1 670.9	23.8	33.0
连 云 港	268 607.1	3 862.0	366.3	112.4	...
扬 州	117 321.4	2 741.0	468.8	14.8	25.5
镇 江	66 824.3	9 832.1	138.8	77.3	28.2
杭 州	966 010.9	8 236.4	196.3	79.7	2.4

重点城市工业废水排放及处理情况（二）（续表）
（2013）

单位：吨

城 市 名 称	工业废水中污染物产生量				
	化学需氧量	氨氮	石油类	挥发酚	氰化物
宁　波	235 946.1	33 270.3	21 093.3	171.5	234.1
温　州	118 814.0	4 940.7	1 065.7	4.1	332.1
湖　州	89 522.5	2 373.8	132.7	…	…
绍　兴	261 744.3	15 547.7	12.3	1.6	0.2
合　肥	71 792.4	4 345.6	1 010.5	0.1	3.0
芜　湖	57 449.9	992.5	467.3	0.3	0.4
马 鞍 山	118 499.8	1 153.1	3 381.2	1 522.1	156.1
福　州	84 056.9	2 549.2	1 055.4	186.2	4.6
厦　门	64 861.5	838.2	288.7	0.5	24.6
泉　州	276 100.6	13 745.6	1 728.5	174.0	41.4
南　昌	99 802.1	9 450.4	469.7	258.2	8.8
九　江	81 670.8	2 035.7	1 170.2	337.6	0.4
济　南	80 686.2	14 299.9	5 032.8	1 263.9	106.4
青　岛	85 447.1	4 083.9	1 022.0	6.1	23.7
淄　博	255 684.2	19 027.7	2 143.8	633.6	18.4
枣　庄	123 921.8	7 670.3	252.1	604.2	9.8
烟　台	117 844.9	3 988.9	538.3	4.7	5.2
潍　坊	412 760.2	9 644.6	1 236.3	226.6	3.4
济　宁	172 371.4	3 111.5	101.9	285.2	6.6
泰　安	44 565.0	1 835.0	345.9	288.1	4.7
日　照	87 673.8	2 584.1	142.3	105.7	…
郑　州	54 412.0	1 680.0	282.0	0.6	5.5
开　封	27 397.9	4 192.3	86.3	0.1	2.4
洛　阳	45 824.2	2 315.4	5 495.3	29.1	2.5
平 顶 山	66 829.6	2 849.5	1 928.6	568.9	27.0
安　阳	72 579.4	2 007.1	1 774.7	1 150.4	82.0
焦　作	202 747.1	5 327.6	1 614.2	1.1	0.6
三 门 峡	54 065.1	5 714.9	344.0	5.1	3.8
武　汉	99 398.0	3 462.2	727.2	547.9	12.3
宜　昌	91 437.1	7 411.3	515.1	16.7	1.3
荆　州	88 670.1	4 458.1	85.1	60.2	22.9
长　沙	43 932.1	960.2	33.1	…	1.1
株　洲	25 167.9	4 182.7	339.5	1.5	8.0
湘　潭	29 613.4	1 407.4	1 660.1	793.8	172.5
岳　阳	119 844.0	23 636.0	875.7	46.0	3.8
常　德	36 755.6	1 674.1	58.4	0.8	…
张 家 界	2 648.5	75.3	11.2	…	…
广　州	196 272.5	6 910.9	1 174.2	23.2	54.3

重点城市工业废水排放及处理情况（二）（续表）
（2013）

单位：吨

城 市 名 称	工业废水中污染物产生量				
	化学需氧量	氨氮	石油类	挥发酚	氰化物
韶　关	10 153.4	826.3	16.0	6.9	0.9
深　圳	69 451.4	6 190.3	19.7	…	102.4
珠　海	32 327.0	1 875.1	146.9	4.7	0.8
汕　头	58 985.6	1 059.8	10.4	…	0.2
湛　江	61 279.8	1 246.8	7.8	3.8	…
南　宁	294 505.2	3 245.7	45.9	22.6	2.6
柳　州	109 045.9	16 091.7	2 071.8	1 654.2	10.8
桂　林	30 919.6	931.7	69.4	…	…
北　海	62 752.4	1 317.7	411.7	241.4	7.6
海　口	2 429.5	84.6	11.8	…	…
重　庆	327 201.4	24 216.3	2 922.9	962.0	91.8
成　都	97 362.9	16 992.5	487.3	…	31.1
自　贡	9 165.2	1 474.1	306.9	6.8	1.8
攀 枝 花	31 807.9	838.6	1 094.1	580.1	7.2
泸　州	36 591.5	37 488.1	21.5	…	…
德　阳	85 170.8	1 400.0	258.8	…	0.3
绵　阳	35 080.5	954.3	24.9	1.1	…
南　充	9 952.3	696.5	225.7	…	…
宜　宾	101 662.7	954.4	125.1	46.2	1.3
贵　阳	21 027.2	1 752.2	513.2	2.5	11.4
遵　义	39 939.5	1 426.3	38.1	0.8	…
昆　明	44 135.9	1 643.2	594.1	659.2	26.9
曲　靖	31 864.8	1 891.9	867.7	1 338.4	57.1
玉　溪	31 790.1	4 673.4	163.2	78.9	8.7
拉　萨	3 066.8	88.4	3.8	…	…
西　安	89 179.1	5 992.2	787.0	5.6	12.6
铜　川	1 665.5	33.1	…	…	…
宝　鸡	105 818.1	11 468.1	234.5	0.7	6.7
咸　阳	35 324.6	2 536.0	172.1	19.0	0.8
渭　南	67 086.7	3 733.3	273.6	0.5	…
延　安	22 156.4	2 165.4	1 026.7	0.6	…
兰　州	33 253.5	9 922.8	1 833.4	173.9	2.6
金　昌	7 130.5	5 675.9	57.0	…	…
西　宁	25 512.6	963.0	594.2	4.6	…
银　川	110 438.8	50 346.4	1 910.5	45.9	0.7
石 嘴 山	31 911.2	3 201.4	307.2	563.6	22.5
乌鲁木齐	50 451.4	38 095.0	3 280.0	1.6	102.0
克拉玛依	217 344.3	63 046.9	4 861.2	359.5	0.7

重点城市工业废水排放及处理情况（三）

（2013）

单位：吨

城 市名 称	工业废水中污染物产生量					
	汞	镉	六价铬	总铬	铅	砷
总　计	9.390	932.650	2 442.753	4 301.805	1 438.843	2 815.671
北　京	…	0.001	11.643	12.328	0.216	3.291
天　津	0.077	0.065	27.576	37.818	2.483	0.125
石 家 庄	…	…	0.469	7.707	0.041	0.003
唐　山	…	…	27.372	47.372	1.185	…
秦 皇 岛	…	…	3.156	3.181	…	0.001
邯　郸	…	0.003	0.026	0.026	0.024	0.043
保　定	…	…	4.515	4.714	4.725	…
太　原	0.009	0.005	0.458	0.482	0.104	0.028
大　同	…	…	0.011	0.011	…	…
阳　泉	0.001	…	…	…	0.031	…
长　治	0.493	…	2.361	2.620	…	…
临　汾	…	12.506	0.018	0.018	9.223	54.810
呼和浩特	0.001	…	…	…	…	…
包　头	0.004	4.008	0.868	0.955	6.175	12.458
赤　峰	0.098	65.055	0.016	0.018	65.012	105.830
沈　阳	0.001	0.003	6.025	8.641	0.965	…
大　连	…	…	1.652	1.768	0.083	…
鞍　山	…	…	3.300	3.300	…	…
抚　顺	0.089	0.002	…	…	0.019	0.123
本　溪	0.001	0.001	0.234	0.234	0.040	0.035
锦　州	…	…	4.799	7.092	…	…
长　春	…	…	3.930	3.936	…	…
吉　林	…	…	…	0.001	…	…
哈 尔 滨	…	…	5.948	5.948	0.361	…
齐齐哈尔	0.540	…	0.618	0.663	…	0.012
牡 丹 江	…	0.001	…	0.002	0.001	…
上　海	0.010	0.070	69.586	89.702	0.740	0.202
南　京	0.004	0.058	1.029	1.301	0.223	0.140
无　锡	…	2.251	37.402	50.247	1.228	…
徐　州	…	…	0.066	3.149	0.274	…
常　州	…	0.011	21.834	22.967	0.119	3.874
苏　州	…	…	29.170	73.002	2.238	…
南　通	0.021	0.030	59.842	66.465	20.681	0.028
连 云 港	…	0.001	0.016	3.000	0.306	…
扬　州	…	0.015	35.821	53.279	2.907	0.917
镇　江	…	…	29.018	83.168	…	0.919
杭　州	…	…	192.046	202.787	1.653	0.377

重点城市工业废水排放及处理情况（三）（续表）
（2013）

单位：吨

城 市 名 称	工业废水中污染物产生量					
	汞	镉	六价铬	总铬	铅	砷
宁　波	…	…	483.157	1 243.964	0.005	…
温　州	…	2.427	681.268	883.469	1.832	1.396
湖　州	…	…	2.194	18.102	1.377	…
绍　兴	…	…	1.273	3.146	0.251	0.034
合　肥	…	0.001	0.061	0.061	0.024	0.041
芜　湖	0.001	0.057	3.006	3.049	1.192	0.228
马 鞍 山	…	…	1.677	1.879	0.085	…
福　州	0.056	…	16.547	16.547	…	…
厦　门	0.004	0.145	266.976	510.788	1.861	0.561
泉　州	0.004	0.045	96.334	181.522	1.942	0.283
南　昌	…	0.075	28.001	34.486	0.207	…
九　江	0.498	89.844	5.113	6.713	22.300	32.770
济　南	…	…	2.503	2.552	…	0.050
青　岛	0.178	…	12.799	23.333	0.027	…
淄　博	0.031	…	0.749	13.323	1.021	1.043
枣　庄	0.001	…	0.049	0.064	0.001	0.037
烟　台	0.698	23.763	12.074	15.737	19.030	209.045
潍　坊	…	…	0.080	0.273	…	…
济　宁	…	…	0.025	0.031	0.227	…
泰　安	0.082	0.001	2.902	3.131	0.790	0.086
日　照	…	0.111	5.595	7.304	0.123	…
郑　州	…	…	3.382	3.382	…	0.021
开　封	0.001	0.336	0.542	14.764	0.714	0.063
洛　阳	0.003	0.595	1.146	1.363	17.278	1.240
平 顶 山	0.167	…	1.240	1.240	0.002	0.574
安　阳	…	61.495	0.274	0.423	22.178	5.103
焦　作	0.342	27.316	0.650	52.631	24.285	4.542
三 门 峡	0.493	73.043	0.068	0.068	145.939	39.711
武　汉	0.125	0.005	17.886	20.272	1.082	82.008
宜　昌	0.011	0.006	4.013	5.063	0.075	0.005
荆　州	0.001	…	1.654	2.112	…	0.098
长　沙	…	0.011	0.199	0.237	0.159	0.005
株　洲	1.626	8.587	1.967	3.040	36.673	4.319
湘　潭	…	0.815	27.537	32.297	5.835	0.681
岳　阳	0.088	10.173	0.098	0.131	16.651	4.122
常　德	…	…	0.053	0.379	…	0.004
张 家 界	…	0.007	0.014	0.014	0.039	0.020
广　州	…	0.046	8.601	36.949	0.587	0.153

重点城市工业废水排放及处理情况（三）（续表）

（2013）

单位：吨

城 市 名 称	工业废水中污染物产生量					
	汞	镉	六价铬	总铬	铅	砷
韶 关	0.001	0.963	0.283	0.597	5.432	2.996
深 圳	0.084	0.061	7.091	76.487	0.308	0.329
珠 海	0.001	0.030	0.152	0.896	0.052	…
汕 头	0.001	0.029	1.973	7.395	0.177	0.065
湛 江	…	…	0.015	0.183	0.001	1.002
南 宁	0.027	0.013	0.309	1.443	0.130	0.265
柳 州	0.115	8.018	3.876	4.347	16.406	3.537
桂 林	0.001	0.054	16.391	32.353	5.319	0.131
北 海	…	…	0.323	5.550	…	…
海 口	…	…	…	15.117	…	…
重 庆	…	0.028	107.302	146.155	7.410	1.646
成 都	…	0.007	3.734	6.426	0.027	0.072
自 贡	…	…	0.009	0.021	0.015	0.015
攀 枝 花	…	0.020	7.592	8.487	0.004	10.086
泸 州	…	…	1.538	1.616	…	…
德 阳	0.413	3.485	0.188	0.348	2.827	65.505
绵 阳	0.108	0.114	5.881	35.868	0.084	185.565
南 充	…	…	…	…	…	…
宜 宾	0.004	0.037	0.006	0.007	0.080	…
贵 阳	…	0.006	0.011	0.024	0.001	1.886
遵 义	…	…	0.032	0.242	0.003	…
昆 明	0.443	129.643	0.091	0.668	170.728	656.236
曲 靖	0.026	27.487	2.905	2.929	34.750	105.287
玉 溪	0.040	60.039	…	…	76.076	131.066
拉 萨	0.058	0.022	…	…	0.309	14.929
西 安	0.042	0.066	7.991	8.766	0.862	…
铜 川	…	…	…	…	…	…
宝 鸡	0.002	1.861	0.652	0.656	2.558	20.931
咸 阳	…	0.039	1.267	2.458	…	…
渭 南	0.006	0.056	…	0.004	0.394	0.018
延 安	…	…	…	0.021	…	…
兰 州	…	…	0.022	0.022	…	…
金 昌	0.171	305.104	0.046	0.091	656.880	937.891
西 宁	0.497	12.383	0.043	0.054	11.850	56.411
银 川	…	0.086	0.367	0.384	1.296	44.297
石 嘴 山	0.407	…	…	…	…	3.969
乌鲁木齐	1.184	0.007	0.128	0.128	0.016	0.075
克拉玛依	…	…	…	0.317	…	…

重点城市工业废水排放及处理情况（四）

（2013）

单位：吨

	工业废水中污染物排放量				
	化学需氧量	氨氮	石油类	挥发酚	氰化物
总　计	1 329 042.2	122 892.7	8 563.0	568.8	93.9
北　京	6 055.0	330.4	48.3	0.1	0.1
天　津	26 214.7	3 338.7	118.2	1.8	1.0
石　家　庄	39 802.2	5 702.1	153.9	12.4	2.0
唐　山	15 330.0	988.5	181.8	4.1	2.4
秦　皇　岛	14 836.8	804.8	35.3	…	…
邯　郸	8 849.0	509.8	82.9	2.1	1.5
保　定	21 043.2	1 109.6	28.3	4.0	…
太　原	4 088.6	294.3	26.6	0.3	0.6
大　同	8 609.5	1 225.8	40.5	32.1	…
阳　泉	288.7	60.1	0.6	…	…
长　治	14 791.4	1 483.6	399.7	285.7	8.3
临　汾	9 067.8	912.1	123.2	26.6	7.9
呼和浩特	10 542.9	631.2	3.3	…	…
包　头	5 634.5	5 650.8	581.9	3.9	1.3
赤　峰	5 388.3	143.4	68.5	…	…
沈　阳	9 158.7	809.0	78.9	0.3	0.1
大　连	19 439.2	2 138.4	466.8	2.0	0.1
鞍　山	4 973.8	447.1	50.5	0.4	1.8
抚　顺	882.6	29.8	21.3	1.0	0.3
本　溪	7 317.6	265.7	39.3	0.5	0.6
锦　州	7 055.0	264.0	50.4	1.1	0.1
长　春	11 669.6	1 383.8	29.8	…	0.3
吉　林	10 799.2	823.4	65.5	0.2	0.3
哈　尔　滨	6 914.4	1 055.3	33.6	0.1	…
齐齐哈尔	18 035.0	2 034.0	44.5	3.3	0.2
牡　丹　江	6 101.6	85.1	12.9	1.0	0.7
上　海	25 503.2	1 934.0	619.6	2.0	3.0
南　京	21 696.9	1 283.4	208.8	5.6	0.7
无　锡	11 070.4	441.6	172.7	0.7	0.4
徐　州	13 339.9	730.2	80.8	0.1	0.4
常　州	6 948.4	321.2	73.7	3.4	1.7
苏　州	45 323.4	3 259.3	92.4	1.6	3.1
南　通	23 954.2	1 093.5	255.5	6.0	1.2
连　云　港	10 621.5	1 584.9	18.8	4.7	…
扬　州	11 918.7	954.2	92.1	1.1	1.2
镇　江	5 876.4	440.7	30.3	1.7	1.8
杭　州	31 947.2	1 373.1	36.5	0.9	0.2

重点城市工业废水排放及处理情况（四）（续表）

（2013）

单位：吨

城 市 名 称	工业废水中污染物排放量				
	化学需氧量	氨氮	石油类	挥发酚	氰化物
宁　波	19 873.0	990.1	28.2	0.1	0.9
温　州	13 802.3	1 218.7	139.2	0.7	3.3
湖　州	9 019.8	417.4	11.1	…	…
绍　兴	28 486.3	1 967.2	2.5	1.5	…
合　肥	7 898.1	386.8	26.0	…	…
芜　湖	7 364.9	97.6	28.1	0.1	…
马 鞍 山	6 622.1	237.8	71.2	3.1	1.9
福　州	5 190.3	405.0	30.9	…	0.1
厦　门	3 727.0	134.0	19.6	…	0.8
泉　州	22 647.5	1 514.1	86.3	…	0.7
南　昌	11 478.8	1 595.7	97.5	2.8	5.3
九　江	11 686.0	850.0	211.0	1.3	0.1
济　南	5 413.0	380.1	60.5	2.2	1.7
青　岛	8 286.2	730.4	46.7	0.5	0.6
淄　博	13 694.5	1 108.8	66.7	2.5	0.2
枣　庄	6 092.2	314.7	34.5	0.2	0.2
烟　台	6 832.5	373.3	28.5	2.3	0.1
潍　坊	20 227.1	2 525.2	20.4	1.8	0.2
济　宁	9 349.1	518.7	10.0	…	…
泰　安	6 412.3	195.3	49.2	1.3	0.1
日　照	4 578.8	290.5	10.5	16.9	…
郑　州	11 977.7	561.8	172.8	0.5	3.6
开　封	9 640.9	1 502.0	56.3	0.1	0.7
洛　阳	9 883.3	623.1	30.6	2.8	0.8
平 顶 山	12 226.2	846.6	292.2	70.7	2.5
安　阳	9 362.8	456.3	41.4	0.1	3.5
焦　作	12 307.0	806.9	48.3	0.3	…
三 门 峡	8 342.5	567.0	75.5	1.3	0.3
武　汉	15 162.9	1 402.3	100.8	0.6	1.7
宜　昌	17 429.3	2 844.7	89.4	0.4	0.3
荆　州	24 878.1	2 088.4	63.8	1.4	0.5
长　沙	13 498.8	455.5	10.3	…	…
株　洲	5 422.2	1 181.1	44.5	0.4	1.7
湘　潭	3 751.8	521.5	108.1	1.6	2.5
岳　阳	25 990.1	12 165.2	49.1	0.1	0.4
常　德	9 980.0	978.7	15.8	…	…
张 家 界	2 623.7	75.0	10.2	…	…
广　州	22 663.5	1 389.4	79.3	1.3	2.0

重点城市工业废水排放及处理情况（四）（续表）

（2013）

单位：吨

城　市 名　称	工业废水中污染物排放量				
	化学需氧量	氨氮	石油类	挥发酚	氰化物
韶　关	4 066.2	515.8	3.5	…	0.9
深　圳	15 731.8	1 423.7	4.8	…	0.5
珠　海	6 424.5	503.3	37.3	0.1	…
汕　头	11 318.2	535.6	1.3	…	…
湛　江	9 377.1	455.8	3.8	0.3	…
南　宁	23 954.0	1 298.0	19.5	9.2	2.1
柳　州	10 164.1	1 271.1	43.6	0.9	1.7
桂　林	8 922.9	517.6	18.1	…	…
北　海	9 013.8	37.0	14.0	0.1	0.1
海　口	857.8	51.1	8.3	…	…
重　庆	51 533.7	3 265.7	335.8	9.5	1.5
成　都	12 321.2	800.6	31.2	…	…
自　贡	2 971.7	455.2	109.7	2.3	0.6
攀 枝 花	1 833.0	131.5	6.7	0.2	…
泸　州	9 073.6	306.6	8.6	…	…
德　阳	6 112.3	512.7	41.3	0.1	…
绵　阳	4 991.2	143.1	7.4	0.1	…
南　充	5 058.8	336.2	10.3	…	…
宜　宾	17 227.6	449.6	56.1	0.1	0.7
贵　阳	6 993.0	292.6	59.8	…	…
遵　义	8 815.2	357.9	17.2	…	…
昆　明	8 114.8	266.1	71.4	0.3	0.4
曲　靖	7 855.2	975.0	39.8	0.2	0.4
玉　溪	11 900.5	125.8	3.0	0.1	0.1
拉　萨	311.6	27.1	0.2	…	…
西　安	21 614.7	1 632.5	184.3	0.1	…
铜　川	1 463.4	24.5	…	…	…
宝　鸡	14 127.0	884.7	40.1	0.2	1.2
咸　阳	14 427.0	1 298.6	53.8	0.3	0.4
渭　南	18 134.6	741.7	10.8	…	…
延　安	5 006.6	248.3	108.0	0.3	…
兰　州	4 445.8	2 723.1	87.9	0.8	…
金　昌	4 340.8	4 297.4	24.9	…	…
西　宁	15 759.3	591.2	62.1	0.5	…
银　川	16 726.1	2 741.0	60.5	4.9	0.6
石 嘴 山	5 139.2	1 230.6	13.1	0.2	0.6
乌鲁木齐	5 949.9	666.2	38.6	0.6	2.5
克拉玛依	2 051.8	104.5	91.5	7.9	…

重点城市工业废水排放及处理情况（五）

（2013）

单位：吨

城 市 名 称	工业废水中污染物排放量					
	汞	镉	六价铬	总铬	铅	砷
总 计	0.325	3.677	37.047	84.055	15.529	20.925
北 京	…	0.001	0.320	0.422	0.054	0.012
天 津	0.006	0.003	0.133	0.356	0.086	0.012
石 家 庄	…	…	0.002	0.351	0.009	…
唐 山	…	…	0.042	0.136	0.004	…
秦 皇 岛	…	…	0.006	0.026	…	…
邯 郸	…	0.003	0.026	0.026	0.024	0.037
保 定	…	…	0.073	0.082	0.177	…
太 原	0.009	0.005	0.132	0.139	0.104	0.028
大 同	…	…	0.003	0.003	…	…
阳 泉	…	…	…	…	0.004	…
长 治	0.002	…	0.003	0.007	…	…
临 汾	…	0.025	0.001	0.001	0.065	0.164
呼和浩特	0.001	…	…	…	…	…
包 头	0.004	…	0.031	0.041	0.180	…
赤 峰	0.034	0.064	…	…	1.008	1.203
沈 阳	0.001	…	0.019	0.070	0.044	…
大 连	…	…	0.051	0.070	0.009	…
鞍 山	…	…	…	…	…	…
抚 顺	0.002	0.002	…	…	0.018	0.022
本 溪	…	…	…	…	…	0.001
锦 州	…	…	0.024	0.291	…	…
长 春	…	…	0.032	0.052	…	…
吉 林	…	…	…	…	…	…
哈 尔 滨	…	…	0.036	0.036	0.013	…
齐齐哈尔	…	…	0.015	0.024	…	0.012
牡 丹 江	…	0.001	…	0.002	0.001	…
上 海	0.008	0.001	1.067	2.381	0.105	0.043
南 京	…	0.001	0.116	0.141	0.003	0.076
无 锡	…	…	0.117	0.429	0.078	…
徐 州	…	…	0.012	0.622	0.019	…
常 州	…	0.005	0.182	0.198	0.018	0.007
苏 州	…	…	0.885	2.682	0.042	…
南 通	0.002	0.007	1.789	2.551	0.330	0.008
连 云 港	…	0.001	0.001	0.008	0.027	…
扬 州	…	…	0.457	1.120	0.164	0.001
镇 江	…	…	0.069	0.241	…	0.091
杭 州	…	…	2.710	3.068	0.005	…

重点城市工业废水排放及处理情况（五）（续表）
（2013）

单位：吨

城　市名　称	工业废水中污染物排放量					
	汞	镉	六价铬	总铬	铅	砷
宁　波	…	…	0.733	2.397	0.001	…
温　州	…	0.175	3.461	4.846	0.082	0.031
湖　州	…	…	0.059	0.352	0.050	…
绍　兴	…	…	0.063	0.210	0.035	0.014
合　肥	…	0.001	0.007	0.007	0.013	0.001
芜　湖	…	0.002	0.054	0.077	0.049	0.012
马 鞍 山	…	…	0.007	0.009	…	…
福　州	0.007	…	0.876	0.876	…	…
厦　门	0.001	0.009	1.054	2.518	0.467	0.137
泉　州	0.001	0.002	0.661	7.012	0.144	0.111
南　昌	…	0.007	16.603	16.698	0.016	…
九　江	0.002	0.095	0.068	0.103	0.342	0.813
济　南	…	…	0.155	0.194	…	…
青　岛	…	…	0.073	0.341	0.001	…
淄　博	…	…	0.006	0.308	0.024	0.022
枣　庄	0.001	…	0.049	0.064	0.001	0.037
烟　台	0.006	0.001	0.154	0.207	0.004	0.064
潍　坊	…	…	0.006	0.148	…	…
济　宁	…	…	0.001	0.003	0.001	…
泰　安	…	…	…	…	0.010	0.022
日　照	…	0.001	0.014	0.024	0.011	…
郑　州	…	…	0.032	0.032	…	0.021
开　封	…	0.269	0.542	0.791	0.714	0.062
洛　阳	0.003	0.284	0.021	0.171	1.042	0.489
平 顶 山	0.002	…	0.014	0.014	0.001	0.061
安　阳	…	…	0.002	0.003	0.166	…
焦　作	0.003	0.010	0.041	21.704	0.244	0.017
三 门 峡	0.003	0.084	0.003	0.003	0.918	0.368
武　汉	0.002	0.005	1.111	1.193	0.123	0.152
宜　昌	…	0.004	0.009	0.063	0.009	0.004
荆　州	0.001	…	0.300	0.524	…	0.098
长　沙	…	0.011	0.045	0.059	0.050	0.005
株　洲	0.140	0.365	0.068	0.090	1.427	0.611
湘　潭	…	0.139	0.089	0.119	0.370	0.110
岳　阳	0.001	0.004	0.005	0.007	0.131	0.325
常　德	…	…	0.006	0.014	…	0.003
张 家 界	…	0.004	0.008	0.008	0.013	0.008
广　州	…	0.013	1.298	2.005	0.041	0.025

重点城市工业废水排放及处理情况（五）（续表）

（2013）

单位：吨

城 市 名 称	工业废水中污染物排放量					
	汞	镉	六价铬	总铬	铅	砷
韶　关	0.001	0.252	0.080	0.194	1.370	0.408
深　圳	0.007	0.019	0.078	0.185	0.110	0.126
珠　海	…	0.001	0.022	0.121	0.015	…
汕　头	0.001	0.001	0.006	0.025	0.011	0.006
湛　江	…	…	0.002	0.019	0.001	0.184
南　宁	0.016	0.007	0.001	0.089	0.057	0.131
柳　州	0.004	0.043	0.051	0.086	0.129	0.094
桂　林	…	0.005	0.080	0.145	0.055	0.045
北　海	…	…	0.006	1.144	…	…
海　口	…	…	…	0.120	…	…
重　庆	…	0.003	0.218	0.423	0.081	0.033
成　都	…	0.001	0.061	0.217	0.006	…
自　贡	…	…	0.001	0.002	0.015	0.015
攀 枝 花	…	0.001	0.001	0.012	0.002	0.010
泸　州	…	…	0.015	0.033	…	…
德　阳	0.002	…	0.001	0.028	…	0.062
绵　阳	0.001	0.003	0.083	0.532	0.002	0.015
南　充	…	…	…	…	…	…
宜　宾	0.003	…	0.001	0.001	…	…
贵　阳	…	0.002	0.004	0.004	0.001	0.001
遵　义	…	…	0.003	0.018	0.002	…
昆　明	…	0.754	…	…	2.685	2.478
曲　靖	…	0.123	0.003	0.017	0.318	0.460
玉　溪	…	0.001	…	…	0.003	0.046
拉　萨	…	…	…	…	…	8.942
西　安	…	0.001	0.039	0.173	0.027	…
铜　川	…	…	…	…	…	…
宝　鸡	…	0.032	0.023	0.023	0.351	0.395
咸　阳	…	0.001	0.090	1.322	…	…
渭　南	0.001	0.005	…	0.004	0.006	0.002
延　安	…	…	…	0.021	…	…
兰　州	…	…	0.011	0.011	…	…
金　昌	0.028	0.544	0.041	0.044	0.924	1.846
西　宁	0.006	0.266	0.007	0.011	0.241	0.171
银　川	…	…	0.022	0.039	0.010	0.037
石 嘴 山	…	…	…	…	…	…
乌鲁木齐	0.012	0.007	0.050	0.050	0.016	0.075
克拉玛依	…	…	…	0.173	…	…

重点城市工业废气排放及处理情况（一）

（2013）

城 市 名 称	工业废气排放总量（标态）/亿米³	工业废气中污染物产生量/万吨			工业废气中污染物排放量/万吨		
		二氧化硫	氮氧化物	烟（粉）尘	二氧化硫	氮氧化物	烟（粉）尘
总　　计	364 825	3 127.6	1 005.6	39 202.5	904.2	796.4	501.5
北　　京	3 692	15.7	9.5	422.9	5.2	7.6	2.7
天　　津	8 080	142.6	35.0	670.3	20.8	25.1	6.3
石 家 庄	7 049	69.8	24.1	802.0	17.6	20.0	10.0
唐　　山	33 110	71.5	31.6	1 635.8	28.3	28.3	47.9
秦 皇 岛	3 022	15.1	6.4	215.8	7.3	5.3	7.8
邯　　郸	14 207	53.2	17.7	981.2	18.5	16.3	21.4
保　　定	1 897	15.7	3.3	132.6	7.9	3.1	3.9
太　　原	4 577	35.6	12.1	488.5	8.9	9.6	3.7
大　　同	2 436	48.8	14.5	364.0	12.4	9.6	5.8
阳　　泉	2 135	21.7	7.3	210.8	9.9	6.5	2.3
长　　治	4 279	39.6	13.6	597.7	13.2	10.8	18.2
临　　汾	3 055	42.8	10.2	417.5	9.7	6.3	9.3
呼和浩特	3 271	40.0	16.7	687.6	9.6	13.2	4.9
包　　头	6 978	55.2	14.1	664.4	19.8	13.5	9.8
赤　　峰	1 731	84.2	6.3	340.0	12.3	6.3	2.5
沈　　阳	2 068	21.9	8.4	216.9	13.1	8.3	6.0
大　　连	2 262	24.3	12.3	422.6	10.3	10.1	4.6
鞍　　山	4 748	13.8	6.7	207.2	11.6	6.7	9.0
抚　　顺	2 470	12.8	5.4	206.3	5.1	4.3	4.8
本　　溪	4 661	8.0	5.2	254.5	7.0	5.2	6.2
锦　　州	787	8.2	2.4	85.8	3.7	1.8	2.0
长　　春	1 863	13.7	11.3	173.5	5.7	9.5	7.3
吉　　林	2 983	10.5	8.7	390.3	7.0	8.1	3.4
哈 尔 滨	2 544	11.8	10.0	394.7	6.6	8.6	8.2
齐齐哈尔	952	7.6	6.9	142.7	5.9	6.8	8.7
牡 丹 江	467	3.6	3.5	174.3	2.1	3.4	4.6
上　　海	13 344	53.9	32.9	660.8	17.3	26.2	6.7
南　　京	7 930	50.9	16.9	541.9	11.1	11.0	6.5
无　　锡	5 979	29.0	15.7	567.8	8.3	11.8	4.4
徐　　州	5 184	46.0	20.6	766.4	13.5	14.9	5.2
常　　州	3 260	11.7	9.0	210.7	3.6	7.3	3.5
苏　　州	14 387	60.0	26.1	743.9	16.5	18.1	6.5
南　　通	2 214	21.5	6.8	231.3	6.3	5.2	3.4
连 云 港	965	10.6	3.7	185.4	4.3	3.0	1.9
扬　　州	1 490	18.7	7.5	166.5	4.6	6.3	1.6
镇　　江	2 723	28.5	10.9	236.6	6.3	7.1	2.1
杭　　州	4 758	17.1	9.0	421.3	8.2	6.7	4.0

重点城市工业废气排放及处理情况（一）（续表）

（2013）

城 市名 称	工业废气排放总量（标态）/亿米³	工业废气中污染物产生量/万吨			工业废气中污染物排放量/万吨		
		二氧化硫	氮氧化物	烟（粉）尘	二氧化硫	氮氧化物	烟（粉）尘
宁　波	6 487	57.7	29.0	647.9	13.5	21.3	2.5
温　州	1 674	11.7	5.6	110.3	3.4	3.5	1.9
湖　州	1 887	8.7	5.3	360.8	3.7	4.9	2.8
绍　兴	1 420	11.9	4.7	136.5	6.0	4.2	3.1
合　肥	2 862	12.3	9.8	481.3	4.1	7.0	4.2
芜　湖	2 644	11.3	7.3	775.7	3.8	6.9	3.9
马 鞍 山	6 877	17.9	12.3	458.4	6.5	9.4	3.2
福　州	3 551	24.0	11.9	296.2	7.6	7.2	4.3
厦　门	1 119	5.4	2.7	46.7	1.9	1.2	0.4
泉　州	3 254	23.6	7.9	159.8	9.1	7.0	5.3
南　昌	1 374	9.0	2.7	78.2	4.1	1.9	1.1
九　江	2 740	25.5	6.9	231.2	9.4	5.9	3.5
济　南	4 087	26.7	9.7	418.8	8.1	7.3	4.7
青　岛	2 129	26.4	13.0	260.1	6.9	6.5	2.8
淄　博	5 275	60.1	13.9	699.7	20.7	12.5	4.7
枣　庄	2 189	22.8	7.0	569.5	6.0	6.7	2.2
烟　台	3 151	44.8	9.0	502.9	8.0	8.3	3.5
潍　坊	2 608	38.1	8.1	492.8	12.8	7.6	3.4
济　宁	3 852	47.4	16.8	852.0	12.3	14.8	6.0
泰　安	1 893	22.7	5.8	236.5	6.0	5.6	1.4
日　照	4 394	12.8	6.7	359.0	5.2	5.5	2.9
郑　州	4 002	30.8	16.9	722.0	10.6	13.4	3.4
开　封	788	11.4	3.4	58.8	4.0	3.3	2.3
洛　阳	6 208	46.0	14.0	687.9	11.7	11.9	5.2
平 顶 山	3 178	24.1	8.0	614.4	9.7	7.0	7.5
安　阳	3 825	37.0	5.9	306.7	11.3	5.8	8.9
焦　作	2 178	11.8	7.3	379.7	6.0	7.3	3.3
三 门 峡	1 657	23.8	7.0	472.8	10.5	7.0	3.0
武　汉	5 642	30.0	11.3	316.0	9.6	9.6	2.0
宜　昌	2 135	92.1	3.8	194.0	5.5	3.6	1.4
荆　州	788	10.5	1.7	50.9	4.4	0.9	1.4
长　沙	623	4.8	2.8	116.1	2.1	1.6	2.0
株　洲	1 674	30.4	3.3	157.5	4.2	2.6	0.8
湘　潭	2 267	18.4	10.3	667.6	4.6	3.7	2.4
岳　阳	1 936	14.8	7.2	124.9	5.8	4.6	2.8
常　德	1 164	10.8	5.8	171.6	3.7	4.9	2.1
张 家 界	154	2.5	0.6	29.4	2.3	0.6	0.3
广　州	3 774	44.7	10.7	335.0	6.6	5.7	1.7

重点城市工业废气排放及处理情况（一）（续表）

（2013）

城 市名 称	工业废气排放总量（标态）/亿米³	工业废气中污染物产生量/万吨			工业废气中污染物排放量/万吨		
		二氧化硫	氮氧化物	烟（粉）尘	二氧化硫	氮氧化物	烟（粉）尘
韶 关	1 206	12.4	3.8	46.3	5.6	3.6	0.5
深 圳	2 067	5.8	2.8	39.0	0.6	1.8	0.2
珠 海	1 328	9.0	4.9	83.2	2.3	4.1	1.0
汕 头	927	10.7	2.9	86.1	2.9	2.4	0.7
湛 江	1 094	11.9	2.9	102.4	2.3	1.7	1.0
南 宁	1 422	12.7	4.7	271.3	3.3	3.5	2.1
柳 州	4 078	19.0	5.2	405.6	4.7	5.0	5.5
桂 林	900	12.6	3.4	37.4	3.6	3.2	1.5
北 海	588	4.1	1.6	54.7	1.2	1.4	0.6
海 口	50	0.2	…	0.2	0.2	…	0.1
重 庆	9 532	140.5	30.2	2 034.7	49.4	24.8	18.0
成 都	3 049	11.6	6.3	236.4	5.2	4.4	2.1
自 贡	498	3.7	0.9	18.8	2.7	0.8	0.3
攀枝花	2 746	15.9	3.6	225.8	10.6	2.9	4.4
泸 州	553	17.0	2.6	107.2	4.2	2.3	0.6
德 阳	769	3.2	1.4	82.0	1.6	1.2	1.1
绵 阳	849	8.2	3.1	224.0	3.5	3.1	0.8
南 充	377	1.0	0.2	1.3	0.8	0.2	0.4
宜 宾	991	45.1	5.9	259.8	10.6	4.5	1.6
贵 阳	1 672	17.5	3.7	310.1	7.1	3.0	2.4
遵 义	1 829	19.8	6.2	291.3	9.4	4.5	1.5
昆 明	3 678	45.6	7.0	315.5	10.3	6.8	5.7
曲 靖	4 109	67.5	11.1	723.7	17.0	10.2	3.7
玉 溪	1 200	12.6	2.0	123.2	4.4	1.9	12.0
拉 萨	59	0.1	0.2	19.8	0.1	0.2	0.1
西 安	844	15.3	4.6	168.0	6.9	3.5	1.6
铜 川	1 359	9.4	15.1	114.3	1.7	4.4	3.6
宝 鸡	1 170	25.0	6.9	276.1	2.9	5.4	1.6
咸 阳	1 368	21.8	10.1	399.7	6.2	7.4	2.4
渭 南	2 192	54.8	12.7	438.3	15.2	10.1	11.2
延 安	175	1.8	0.5	11.8	1.7	0.5	0.6
兰 州	4 068	16.8	9.0	382.1	7.2	8.0	4.0
金 昌	801	152.3	2.5	93.2	10.4	2.2	0.7
西 宁	3 791	14.1	5.4	305.6	7.2	5.3	5.3
银 川	2 168	35.1	9.1	456.8	9.2	8.4	2.7
石嘴山	1 601	34.6	19.8	450.4	9.0	8.1	8.7
乌鲁木齐	3 527	22.3	11.4	337.1	7.4	11.4	5.2
克拉玛依	1 169	16.6	3.7	57.6	4.5	3.7	0.5

重点城市工业废气排放及处理情况（二）

（2013）

城 市 名 称	废气治理设施数/ 套	废气治理设施处理能力 （标态）/（万米³/小时）	废气治理设施运行费用/ 万元
总　计	130 563	807 516	8 939 035
北　京	3 316	10 729	90 066
天　津	4 262	21 118	256 506
石 家 庄	2 569	16 229	121 640
唐　山	3 620	66 064	484 565
秦 皇 岛	3 623	7 420	36 805
邯　郸	1 752	38 116	306 463
保　定	1 307	4 751	20 964
太　原	1 477	13 269	176 423
大　同	1 647	5 365	73 841
阳　泉	1 140	4 486	39 799
长　治	934	7 009	92 242
临　汾	974	8 430	90 381
呼 和 浩 特	1 080	10 281	48 634
包　头	858	11 431	133 541
赤　峰	730	5 377	42 147
沈　阳	2 107	9 089	35 366
大　连	1 869	7 085	58 210
鞍　山	1 528	13 868	88 005
抚　顺	486	4 837	34 612
本　溪	773	7 716	72 801
锦　州	506	5 662	15 293
长　春	885	5 784	61 347
吉　林	876	5 762	65 214
哈 尔 滨	1 317	8 452	32 707
齐 齐 哈 尔	719	3 338	18 177
牡 丹 江	303	1 612	8 023
上　海	4 896	29 883	471 487
南　京	1 685	11 924	238 355
无　锡	2 704	16 550	143 703
徐　州	1 454	9 526	172 345
常　州	1 015	10 991	66 528
苏　州	4 642	26 474	291 255
南　通	1 593	5 730	58 358
连 云 港	759	3 139	31 573
扬　州	553	4 609	62 516
镇　江	623	3 588	87 723
杭　州	2 575	9 332	104 910

重点城市工业废气排放及处理情况（二）（续表）

（2013）

城 市 名 称	废气治理设施数/ 套	废气治理设施处理能力（标态）/ （万米³/小时）	废气治理设施运行费用/ 万元
宁 波	2 673	11 734	251 300
温 州	2 273	4 293	79 650
湖 州	1 351	5 146	109 280
绍 兴	2 317	3 856	45 840
合 肥	603	2 353	43 063
芜 湖	791	3 996	45 912
马 鞍 山	673	9 939	182 600
福 州	800	8 824	130 159
厦 门	1 011	2 719	39 722
泉 州	1 238	5 336	77 800
南 昌	646	2 805	37 634
九 江	689	4 391	52 369
济 南	1 086	16 346	142 439
青 岛	1 280	7 238	98 042
淄 博	3 209	14 032	147 350
枣 庄	864	4 678	41 583
烟 台	1 245	8 937	78 355
潍 坊	1 585	7 735	78 222
济 宁	1 107	12 756	108 136
泰 安	686	4 740	38 915
日 照	951	8 565	170 042
郑 州	1 187	7 679	72 179
开 封	246	881	7 225
洛 阳	1 067	8 033	73 472
平 顶 山	471	4 027	23 390
安 阳	978	7 559	85 784
焦 作	1 048	4 602	31 313
三 门 峡	558	2 783	27 149
武 汉	1 039	8 882	125 055
宜 昌	932	6 729	30 325
荆 州	342	739	13 973
长 沙	333	630	6 566
株 洲	783	2 010	42 634
湘 潭	258	1 635	24 206
岳 阳	400	2 374	42 479
常 德	762	4 122	31 111
张 家 界	44	83	954
广 州	2 305	9 671	206 143

重点城市工业废气排放及处理情况（二）（续表）

（2013）

城 市 名 称	废气治理设施数/ 套	废气治理设施处理能力 （标态）/（万米³/小时）	废气治理设施运行费用/ 万元
韶 关	887	5 853	56 550
深 圳	1 226	2 612	40 847
珠 海	1 231	4 632	51 118
汕 头	552	2 656	21 739
湛 江	287	2 328	45 638
南 宁	835	2 722	33 838
柳 州	964	8 011	84 457
桂 林	532	1 200	10 817
北 海	128	1 155	13 690
海 口	61	103	2 090
重 庆	4 439	23 084	231 696
成 都	2 558	6 552	62 816
自 贡	150	531	6 244
攀 枝 花	696	6 262	95 847
泸 州	236	1 452	32 414
德 阳	664	2 634	12 513
绵 阳	663	2 627	30 321
南 充	35	337	1 070
宜 宾	529	4 527	41 356
贵 阳	469	3 222	46 556
遵 义	721	2 742	70 545
昆 明	1 616	7 252	117 093
曲 靖	1 013	8 576	85 195
玉 溪	1 049	2 257	36 426
拉 萨	150	124	1 274
西 安	661	2 817	27 281
铜 川	245	476	12 245
宝 鸡	539	2 599	54 973
咸 阳	427	5 367	27 699
渭 南	416	4 338	62 891
延 安	278	286	5 451
兰 州	902	5 671	63 192
金 昌	227	1 173	71 215
西 宁	718	5 114	80 864
银 川	436	4 111	88 174
石 嘴 山	567	3 639	49 594
乌鲁木齐	293	9 174	74 840
克拉玛依	175	1 417	13 574

重点城市工业固体废物产生及处置利用情况

（2013）

单位：万吨

城 市 名 称	一般工业固体废物产生量	一般工业固体废物综合利用量	一般工业固体废物处置量	一般工业固体废物贮存量	一般工业固体废物倾倒丢弃量	一般工业固体废弃物综合利用率/%
总　计	142 948	96 386	38 461	9 536	21	66.9
北　京	1 044	904	140	0	0	86.6
天　津	1 592	1 582	10	0	0	99.4
石 家 庄	1 558	1 534	22	63	0	94.7
唐　山	10 222	7 611	2 572	244	0	73.3
秦 皇 岛	1 761	868	694	198	0	49.3
邯　郸	3 505	3 405	23	141	0	95.4
保　定	698	626	72	0	0	89.6
太　原	2 632	1 436	1 021	179	0	54.5
大　同	3 801	1 641	1 812	349	0	43.2
阳　泉	1 974	461	1 425	89	0	23.3
长　治	2 618	1 756	556	324	0	66.6
临　汾	2 958	2 392	323	329	0	79.4
呼和浩特	876	408	461	7	0	46.6
包　头	2 958	1 423	1 221	315	0	48.1
赤　峰	1 950	502	1 264	186	0	25.7
沈　阳	791	733	133	51	0	92.7
大　连	518	468	50	0	0	90.3
鞍　山	5 178	1 350	1 277	2 550	0	26.1
抚　顺	2 913	727	1 866	320	0	24.9
本　溪	6 464	1 050	5 122	284	8	16.2
锦　州	252	235	2	14	0	93.3
长　春	606	605	1	0	0	99.8
吉　林	1 189	973	39	336	0	72.3
哈 尔 滨	575	539	35	0	0	93.9
齐齐哈尔	431	351	27	53	0	81.5
牡 丹 江	251	183	4	64	0	73.1
上　海	2 054	1 995	58	5	0	96.9
南　京	1 697	1 555	21	129	0	91.2
无　锡	1 052	957	60	35	0	91.0
徐　州	1 608	1 689	29	0	0	99.2
常　州	667	655	12	0	0	98.2
苏　州	2 486	2 435	51	0	0	97.9
南　通	439	430	8	1	0	98.0
连 云 港	493	478	1	23	0	95.3
扬　州	315	308	6	1	0	97.7
镇　江	723	710	5	8	0	98.1
杭　州	687	648	38	3	0	94.1

重点城市工业固体废物产生及处置利用情况（续表）
（2013）

单位：万吨

城　市名　称	一般工业固体废物产生量	一般工业固体废物综合利用量	一般工业固体废物处置量	一般工业固体废物贮存量	一般工业固体废物倾倒丢弃量	一般工业固体废弃物综合利用率/%
宁　波	1 253	1 168	49	37	0	93.2
温　州	214	212	2	0	0	99.0
湖　州	185	178	7	0	0	96.3
绍　兴	360	334	23	4	0	92.7
合　肥	1 024	955	8	62	0	93.2
芜　湖	329	323	11	0	0	96.7
马 鞍 山	2 381	1 672	524	186	0	70.2
福　州	810	764	44	5	0	94.3
厦　门	103	97	9	2	0	90.3
泉　州	806	775	15	24	0	95.2
南　昌	224	219	4	0	1	97.7
九　江	989	474	90	442	0	47.3
济　南	932	920	11	0	0	98.7
青　岛	822	791	22	20	0	94.9
淄　博	1 688	1 614	5	77	0	95.2
枣　庄	767	775	0	0	0	100.0
烟　台	2 489	2 083	370	36	0	83.7
潍　坊	888	812	22	66	0	91.5
济　宁	2 217	2 022	140	63	0	90.9
泰　安	1 167	1 251	0	18	0	98.5
日　照	967	957	6	4	0	98.9
郑　州	1 549	1 139	377	33	0	73.5
开　封	132	128	7	1	0	95.0
洛　阳	3 294	1 964	1 321	10	0	59.6
平 顶 山	2 432	2 299	1	184	0	92.5
安　阳	1 222	1 074	149	0	0	87.8
焦　作	967	570	403	23	0	57.2
三 门 峡	1 795	608	1 122	84	0	33.5
武　汉	1 384	1 409	8	10	0	98.7
宜　昌	1 658	795	861	3	0	47.9
荆　州	285	99	0	186	0	34.6
长　沙	101	87	10	4	0	85.7
株　洲	366	334	21	22	0	88.6
湘　潭	757	731	1	25	0	96.5
岳　阳	373	298	6	85	0	76.7
常　德	300	294	1	5	0	97.9
张 家 界	47	45	0	1	0	97.2
广　州	556	529	22	5	0	95.2

重点城市工业固体废物产生及处置利用情况（续表）

（2013）

单位：万吨

城　市 名　称	一般工业固体 废物产生量	一般工业固体废 物综合利用量	一般工业固体 废物处置量	一般工业固体 废物贮存量	一般工业固体 物倾倒丢弃量	一般工业固体 废弃物综合利 用率/%
韶　关	758	606	21	135	0	79.5
深　圳	111	63	46	2	0	56.8
珠　海	265	246	17	2	0	92.8
汕　头	159	155	2	2	0	97.6
湛　江	262	258	5	1	0	97.7
南　宁	396	376	20	1	0	94.7
柳　州	1 187	1 129	37	42	0	93.5
桂　林	210	186	1	23	0	88.3
北　海	229	229	0	0	0	99.9
海　口	6	6	0	0	0	93.6
重　庆	3 162	2 695	415	79	11	84.2
成　都	533	525	8	0	0	98.4
自　贡	55	50	5	0	0	90.6
攀 枝 花	6 037	1 338	4 659	64	0	22.1
泸　州	235	213	14	8	0	90.5
德　阳	366	221	172	0	0	56.2
绵　阳	250	245	3	2	0	97.9
南　充	16	16	0	0	0	98.8
宜　宾	546	501	41	5	0	91.7
贵　阳	1 104	516	595	5	0	46.2
遵　义	652	380	265	8	0	58.2
昆　明	3 319	1 413	1 841	77	0	42.6
曲　靖	2 256	1 606	166	486	0	71.1
玉　溪	2 823	1 551	1 213	72	0	54.9
拉　萨	279	5	25	264	0	2.0
西　安	255	244	9	2	0	95.8
铜　川	168	166	2	0	0	98.8
宝　鸡	436	256	170	10	0	58.8
咸　阳	593	551	20	25	0	92.5
渭　南	1 826	587	1 237	2	0	32.1
延　安	101	80	20	0	0	79.8
兰　州	625	608	15	2	0	97.4
金　昌	1 245	263	854	127	0	21.1
西　宁	538	556	4	9	0	97.8
银　川	686	581	59	45	0	84.8
石 嘴 山	682	485	216	7	0	70.7
乌鲁木齐	1 128	988	139	0	0	87.6
克拉玛依	79	71	8	0	0	90.3

重点城市工业危险废物产生及处置利用情况

（2013）

单位：万吨

城 市 名 称	危险废物 产生量	危险废物综合 利用量	危险废物 处置量	危险废物 贮存量	危险废物倾倒 丢弃量	危险废物处置 利用率/%
总　计	1 543.43	1 013.26	506.55	54.39	0	96.5
北　京	13.22	5.79	6.81	0.63	0	95.3
天　津	11.81	3.53	8.29	…	0	100.0
石 家 庄	24.57	7.43	17.09	0.05	0	99.8
唐　山	28.99	27.64	1.35	…	0	100.0
秦 皇 岛	1.73	0.06	1.67	…	0	100.0
邯　郸	1.03	0.59	0.27	0.17	0	83.4
保　定	1.85	0.64	1.21	…	0	100.0
太　原	4.57	1.54	3.03	…	0	99.9
大　同	0.24	0.20	0.04	…	0	99.7
阳　泉	0.15	0.01	0.02	0.12	0	20.6
长　治	3.39	3.13	0.27	…	0	99.9
临　汾	4.79	4.49	0.29	0.02	0	99.6
呼和浩特	2.16	1.96	0.21	…	0	100.0
包　头	3.92	2.42	1.14	0.37	0	90.7
赤　峰	36.86	15.83	19.31	1.73	0	95.3
沈　阳	4.42	1.71	2.69	0.02	0	99.5
大　连	13.13	2.40	10.73	…	0	100.0
鞍　山	2.04	1.67	0.37	…	0	100.0
抚　顺	8.56	3.65	4.69	0.27	0	96.9
本　溪	10.90	10.72	0.18	…	0	100.0
锦　州	32.88	44.74	0.21	5.00	0	90.0
长　春	3.27	0.44	2.83	…	0	100.0
吉　林	69.57	38.08	31.49	…	0	100.0
哈 尔 滨	2.30	1.71	0.59	…	0	100.0
齐齐哈尔	0.80	0.18	0.06	0.56	0	29.9
牡 丹 江	0.37	0.15	0.22	0.01	0	98.7
上　海	54.31	28.76	25.64	0.39	0	99.3
南　京	37.71	22.29	15.75	0.56	0	98.6
无　锡	54.96	30.64	24.16	0.18	0	99.7
徐　州	0.99	0.15	0.53	0.31	0	68.7
常　州	13.42	6.00	7.26	0.17	0	98.7
苏　州	62.94	22.66	39.98	0.36	0	99.4
南　通	12.70	6.83	5.67	0.54	0	95.9
连 云 港	2.64	1.78	0.86	…	0	100.0
扬　州	10.47	7.19	3.21	0.66	0	94.0
镇　江	9.92	2.32	7.58	0.03	0	99.7
杭　州	14.64	5.08	9.15	0.69	0	95.4

重点城市工业危险废物产生及处置利用情况（续表）

（2013）

单位：万吨

城　市 名　称	危险废物 产生量	危险废物 综合利用量	危险废物 处置量	危险废物 贮存量	危险废物倾 倒丢弃量	危险废物处置 利用率/%
宁　波	46.47	15.10	31.66	1.33	0	97.2
温　州	4.82	1.94	2.86	0.01	0	99.7
湖　州	7.81	1.06	6.57	0.21	0	97.3
绍　兴	12.78	1.84	10.89	0.30	0	97.7
合　肥	3.06	1.16	1.90	0.01	0	99.8
芜　湖	3.25	1.31	1.93	0.03	0	99.0
马　鞍　山	22.22	21.91	0.19	0.12	0	99.4
福　州	3.22	1.16	1.35	0.71	0	77.8
厦　门	2.59	0.49	1.98	0.14	0	94.7
泉　州	8.11	0.25	4.56	3.30	0	59.3
南　昌	12.42	11.92	0.50	…	0	100.0
九　江	11.72	11.13	0.59	0.04	0	99.7
济　南	15.79	2.99	12.77	0.06	0	99.6
青　岛	3.25	1.48	1.75	0.05	0	98.5
淄　博	63.33	49.10	14.36	0.21	0	99.7
枣　庄	1.43	1.32	0.11	…	0	100.0
烟　台	192.17	181.37	6.20	7.97	0	95.9
潍　坊	8.51	6.95	1.27	0.35	0	95.9
济　宁	5.17	4.91	0.27	0.01	0	99.7
泰　安	5.85	5.36	0.46	0.06	0	98.9
日　照	0.40	0.02	0.39	…	0	99.7
郑　州	1.34	0.91	0.43	0.01	0	99.3
开　封	0.16	0.15	0.01	…	0	98.1
洛　阳	2.69	1.65	1.04	…	0	100.0
平　顶　山	0.13	0.13	…	…	0	100.0
安　阳	7.01	6.95	0.06	0.01	0	99.9
焦　作	0.73	0.38	0.34	0.01	0	99.0
三　门　峡	24.58	16.96	7.62	…	0	100.0
武　汉	15.09	3.01	2.55	9.53	0	36.8
宜　昌	0.73	0.61	0.12	0.02	0	97.2
荆　州	1.33	0.92	0.41	…	0	100.0
长　沙	0.50	0.31	0.18	0.05	0	91.1
株　洲	22.07	9.34	6.26	6.47	0	70.7
湘　潭	7.94	7.50	0.41	0.03	0	99.6
岳　阳	129.31	118.43	10.44	0.44	0	99.7
常　德	2.93	2.16	0.70	0.64	0	81.7
张　家　界	0.12	0.12	…	…	0	100.0
广　州	31.02	16.14	14.88	…	0	100.0

重点城市工业固体废物产生及处置利用情况（续表）
（2013）

单位：万吨

城 市 名 称	危险废物 产生量	危险废物 综合利用量	危险废物 处置量	危险废物 贮存量	危险废物倾倒 丢弃量	危险废物处置 利用率/%
韶　关	15.78	9.90	5.76	0.34	0	97.8
深　圳	31.58	16.46	15.12	…	0	100.0
珠　海	9.48	4.46	5.02	…	0	100.0
汕　头	0.78	0.61	0.16	…	0	99.4
湛　江	0.28	…	0.22	0.06	0	78.0
南　宁	1.10	0.07	0.67	0.35	0	67.7
柳　州	2.16	1.03	0.89	0.28	0	87.4
桂　林	0.10	0.09	…	…	0	99.6
北　海	0.04	…	0.04	…	0	100.0
海　口	0.74	…	0.74	…	0	100.0
重　庆	46.68	32.78	13.23	1.03	0	97.8
成　都	15.46	0.31	15.04	0.17	0	98.9
自　贡	0.37	0.19	0.17	…	0	99.0
攀 枝 花	7.33	6.83	0.48	0.01	0	99.9
泸　州	0.04	…	0.04	…	0	100.0
德　阳	0.66	…	0.66	…	0	99.8
绵　阳	9.98	2.41	7.51	0.06	0	99.4
南　充	0.17	0.17	…	…	0	100.0
宜　宾	0.78	0.35	0.41	0.03	0	96.5
贵　阳	1.80	1.43	0.40	…	0	100.0
遵　义	0.22	0.09	0.12	…	0	99.9
昆　明	74.44	65.53	8.91	…	0	100.0
曲　靖	8.23	3.24	2.90	2.47	0	71.3
玉　溪	3.53	2.33	…	1.23	0	65.6
拉　萨	…	…	…	…	0	—
西　安	0.96	0.01	0.95	…	0	99.8
铜　川	…	…	…	…	0	—
宝　鸡	1.17	0.03	1.20	…	0	100.0
咸　阳	0.28	0.04	0.24	…	0	99.8
渭　南	3.81	3.38	0.43	…	0	100.0
延　安	7.45	1.55	5.31	0.66	0	91.2
兰　州	10.15	0.20	9.95	…	0	100.0
金　昌	0.96	0.97	…	0.12	0	89.0
西　宁	17.08	18.93	0.15	0.95	0	95.2
银　川	3.97	2.89	0.19	0.89	0	77.7
石 嘴 山	0.73	0.72	0.01	…	0	100.0
乌鲁木齐	13.02	11.89	1.13	…	0	100.0
克拉玛依	7.87	1.48	5.65	0.74	0	90.6

重点城市汇总工业企业概况（一）

（2013）

城 市 名 称	汇总工业 企业数/ 个	工业总产值 （现价）/ 亿元	工业炉窑数/ 台	工业锅炉 数/台	35 蒸吨及 以上	20（含）～ 35 蒸吨	10（含）～ 20 蒸吨	10 蒸吨以下
总 计	68 783	249 768.3	43 300	53 671	7 975	4 191	7 641	33 864
北 京	932	6 896.7	242	1 962	329	239	402	992
天 津	2 657	13 003.0	866	2 484	424	302	369	1 389
石 家 庄	1 640	2 595.7	1 051	1 246	141	40	99	966
唐 山	1 450	6 391.2	1 352	602	146	70	158	228
秦 皇 岛	327	1 264.3	139	392	46	34	56	256
邯 郸	652	2 710.3	722	409	110	28	58	213
保 定	1 439	1 440.8	521	974	22	26	156	770
太 原	302	1 800.0	431	401	107	51	74	169
大 同	654	946.3	320	963	62	56	99	746
阳 泉	306	438.1	405	228	28	7	25	168
长 治	530	1 491.4	359	551	77	52	72	350
临 汾	481	1 394.3	250	467	67	51	54	295
呼 和 浩 特	557	830.6	69	1 061	143	79	255	584
包 头	121	1 727.5	334	200	52	14	39	95
赤 峰	490	456.8	250	473	92	50	105	226
沈 阳	971	3 194.8	289	1 593	281	187	374	751
大 连	833	3 741.7	418	1 153	204	211	210	528
鞍 山	646	2 830.8	2 160	616	134	54	142	286
抚 顺	270	1 167.7	393	338	45	57	53	183
本 溪	401	1 978.8	382	383	57	19	50	257
锦 州	217	750.4	316	343	74	65	52	152
长 春	309	2 130.6	108	701	219	94	125	263
吉 林	250	1 449.6	156	322	105	32	32	153
哈 尔 滨	352	1 322.1	113	711	221	117	135	238
齐 齐 哈 尔	340	348.8	327	652	73	72	133	374
牡 丹 江	103	163.6	23	232	43	20	46	123
上 海	2 089	18 255.7	1 179	2 019	153	65	153	1 648
南 京	727	7 384.1	465	449	77	23	34	315
无 锡	1 324	6 173.1	803	883	147	12	73	651
徐 州	589	2 389.5	242	367	88	25	54	200
常 州	1 000	3 112.2	463	529	56	14	25	434
苏 州	1 873	9 927.0	652	1 713	263	92	242	1 116
南 通	1 326	2 109.1	543	627	79	12	54	482
连 云 港	295	1 113.7	37	262	32	16	56	158
扬 州	536	1 727.6	158	245	43	5	14	183
镇 江	610	1 910.6	149	306	36	7	19	244
杭 州	1 377	4 348.5	469	917	155	54	147	561

重点城市汇总工业企业概况（一）（续表）

（2013）

城 市 名 称	汇总工业企业数/个	工业总产值（现价）/亿元	工业炉窑数/台	工业锅炉数/台	35蒸吨及以上	20（含）～35蒸吨	10（含）～20蒸吨	10蒸吨以下
宁　波	1 355	6 596.9	715	792	108	45	67	572
温　州	1 950	1 268.4	661	1 033	11	13	85	924
湖　州	1 132	1 404.5	371	586	45	16	47	478
绍　兴	992	3 265.4	267	1 091	125	58	158	750
合　肥	913	2 199.2	408	356	42	12	37	265
芜　湖	576	1 476.5	284	223	10	16	24	173
马 鞍 山	568	1 600.7	426	145	44	11	20	70
福　州	461	1 896.8	273	398	18	17	81	282
厦　门	375	2 469.5	83	194	17	8	34	135
泉　州	1 356	2 023.1	909	783	43	39	133	568
南　昌	338	1 113.3	162	290	7	21	51	211
九　江	594	915.3	457	207	29	8	34	136
济　南	624	2 029.3	639	378	136	30	94	118
青　岛	738	3 970.1	235	608	172	67	66	303
淄　博	1 022	3 328.4	1 189	513	198	59	62	194
枣　庄	206	513.2	87	224	73	6	22	123
烟　台	587	4 363.6	207	573	129	39	71	334
潍　坊	860	4 520.9	250	820	244	65	104	407
济　宁	446	2 471.9	395	437	149	22	26	240
泰　安	258	940.2	102	303	82	13	51	157
日　照	160	1 532.8	129	172	63	6	13	90
郑　州	619	1 827.5	628	439	56	32	55	296
开　封	384	351.1	277	146	15	7	27	97
洛　阳	644	1 686.0	587	266	54	22	23	167
平 顶 山	237	811.1	153	134	48	9	17	60
安　阳	456	1 293.7	362	254	26	22	21	185
焦　作	432	814.3	363	437	59	23	22	333
三 门 峡	373	747.4	449	169	44	5	25	95
武　汉	385	5 077.4	256	330	58	18	37	217
宜　昌	341	993.0	350	302	33	38	39	192
荆　州	482	586.8	135	313	18	15	16	264
长　沙	430	1 333.2	108	280	7	10	12	251
株　洲	523	1 016.4	574	199	17	9	19	154
湘　潭	195	743.8	162	140	17	1	14	108
岳　阳	291	2 839.1	121	239	26	10	32	171
常　德	215	948.7	73	182	26	12	25	119
张 家 界	72	15.1	43	22	2	0	1	19
广　州	1 182	7 218.0	243	826	88	37	123	578

重点城市汇总工业企业概况（一）（续表）

（2013）

城市名称	汇总工业企业数/个	工业总产值（现价）/亿元	工业炉窑数/台	工业锅炉数/台	35 蒸吨及以上	20（含）～35 蒸吨	10（含）～20 蒸吨	10 蒸吨以下
韶　关	366	773.2	184	127	16	9	25	77
深　圳	1 809	7 962.2	18	429	30	10	43	346
珠　海	516	2 546.9	77	331	20	8	36	267
汕　头	568	478.8	25	520	12	10	96	402
湛　江	338	911.1	101	255	51	37	26	141
南　宁	420	600.3	83	294	43	35	32	184
柳　州	346	1 816.4	247	211	49	21	12	129
桂　林	466	397.9	272	175	4	6	23	142
北　海	84	545.0	61	64	15	7	20	22
海　口	112	343.4	16	80	0	3	6	71
重　庆	3 145	7 667.5	1 768	1 455	116	62	120	1 157
成　都	1 350	3 119.4	2 251	879	34	37	115	693
自　贡	381	355.3	136	114	9	4	6	95
攀枝花	232	817.6	401	102	27	8	13	54
泸　州	599	471.1	238	266	22	10	11	223
德　阳	360	1 477.6	635	231	15	19	34	163
绵　阳	506	1 199.1	311	302	9	13	31	249
南　充	474	147.4	268	101	6	1	7	87
宜　宾	320	787.0	205	203	24	29	19	131
贵　阳	170	1 192.8	1 063	126	34	9	26	57
遵　义	139	762.6	493	171	16	30	36	89
昆　明	596	2 024.8	330	323	33	21	43	226
曲　靖	490	956.4	256	198	52	15	29	102
玉　溪	272	1 192.4	144	128	16	12	13	87
拉　萨	50	34.9	8	26	0	0	1	25
西　安	436	1 923.7	224	520	85	83	82	270
铜　川	115	146.0	45	60	2	0	19	39
宝　鸡	209	836.4	105	271	46	32	81	112
咸　阳	498	909.2	271	339	42	19	51	227
渭　南	349	502.2	367	219	30	10	42	137
延　安	166	645.9	126	341	23	24	76	218
兰　州	308	1 036.9	909	617	43	66	214	294
金　昌	39	828.0	80	102	30	13	4	55
西　宁	133	593.4	264	187	24	15	24	124
银　川	125	728.4	82	363	93	71	79	120
石嘴山	269	489.2	613	338	42	14	55	227
乌鲁木齐	211	1 276.7	174	184	52	12	17	103
克拉玛依	68	2 151.8	160	441	40	223	42	136

重点城市汇总工业企业概况（二）

（2013）

单位：万吨

城 市 名 称	工业用水 总量	取水量	重 复 用水量	工业煤炭 消耗量	燃料煤
总　计	24 120 676	3 004 113	21 116 564	207 004	153 728
北　京	307 114	19 460	287 654	1 535	1 462
天　津	932 262	40 010	892 253	4 879	4 588
石 家 庄	697 923	41 735	656 188	4 080	3 474
唐　山	1 458 094	67 040	1 391 053	8 278	3 622
秦 皇 岛	45 699	9 277	36 421	1 091	1 032
邯　郸	459 862	18 316	441 546	6 049	2 786
保　定	143 002	20 676	122 326	744	713
太　原	321 521	12 307	309 214	3 651	1 609
大　同	179 054	13 057	165 997	2 617	2 387
阳　泉	32 664	4 417	28 248	897	897
长　治	191 608	15 888	175 719	6 304	2 228
临　汾	202 243	22 093	180 150	6 594	2 091
呼和浩特	101 384	9 688	91 695	2 973	2 900
包　头	613 719	22 602	591 117	3 488	2 467
赤　峰	54 630	6 842	47 788	1 651	1 459
沈　阳	513 822	13 519	500 304	1 955	1 924
大　连	156 844	88 718	68 125	2 162	2 082
鞍　山	263 927	34 662	229 266	2 102	958
抚　顺	243 098	10 365	232 734	975	866
本　溪	231 378	17 267	214 110	1 562	432
锦　州	75 397	7 424	67 973	546	538
长　春	245 905	13 111	232 795	2 669	2 071
吉　林	338 166	39 818	298 347	2 191	1 724
哈 尔 滨	170 801	10 080	160 720	1 937	1 828
齐齐哈尔	138 788	73 180	65 608	1 241	1 100
牡 丹 江	55 974	6 321	49 652	693	673
上　海	798 569	71 480	727 088	5 362	4 111
南　京	658 702	195 211	463 491	3 548	2 691
无　锡	150 393	36 811	113 582	2 768	2 688
徐　州	441 556	26 629	414 928	4 182	3 273
常　州	157 578	17 337	140 240	1 367	1 367
苏　州	914 662	96 294	818 368	6 180	4 831
南　通	150 801	23 402	127 399	1 796	1 786
连 云 港	129 923	12 106	117 818	652	614
扬　州	248 557	20 060	228 497	1 194	1 193
镇　江	114 718	14 064	100 654	2 305	2 238
杭　州	342 678	55 646	287 032	1 522	1 334

重点城市汇总工业企业概况（二）（续表）

（2013）

单位：万吨

城　市 名　称	工业用水 总量	取水量	重　复 用水量	工业煤炭 消耗量	燃料煤
宁　波	1 068 300	573 017	495 284	4 564	4 336
温　州	34 408	10 524	23 884	1 028	1 028
湖　州	68 638	13 503	55 136	777	574
绍　兴	174 819	35 704	139 115	1 154	1 093
合　肥	209 204	11 907	197 297	1 511	1 455
芜　湖	43 086	6 920	36 166	1 377	889
马 鞍 山	309 707	12 913	296 794	2 285	1 327
福　州	52 324	10 573	41 751	1 949	1 847
厦　门	102 069	7 515	94 554	499	499
泉　州	230 831	28 146	202 685	1 537	1 424
南　昌	66 868	13 460	53 408	555	374
九　江	91 883	13 754	78 129	760	595
济　南	293 884	13 428	280 456	1 828	1 305
青　岛	145 620	25 309	120 311	1 381	1 310
淄　博	533 047	30 926	502 120	3 302	2 740
枣　庄	157 717	14 038	143 679	1 916	917
烟　台	51 275	15 649	35 626	2 110	1 988
潍　坊	179 782	36 746	143 036	2 224	1 868
济　宁	591 502	29 468	562 034	4 245	3 525
泰　安	104 343	13 148	91 194	1 423	1 027
日　照	221 292	13 820	207 472	1 309	803
郑　州	194 553	20 233	174 320	2 861	2 811
开　封	161 042	10 489	150 553	660	498
洛　阳	264 927	19 356	245 572	2 674	2 538
平 顶 山	247 042	14 372	232 671	5 385	2 222
安　阳	220 809	15 362	205 447	1 484	727
焦　作	122 507	20 195	102 312	1 504	897
三 门 峡	157 223	8 302	148 921	1 513	952
武　汉	638 112	123 700	514 412	2 565	1 590
宜　昌	209 201	26 726	182 475	1 065	678
荆　州	35 932	13 862	22 069	350	322
长　沙	9 568	5 590	3 978	462	448
株　洲	63 034	8 229	54 804	388	347
湘　潭	168 662	7 020	161 642	968	380
岳　阳	174 420	14 798	159 622	1 661	1 612
常　德	107 129	13 799	93 330	732	584
张 家 界	275	203	72	78	63
广　州	100 332	32 727	67 606	1 765	1 719

重点城市汇总工业企业概况（二）（续表）

（2013）

单位：万吨

城　市 名　称	工业用水 总量	取水量	重　复 用水量	工业煤炭 消耗量	燃料煤
韶　关	177 039	10 396	166 643	890	514
深　圳	23 331	17 294	6 037	396	396
珠　海	51 646	10 961	40 685	655	627
汕　头	11 399	7 716	3 683	1 283	1 251
湛　江	49 192	8 187	41 005	677	673
南　宁	159 144	77 938	81 205	620	615
柳　州	356 777	15 277	341 501	1 300	385
桂　林	60 322	6 052	54 271	312	230
北　海	52 448	2 144	50 305	207	206
海　口	3 096	1 083	2 012	2	2
重　庆	626 175	130 733	495 442	4 505	3 700
成　都	147 556	16 394	131 162	723	697
自　贡	25 577	2 414	23 162	132	111
攀枝花	100 880	14 580	86 300	1 065	587
泸　州	167 829	7 845	159 983	413	406
德　阳	56 148	9 031	47 118	191	172
绵　阳	67 680	9 208	58 472	444	334
南　充	8 733	4 046	4 687	75	54
宜　宾	50 662	12 431	38 230	1 083	1 024
贵　阳	176 929	8 306	168 623	703	388
遵　义	192 192	7 277	184 914	903	887
昆　明	182 868	12 958	169 909	1 631	1 162
曲　靖	216 423	11 011	205 412	2 709	2 071
玉　溪	34 198	5 556	28 641	373	251
拉　萨	1 849	734	1 115	21	20
西　安	42 773	12 712	30 061	879	852
铜　川	17 919	784	17 136	835	835
宝　鸡	39 454	6 221	33 232	1 181	1 012
咸　阳	13 958	7 036	6 922	3 381	1 275
渭　南	244 107	9 870	234 237	2 165	1 629
延　安	6 545	3 959	2 585	151	141
兰　州	91 427	11 126	80 302	1 335	1 322
金　昌	71 725	7 154	64 571	477	429
西　宁	99 004	7 485	91 519	883	612
银　川	108 991	9 619	99 371	2 199	2 167
石嘴山	77 567	8 732	68 834	1 986	1 255
乌鲁木齐	221 929	12 231	209 697	2 201	1 657
克拉玛依	196 804	5 262	191 542	430	430

重点城市汇总工业企业概况（三）

（2013）

城 市名 称	燃料油消耗量/万吨	焦炭消耗量/万吨	天然气消耗量/亿米³	其他燃料（折标煤）消耗量/万吨
总 计	1 000	19 694	639	6 309
北 京	2	0	43	113
天 津	9	577	19	184
石 家 庄	4	475	1	222
唐 山	0	3 207	4	520
秦 皇 岛	5	320	2	0
邯 郸	0	1 251	28	105
保 定	0	61	1	3
太 原	1	272	2	79
大 同	0	55	0	65
阳 泉	0	4	1	21
长 治	0	235	0	223
临 汾	1	323	0	1
呼和浩特	1	17	8	19
包 头	0	468	2	118
赤 峰	0	105	0	10
沈 阳	6	1	4	5
大 连	41	4	2	37
鞍 山	7	667	1	353
抚 顺	16	158	3	1
本 溪	0	850	0	311
锦 州	4	20	2	17
长 春	1	2	1	24
吉 林	11	174	6	39
哈 尔 滨	3	47	1	18
齐齐哈尔	0	35	1	16
牡 丹 江	0	0	0	2
上 海	47	634	50	193
南 京	38	526	15	352
无 锡	6	337	5	23
徐 州	498	46	2	5
常 州	1	120	11	11
苏 州	38	1 098	39	31
南 通	3	4	1	28
连 云 港	1	114	2	20
扬 州	0	3	7	40
镇 江	1	40	2	17
杭 州	3	111	23	45

重点城市汇总工业企业概况（三）（续表）

（2013）

城　市 名　称	燃料油消耗量/ 万吨	焦炭消耗量/ 万吨	天然气消耗量/ 亿米³	其他燃料（折标煤）消耗量/ 万吨
宁　波	44	141	14	41
温　州	5	2	0	3
湖　州	11	0	3	48
绍　兴	8	2	2	0
合　肥	1	59	1	44
芜　湖	1	129	3	11
马 鞍 山	0	709	2	425
福　州	4	255	3	5
厦　门	8	0	5	3
泉　州	14	93	18	2
南　昌	1	116	0	20
九　江	1	188	1	36
济　南	3	378	3	65
青　岛	7	127	2	58
淄　博	23	167	14	62
枣　庄	0	8	0	20
烟　台	2	18	2	34
潍　坊	9	220	2	79
济　宁	0	0	1	31
泰　安	0	115	7	12
日　照	8	509	3	0
郑　州	2	1	10	1
开　封	0	1	0	1
洛　阳	1	12	2	31
平 顶 山	0	36	2	19
安　阳	0	588	1	6
焦　作	2	2	2	55
三 门 峡	0	2	1	0
武　汉	3	0	10	17
宜　昌	0	14	4	5
荆　州	0	0	0	50
长　沙	0	1	1	2
株　洲	2	14	2	27
湘　潭	0	236	0	20
岳　阳	1	3	3	22
常　德	0	1	0	18
张 家 界	0	0	0	0
广　州	15	92	8	112

重点城市汇总工业企业概况（三）（续表）

（2013）

城市 名称	燃料油消耗量/ 万吨	焦炭消耗量/ 万吨	天然气消耗量/ 亿米^3	其他燃料（折标煤）消耗量/ 万吨
韶 关	1	257	0	190
深 圳	9	0	22	155
珠 海	12	79	4	7
汕 头	1	0	0	1
湛 江	3	0	0	99
南 宁	5	0	0	74
柳 州	1	487	0	419
桂 林	0	23	0	17
北 海	0	54	1	14
海 口	0	0	0	1
重 庆	1	298	38	29
成 都	0	69	18	88
自 贡	0	0	2	1
攀 枝 花	0	288	0	44
泸 州	0	0	8	14
德 阳	0	3	9	1
绵 阳	0	8	2	4
南 充	0	1	2	3
宜 宾	0	0	1	8
贵 阳	1	11	2	19
遵 义	0	0	4	0
昆 明	8	219	27	13
曲 靖	0	192	0	22
玉 溪	1	384	0	7
拉 萨	13	0	0	0
西 安	1	1	2	0
铜 川	0	0	1	0
宝 鸡	0	8	0	5
咸 阳	1	0	5	1
渭 南	0	0	0	0
延 安	7	0	1	1
兰 州	1	178	6	53
金 昌	2	26	0	18
西 宁	0	105	6	15
银 川	2	11	2	14
石 嘴 山	0	29	1	4
乌鲁木齐	1	362	16	234
克拉玛依	0	0	35	82

重点城市工业污染防治投资情况（一）

（2013）

单位：个

城　市 名　称	汇总工业 企业数	本年施工 项目总数	工业废水 治理项目	工业废气 治理项目	脱硫治理 项目	脱硝治理 项目	工业固体 废物治理 项目	噪声治理 项目	其他治理 项目
总　　计	3 055	3 008	775	1 521	330	428	159	56	497
北　京	38	45	7	24	1	8	0	1	13
天　津	81	66	14	34	16	3	0	0	18
石　家　庄	76	129	5	72	7	7	49	1	2
唐　山	48	51	3	45	22	5	0	0	3
秦　皇　岛	9	15	0	15	5	3	0	0	0
邯　郸	36	39	7	17	6	4	3	1	11
保　定	14	13	10	3	0	1	0	0	0
太　原	33	42	9	18	4	7	7	0	8
大　同	26	53	2	16	8	7	13	0	22
阳　泉	11	12	1	5	1	0	4	1	1
长　治	39	34	5	15	5	8	0	2	12
临　汾	17	9	2	5	1	3	0	0	2
呼和浩特	10	25	8	11	0	5	4	0	2
包　头	16	23	2	18	6	8	1	0	2
赤　峰	16	19	5	12	4	7	0	0	2
沈　阳	12	6	0	4	2	1	0	0	2
大　连	13	7	0	4	0	1	1	1	1
鞍　山	2	0	0	0	0	0	0	0	0
抚　顺	8	18	1	16	2	4	0	0	1
本　溪	7	6	0	5	2	2	0	0	1
锦　州	11	3	0	3	2	0	0	0	0
长　春	8	7	1	6	3	1	0	0	0
吉　林	7	10	2	6	2	1	0	0	2
哈　尔　滨	16	19	1	18	3	13	0	0	0
齐齐哈尔	13	16	1	15	7	2	0	0	0
牡　丹　江	5	3	0	3	1	1	0	0	0
上　海	116	112	30	39	2	0	2	6	35
南　京	66	65	15	28	5	7	0	3	19
无　锡	76	79	56	21	2	10	0	0	2
徐　州	19	20	1	17	6	4	0	0	2
常　州	19	18	0	18	2	3	0	0	0
苏　州	54	54	8	36	6	11	1	2	7
南　通	90	76	53	16	0	4	0	1	6
连　云　港	41	31	18	11	3	0	0	1	1
扬　州	12	9	2	7	1	4	0	0	0
镇　江	20	8	6	2	0	1	0	0	0
杭　州	78	88	39	35	2	17	1	3	10

重点城市工业污染防治投资情况（一）（续表）

（2013）

单位：个

城 市 名 称	汇总工业 企业数	本年施工 项目总数	工业废水 治理项目	工业废气 治理项目	脱硫治理 项目	脱硝治理 项目	工业固体 废物治理 项目	噪声治理 项目	其他治理 项目
宁 波	59	81	23	42	3	9	0	1	15
温 州	7	4	0	3	0	1	0	0	1
湖 州	55	47	17	23	2	2	0	0	7
绍 兴	156	194	69	65	1	2	2	0	58
合 肥	29	22	6	15	2	8	0	0	1
芜 湖	15	19	4	12	1	8	0	0	3
马 鞍 山	8	14	0	13	4	7	0	0	1
福 州	11	11	2	4	1	2	0	0	5
厦 门	38	18	8	5	1	0	0	1	4
泉 州	106	111	26	30	6	2	42	2	11
南 昌	10	15	2	10	3	1	2	0	1
九 江	9	12	0	11	3	5	0	0	1
济 南	33	35	2	31	12	5	0	1	1
青 岛	21	27	3	21	8	5	1	1	1
淄 博	56	72	13	46	15	7	2	1	10
枣 庄	40	45	1	11	1	6	0	0	33
烟 台	37	53	9	28	5	8	2	1	13
潍 坊	85	88	18	68	16	17	0	0	2
济 宁	29	38	4	19	3	10	1	0	14
泰 安	26	34	4	29	13	7	0	0	1
日 照	13	10	1	9	2	1	0	0	0
郑 州	28	24	1	21	1	11	1	0	1
开 封	2	1	0	1	1	0	0	0	0
洛 阳	14	17	1	10	4	4	1	0	5
平 顶 山	3	6	0	5	1	3	1	0	0
安 阳	7	9	2	7	1	2	0	0	0
焦 作	24	16	7	7	2	4	0	0	2
三 门 峡	4	3	0	3	1	1	0	0	0
武 汉	22	14	4	6	0	4	0	1	3
宜 昌	29	21	6	13	6	3	0	0	2
荆 州	3	2	0	1	1	0	0	0	1
长 沙	22	9	3	4	0	1	0	0	2
株 洲	21	22	7	12	0	0	0	1	2
湘 潭	7	7	2	5	1	2	0	0	0
岳 阳	6	2	0	2	1	0	0	0	0
常 德	14	9	4	5	1	3	0	0	0
张 家 界	1	0	0	0	0	0	0	0	0
广 州	39	27	7	10	3	0	0	1	9

重点城市工业污染防治投资情况（一）（续表）

（2013）

单位：个

城　市 名　称	汇总工业 企业数	本年施工 项目总数	工业废水 治理项目	工业废气 治理项目	脱硫治理 项目	脱硝治理 项目	工业固体 废物治理 项目	噪声治理 项目	其他治理 项目
韶　关	35	24	6	4	2	1	1	0	13
深　圳	31	29	18	7	1	2	1	0	3
珠　海	48	20	6	12	1	0	0	2	0
汕　头	22	23	7	15	10	2	0	1	0
湛　江	21	18	15	3	0	0	0	0	0
南　宁	9	12	4	7	0	3	0	0	1
柳　州	31	26	8	8	2	3	1	3	6
桂　林	6	6	0	4	1	2	2	0	0
北　海	9	7	4	1	0	1	0	0	2
海　口	2	0	0	0	0	0	0	0	0
重　庆	108	51	19	25	3	5	0	1	6
成　都	64	46	16	19	6	7	1	2	8
自　贡	4	3	1	2	0	2	0	0	0
攀枝花	33	41	5	17	3	2	2	2	15
泸　州	25	23	21	1	0	1	0	0	1
德　阳	12	7	3	3	0	3	0	0	1
绵　阳	11	10	3	6	1	5	1	0	0
南　充	1	0	0	0	0	0	0	0	0
宜　宾	9	3	2	1	0	1	0	0	0
贵　阳	17	13	3	9	1	6	0	0	1
遵　义	19	14	2	9	0	8	1	1	1
昆　明	40	53	12	22	6	5	3	4	12
曲　靖	42	40	14	21	5	12	0	0	5
玉　溪	14	7	0	6	2	1	0	0	1
拉　萨	26	9	5	1	0	0	0	1	2
西　安	20	11	4	7	1	1	0	0	0
铜　川	4	3	0	3	0	3	0	0	0
宝　鸡	17	13	3	9	0	7	0	1	0
咸　阳	21	30	10	10	5	3	0	2	8
渭　南	12	8	2	6	1	4	0	0	0
延　安	5	3	1	0	0	0	0	1	1
兰　州	6	3	0	3	0	0	0	0	0
金　昌	7	6	2	1	0	1	0	0	3
西　宁	24	19	1	13	1	4	0	0	5
银　川	30	31	6	22	8	9	0	1	2
石嘴山	10	17	2	7	1	4	5	0	3
乌鲁木齐	6	7	0	7	3	4	0	0	0
克拉玛依	2	3	0	3	1	2	0	0	0

重点城市工业污染防治投资情况（二）

（2013）

单位：个

城 市 名 称	本年竣工项目总数	工业废水治理项目	工业废气治理项目	脱硫治理项目	脱硝治理项目	工业固体废物治理项目	噪声治理项目	其他治理项目
总 计	3 212	849	1 634	376	430	160	54	515
北 京	45	9	25	1	7	0	1	10
天 津	66	13	39	18	5	0	0	14
石 家 庄	116	3	62	3	1	49	1	1
唐 山	65	3	59	29	8	0	0	3
秦 皇 岛	15	0	14	5	3	0	0	1
邯 郸	40	8	15	5	4	3	1	13
保 定	13	11	2	0	1	0	0	0
太 原	45	9	21	6	6	5	0	10
大 同	60	2	23	11	10	13	0	22
阳 泉	21	3	10	2	4	4	1	3
长 治	28	2	13	6	5	0	1	12
临 汾	17	3	12	4	3	1	0	1
呼 和 浩 特	25	9	9	0	3	5	0	2
包 头	20	2	16	7	6	1	0	1
赤 峰	16	4	10	3	6	0	0	2
沈 阳	13	0	11	2	1	0	0	2
大 连	13	1	10	1	6	1	0	1
鞍 山	3	0	2	1	0	1	0	0
抚 顺	16	1	14	2	2	0	0	1
本 溪	7	1	5	0	3	0	0	1
锦 州	12	3	7	5	0	0	0	2
长 春	9	2	7	3	1	0	0	0
吉 林	9	0	6	2	1	0	0	3
哈 尔 滨	20	1	19	4	13	0	0	0
齐 齐 哈 尔	15	1	14	6	2	0	0	0
牡 丹 江	3	1	2	1	1	0	0	0
上 海	115	28	40	5	1	3	6	38
南 京	79	18	38	7	7	0	3	20
无 锡	68	44	22	2	10	0	0	2
徐 州	27	2	22	8	7	0	0	3
常 州	24	1	23	2	3	0	0	0
苏 州	64	16	37	6	10	1	3	7
南 通	73	48	16	0	4	2	1	6
连 云 港	32	18	13	2	1	0	0	1
扬 州	10	1	8	1	4	0	0	1
镇 江	12	5	7	1	3	0	0	0
杭 州	93	40	36	2	16	1	3	13

重点城市工业污染防治投资情况（二）（续表）
（2013）

单位：个

城　市名　称	本年竣工项目总数	工业废水治理项目	工业废气治理项目	脱硫治理项目	脱硝治理项目	工业固体废物治理项目	噪声治理项目	其他治理项目
宁　波	65	18	34	0	9	0	1	12
温　州	9	2	6	1	2	0	0	1
湖　州	49	18	24	2	2	0	0	7
绍　兴	194	71	65	4	4	2	0	56
合　肥	25	9	15	2	9	0	0	1
芜　湖	21	3	15	0	11	0	0	3
马鞍山	15	0	14	4	6	0	0	1
福　州	17	2	8	2	5	0	0	7
厦　门	43	23	12	1	2	0	0	8
泉　州	74	6	14	6	2	41	5	8
南　昌	19	5	10	3	1	2	0	2
九　江	13	1	11	2	5	0	0	1
济　南	36	5	31	13	5	0	0	0
青　岛	24	3	19	6	5	0	1	1
淄　博	65	11	41	13	2	2	0	11
枣　庄	48	2	11	1	6	0	1	34
烟　台	57	12	28	9	5	2	1	14
潍　坊	70	17	52	9	7	0	0	1
济　宁	35	4	17	6	6	1	0	13
泰　安	29	2	27	12	7	0	0	0
日　照	13	4	9	3	1	0	0	0
郑　州	34	3	28	2	12	2	0	1
开　封	2	1	1	1	0	0	0	0
洛　阳	11	0	5	0	3	1	0	5
平顶山	4	0	3	0	2	1	0	0
安　阳	15	3	11	3	2	0	0	1
焦　作	19	12	4	1	2	0	0	3
三门峡	5	0	4	2	1	1	0	0
武　汉	22	6	10	3	4	0	0	6
宜　昌	25	7	15	8	1	1	0	2
荆　州	3	0	2	2	0	0	0	1
长　沙	19	5	12	1	4	0	0	2
株　洲	26	10	13	0	1	0	1	2
湘　潭	4	1	3	0	2	0	0	0
岳　阳	4	1	3	2	0	0	0	0
常　德	12	3	9	3	3	0	0	0
张家界	1	0	1	0	1	0	0	0
广　州	42	11	20	5	2	0	1	10

重点城市工业污染防治投资情况（二）（续表）

（2013）

单位：个

城 市 名 称	本年竣工项目总数	工业废水治理项目	工业废气治理项目	脱硫治理项目	脱硝治理项目	工业固体废物治理项目	噪声治理项目	其他治理项目
韶　关	26	9	6	4	1	1	0	10
深　圳	32	21	7	1	2	1	0	3
珠　海	28	9	17	2	1	0	2	0
汕　头	26	8	17	11	3	0	1	0
湛　江	20	17	3	0	0	0	0	0
南　宁	13	5	7	0	2	0	0	1
柳　州	32	11	12	1	3	1	3	5
桂　林	5	1	4	1	2	0	0	0
北　海	7	5	1	0	1	0	0	1
海　口	2	0	2	2	0	0	0	0
重　庆	70	27	35	2	9	0	1	7
成　都	52	17	20	9	5	2	3	10
自　贡	4	2	2	0	2	0	0	0
攀枝花	39	5	17	4	2	1	1	15
泸　州	22	20	1	0	1	0	0	1
德　阳	12	9	3	0	3	0	0	0
绵　阳	9	2	6	1	5	0	0	1
南　充	0	0	0	0	0	0	0	0
宜　宾	6	4	2	0	1	0	0	0
贵　阳	15	3	11	2	6	0	0	1
遵　义	18	4	9	0	9	1	2	2
昆　明	56	12	24	6	5	3	5	12
曲　靖	38	11	18	3	10	1	0	8
玉　溪	16	1	13	6	2	0	0	2
拉　萨	27	16	7	3	0	1	1	2
西　安	17	9	8	1	3	0	0	0
铜　川	3	0	3	0	3	0	0	0
宝　鸡	20	7	12	0	9	0	1	0
咸　阳	27	9	9	3	5	0	2	7
渭　南	8	2	5	1	4	0	0	1
延　安	3	1	1	1	0	0	0	1
兰　州	7	0	5	0	2	0	0	2
金　昌	8	2	2	0	2	0	0	4
西　宁	24	1	18	2	6	0	0	5
银　川	20	3	16	7	6	0	0	1
石嘴山	16	1	10	2	5	2	0	3
乌鲁木齐	6	1	5	5	0	0	0	0
克拉玛依	0	0	0	0	0	0	0	0

重点城市工业污染治理项目建设情况（三）

（2013）

城 市 名 称	本年竣工项目新增设计处理能力		
	治理废水/ （万吨/日）	治理废气（标态）/ （万米³/小时）	治理固废/ （万吨/日）
总　计	150	60 239	5
北　京	3	220	0
天　津	3	1 221	0
石 家 庄	0	693	2
唐　山	*1	6 067	0
秦 皇 岛	0	358	0
邯　郸	0	386	0
保　定	1	0	0
太　原	1	507	0
大　同	0	801	0
阳　泉	0	19	0
长　治	0	1 353	0
临　汾	1	159	0
呼 和 浩 特	0	1 058	0
包　头	1	1 870	0
赤　峰	1	967	0
沈　阳	0	85	0
大　连	0	265	0
鞍　山	0	2	0
抚　顺	0	404	0
本　溪	0	995	0
锦　州	3	42	0
长　春	0	21	0
吉　林	0	5	0
哈 尔 滨	0	514	0
齐 齐 哈 尔	1	649	0
牡 丹 江	0	107	0
上　海	3	460	0
南　京	4	886	0
无　锡	1	697	0
徐　州	1	381	0
常　州	0	1 218	0
苏　州	2	1 432	0
南　通	5	1 549	0
连 云 港	4	887	0
扬　州	0	493	0
镇　江	2	128	0
杭　州	4	693	0

重点城市工业污染治理项目建设情况（三）（续表）

（2013）

城 市 名 称	本年竣工项目新增设计处理能力		
	治理废水/ （万吨/日）	治理废气（标态）/ （万米³/小时）	治理固废/ （万吨/日）
宁 波	2	266	0
温 州	0	315	0
湖 州	1	557	0
绍 兴	10	310	0
合 肥	5	1 032	0
芜 湖	1	708	0
马 鞍 山	0	1 542	0
福 州	2	1 049	0
厦 门	1	583	0
泉 州	5	221	0
南 昌	0	0	0
九 江	0	1 125	0
济 南	0	837	0
青 岛	0	222	0
淄 博	3	1 871	0
枣 庄	3	602	0
烟 台	3	700	0
潍 坊	2	791	0
济 宁	4	1 011	0
泰 安	2	783	0
日 照	1	751	0
郑 州	0	1 399	0
开 封	0	0	0
洛 阳	0	49	0
平 顶 山	0	757	0
安 阳	1	1 023	0
焦 作	3	203	0
三 门 峡	0	129	0
武 汉	0	258	0
宜 昌	1	239	0
荆 州	0	316	0
长 沙	1	95	0
株 洲	2	108	0
湘 潭	1	312	0
岳 阳	0	240	0
常 德	2	458	0
张 家 界	0	50	0
广 州	1	592	0

重点城市工业污染治理项目建设情况（三）（续表）

（2013）

城 市 名 称	本年竣工项目新增设计处理能力		
	治理废水/ （万吨/日）	治理废气（标态）/ （万米³/小时）	治理固废/ （万吨/日）
韶　关	4	576	0
深　圳	0	17	0
珠　海	1	25	0
汕　头	1	11	0
湛　江	4	5	0
南　宁	3	39	0
柳　州	2	253	0
桂　林	0	732	0
北　海	0	104	0
海　口	0	4	0
重　庆	1	2 334	0
成　都	7	530	0
自　贡	0	56	0
攀 枝 花	1	702	0
泸　州	1	215	0
德　阳	3	63	0
绵　阳	0	420	0
南　充	0	0	0
宜　宾	5	0	0
贵　阳	0	106	0
遵　义	0	430	0
昆　明	1	366	0
曲　靖	8	245	0
玉　溪	0	88	0
拉　萨	1	0	0
西　安	0	334	0
铜　川	0	487	0
宝　鸡	2	528	0
咸　阳	2	186	0
渭　南	1	727	0
延　安	0	300	0
兰　州	0	64	0
金　昌	0	230	0
西　宁	0	229	0
银　川	2	494	0
石 嘴 山	0	246	0
乌鲁木齐	1	3	0
克拉玛依	0	1	0

重点城市农业污染排放情况

（2013）

单位：吨

城市名称	农业污染物排放（流失）总量			
	化学需氧量	氨氮	总氮	总磷
总　计	4 852 744	314 245	1 694 705	209 092
北　京	74 717	4 504	25 809	3 496
天　津	109 934	5 633	28 419	3 255
石 家 庄	159 511	5 522	44 503	5 674
唐　山	129 445	6 252	64 085	8 166
秦 皇 岛	38 711	2 497	12 136	1 790
邯　郸	72 592	4 267	33 478	4 096
保　定	84 726	4 984	36 875	3 822
太　原	9 258	912	3 423	351
大　同	15 836	942	5 892	523
阳　泉	3 362	292	1 440	276
长　治	14 878	1 214	6 947	900
临　汾	19 051	1 333	9 921	945
呼和浩特	100 842	1 169	23 509	2 019
包　头	63 039	1 012	13 472	1 250
赤　峰	88 907	1 956	19 633	2 061
沈　阳	211 031	6 683	45 821	5 645
大　连	99 220	3 742	19 731	3 847
鞍　山	50 384	1 930	11 247	1 939
抚　顺	21 486	811	4 720	725
本　溪	15 895	722	3 721	560
锦　州	81 202	4 452	21 938	2 320
长　春	132 856	4 905	29 878	4 557
吉　林	87 866	2 612	19 692	2 387
哈 尔 滨	206 213	6 056	57 200	6 051
齐齐哈尔	173 438	4 314	33 982	3 247
牡 丹 江	22 954	911	5 036	569
上　海	31 112	3 204	15 403	1 976
南　京	18 209	1 809	8 527	977
无　锡	8 662	1 133	6 081	723
徐　州	48 053	3 337	23 753	2 438
常　州	14 948	1 417	8 477	1 367
苏　州	22 054	2 201	10 213	1 404
南　通	51 369	6 497	24 058	2 434
连 云 港	47 736	2 832	13 526	1 559
扬　州	11 257	1 776	7 797	639
镇　江	7 749	1 083	5 648	815
杭　州	28 617	3 574	13 720	1 821

重点城市农业污染排放情况（续表）

（2013）

单位：吨

城 市 名 称	农业污染物排放（流失）总量			
	化学需氧量	氨氮	总氮	总磷
宁 波	16 375	2 738	10 315	1 292
温 州	14 252	1 849	6 988	898
湖 州	21 055	1 971	7 946	1 043
绍 兴	15 106	2 276	7 074	744
合 肥	62 769	3 432	14 164	2 021
芜 湖	9 930	1 480	6 357	846
马 鞍 山	5 156	796	4 017	386
福 州	33 229	5 848	11 818	1 834
厦 门	13 254	1 811	3 482	420
泉 州	14 706	2 129	8 132	1 020
南 昌	35 123	3 214	10 021	1 221
九 江	14 950	2 085	7 774	811
济 南	72 900	3 092	26 589	2 895
青 岛	110 934	5 361	41 694	6 934
淄 博	35 063	1 815	12 436	1 429
枣 庄	29 951	1 876	12 861	1 610
烟 台	107 701	5 636	33 093	5 431
潍 坊	133 336	8 663	54 344	7 301
济 宁	92 852	7 087	33 833	4 308
泰 安	78 501	3 682	26 734	3 181
日 照	27 153	1 674	8 505	1 374
郑 州	53 376	3 019	17 593	2 329
开 封	46 765	3 270	27 642	3 230
洛 阳	40 125	2 209	15 838	1 758
平 顶 山	26 157	1 989	12 695	1 377
安 阳	40 079	3 500	19 678	2 911
焦 作	34 534	1 609	15 524	1 994
三 门 峡	10 026	810	4 733	565
武 汉	39 408	3 523	18 875	2 851
宜 昌	30 952	4 378	15 616	1 915
荆 州	94 172	5 611	26 864	3 442
长 沙	47 122	4 522	15 431	1 975
株 洲	26 485	3 422	9 142	986
湘 潭	38 566	3 943	11 514	1 184
岳 阳	64 866	7 547	21 283	2 433
常 德	54 692	5 796	24 513	3 013
张 家 界	5 030	566	2 602	260
广 州	46 748	3 871	11 855	1 933

重点城市农业污染排放情况（续表）

（2013）

单位：吨

城 市 名 称	农业污染物排放（流失）总量			
	化学需氧量	氨氮	总氮	总磷
韶 关	29 072	3 120	7 314	994
深 圳	5 648	218	657	96
珠 海	10 850	939	2 856	405
汕 头	12 828	1 220	3 730	511
湛 江	46 443	5 715	21 228	3 206
南 宁	36 112	4 103	18 617	2 383
柳 州	17 219	2 069	9 265	1 062
桂 林	32 106	3 171	13 403	1 866
北 海	9 541	821	4 277	509
海 口	11 100	1 072	5 135	729
重 庆	121 034	12 517	52 116	6 553
成 都	73 766	7 963	27 313	3 167
自 贡	18 372	1 642	5 288	581
攀 枝 花	2 887	311	1 615	208
泸 州	24 038	2 673	9 130	970
德 阳	25 088	2 156	9 928	1 199
绵 阳	30 987	3 786	15 224	1 758
南 充	50 926	5 044	17 135	1 897
宜 宾	25 897	3 128	10 981	1 121
贵 阳	8 199	608	4 157	540
遵 义	10 898	2 022	8 624	951
昆 明	14 454	1 750	7 513	1 031
曲 靖	14 413	2 085	10 273	970
玉 溪	5 174	366	3 395	626
拉 萨	1 274	98	1 407	92
西 安	27 677	1 044	14 485	1 310
铜 川	5 941	271	1 297	157
宝 鸡	21 680	962	10 530	900
咸 阳	34 003	1 480	15 230	1 208
渭 南	29 900	2 312	14 848	1 309
延 安	6 659	822	4 782	406
兰 州	6 922	441	3 034	279
金 昌	7 526	165	2 177	169
西 宁	9 898	378	2 501	289
银 川	28 113	574	10 026	1 253
石 嘴 山	6 625	131	2 436	244
乌鲁木齐	7 998	386	2 997	332
克拉玛依	956	94	493	36

重点城市城镇生活污染排放及处理情况（一）

（2013）

年 份 地 区	城镇人口/ 万人	生活煤炭消费量/ 万吨	生活天然气消费量/ 万米^3	生活用水总量/ 万吨	居民家庭用水 总量	公共服务用水 总量
总 计	41 422	8 834	2 555 022	3 401 259	2 789 989	611 270
北 京	1 825	412	625 243	158 674	90 700	67 974
天 津	1 214	204	124 829	73 554	59 554	14 000
石 家 庄	549	107	9 750	38 632	29 040	9 592
唐 山	425	74	5 800	28 319	25 173	3 147
秦 皇 岛	163	56	2 674	12 481	7 027	5 454
邯 郸	447	161	7 138	26 696	24 127	2 569
保 定	437	96	11 886	29 754	22 315	7 438
太 原	386	318	47 863	25 006	18 442	6 564
大 同	201	73	14 458	11 454	8 638	2 816
阳 泉	89	47	28 000	6 855	5 191	1 664
长 治	160	227	1 240	11 999	10 625	1 374
临 汾	194	224	23 834	10 959	10 132	827
呼 和 浩 特	199	33	45 650	14 914	10 302	4 612
包 头	227	141	32 450	10 095	8 480	1 615
赤 峰	193	106	1	12 400	11 100	1 300
沈 阳	662	191	11 771	45 078	36 964	8 114
大 连	540	120	5 575	36 860	33 706	3 154
鞍 山	257	65	471	17 091	12 628	4 463
抚 顺	157	40	0	9 709	6 472	3 237
本 溪	134	45	14	5 287	3 413	1 873
锦 州	163	52	2 132	13 273	10 323	2 950
长 春	391	80	7 202	24 259	7 123	17 136
吉 林	277	75	3 100	16 304	14 886	1 418
哈 尔 滨	683	1 347	11 229	42 605	39 615	2 990
齐 齐 哈 尔	265	262	18 224	14 245	13 082	1 163
牡 丹 江	141	137	443	11 209	9 040	2 169
上 海	2 156	230	155 000	196 900	181 148	15 752
南 京	681	10	70 229	62 010	39 892	22 118
无 锡	490	4	3	47 223	34 472	12 750
徐 州	492	18	3 439	37 175	34 660	2 514
常 州	325	5	28 000	27 734	26 001	1 733
苏 州	775	6	36 179	79 686	68 167	11 519
南 通	435	12	3 909	34 962	32 737	2 225
连 云 港	243	37	9 211	19 683	18 176	1 507
扬 州	271	23	8 738	21 153	9 565	11 588
镇 江	209	16	4 796	16 941	15 720	1 221
杭 州	669	4	17 692	63 716	58 716	5 000

重点城市城镇生活污染排放及处理情况（一）（续表）

（2013）

年 份 地 区	城镇人口/ 万人	生活煤炭消费量/ 万吨	生活天然气消费量/ 万米³	生活用水总量/ 万吨		
					居民家庭用水 总量	公共服务用水 总量
宁　波	541	34	9 329	51 459	45 097	6 362
温　州	614	5	1 801	50 448	46 188	4 260
湖　州	167	12	1 457	14 008	6 117	7 890
绍　兴	302	7	8 057	24 898	17 460	7 439
合　肥	506	32	11 060	48 270	43 395	4 875
芜　湖	221	20	3 635	17 453	16 001	1 452
马 鞍 山	129	10	4 706	10 431	9 031	1 400
福　州	468	7	4 059	37 903	35 171	2 732
厦　门	331	2	4 933	27 168	21 801	5 367
泉　州	515	22	608	41 729	38 721	3 007
南　昌	342	3	42	41 005	37 076	3 930
九　江	237	24	12	17 073	12 050	5 023
济　南	462	140	22 000	37 236	33 512	3 724
青　岛	607	124	24 280	50 484	45 423	5 061
淄　博	300	52	8 608	21 341	17 698	3 642
枣　庄	192	55	4 428	11 533	8 739	2 794
烟　台	404	57	7 008	25 661	22 897	2 763
潍　坊	477	77	3 613	31 369	26 141	5 228
济　宁	397	142	3 279	31 591	28 832	2 760
泰　安	299	199	15 300	21 257	19 122	2 135
日　照	146	82	0	8 390	7 400	990
郑　州	591	179	29 749	55 807	40 590	15 217
开　封	173	90	5 782	13 220	12 100	1 120
洛　阳	326	120	1 500	25 722	23 978	1 744
平 顶 山	196	4	3 000	17 865	10 065	7 800
安　阳	197	49	7 438	14 027	13 239	788
焦　作	183	60	518	14 900	13 925	975
三 门 峡	110	170	416	7 899	7 190	710
武　汉	810	100	2 500	78 052	73 321	4 730
宜　昌	224	61	3 223	16 352	14 114	2 238
荆　州	275	42	3 202	18 894	17 590	1 304
长　沙	539	28	24 911	49 328	45 875	3 453
株　洲	231	18	22 794	20 243	16 869	3 374
湘　潭	163	11	3 630	13 664	9 565	4 099
岳　阳	290	44	3 805	21 071	18 210	2 860
常　德	256	51	5 784	18 802	15 981	2 821
张 家 界	66	8	538	4 835	4 035	800
广　州	1 122	5	22 000	141 967	100 640	41 327

重点城市城镇生活污染排放及处理情况（一）（续表）

（2013）

年　份 地　区	城镇人口/ 万人	生活煤炭消费量/ 万吨	生活天然气消费量/ 万米³	生活用水总量/ 万吨	居民家庭用水 总量	公共服务用水 总量
韶　　关	155	11	0	12 566	11 681	885
深　　圳	1 119	2	20 000	140 019	139 899	120
珠　　海	140	0	5 100	18 282	14 278	4 004
汕　　头	382	1	400	27 583	24 915	2 668
湛　　江	280	13	5 089	25 772	23 014	2 758
南　　宁	395	51	4 782	29 653	23 542	6 111
柳　　州	230	20	768	23 405	18 431	4 973
桂　　林	215	22	1 670	17 024	15 173	1 850
北　　海	84	6	2 200	6 538	5 852	686
海　　口	165	0	2 027	13 900	9 655	4 245
重　　庆	1 754	146	194 906	127 455	114 709	12 745
成　　都	992	33	180 983	124 825	78 667	46 158
自　　贡	125	15	9 057	7 826	6 961	864
攀 枝 花	78	6	0	4 024	2 639	1 385
泸　　州	184	40	13 108	11 152	9 529	1 623
德　　阳	162	8	9 781	11 517	11 017	500
绵　　阳	211	18	42 232	15 926	14 686	1 240
南　　充	258	8	18 376	14 829	11 351	3 478
宜　　宾	190	186	15 006	11 346	9 303	2 043
贵　　阳	314	103	10 322	30 920	18 890	12 030
遵　　义	253	130	458	11 863	9 704	2 159
昆　　明	465	30	0	57 508	46 054	11 454
曲　　靖	241	70	288	14 785	12 905	1 880
玉　　溪	101	3	0	5 915	5 191	724
拉　　萨	38	2	4	2 489	2 086	403
西　　安	619	187	100 924	45 242	32 822	12 420
铜　　川	52	12	3 100	2 657	2 057	600
宝　　鸡	175	100	20 760	11 490	10 456	1 034
咸　　阳	230	105	32 800	13 537	9 060	4 477
渭　　南	180	23	8 618	8 664	7 942	722
延　　安	119	49	6 723	6 561	5 716	845
兰　　州	296	45	31 000	18 624	15 590	3 034
金　　昌	31	20	132	2 209	1 953	256
西　　宁	158	60	27 580	8 994	5 801	3 193
银　　川	151	34	70 796	15 574	12 886	2 688
石 嘴 山	54	21	3 164	3 942	2 267	1 675
乌鲁木齐	263	49	55 134	23 837	17 603	6 234
克拉玛依	29	2	553	4 564	3 192	1 372

重点城市城镇生活污染排放及处理情况（二）

（2013）

城　市　名　称	城镇生活污水排放量/万吨	城镇生活化学需氧量产生量/吨	城镇生活氨氮产生量/吨	城镇生活化学需氧量排放量/吨	城镇生活氨氮排放量/吨
总　计	2 917 381	10 336 627.6	1 308 355.1	3 958 510.7	697 402.5
北　京	134 991	527 173.0	64 729.0	89 868.0	14 189.0
天　津	65 469	304 702.3	41 599.1	84 885.5	15 670.8
石　家　庄	34 769	120 165.8	16 022.1	6 830.8	2 364.3
唐　山	23 517	93 239.5	12 742.7	16 271.0	5 922.0
秦　皇　岛	11 538	30 082.1	4 533.6	5 512.8	1 761.2
邯　郸	25 700	98 827.0	13 437.0	45 497.0	8 482.0
保　定	26 778	85 487.4	11 045.9	23 898.2	5 464.8
太　原	18 315	90 164.9	11 975.0	11 333.1	3 243.5
大　同	8 748	37 985.9	5 345.3	14 861.9	2 765.5
阳　泉	5 827	20 740.6	2 767.3	7 457.9	1 367.4
长　治	10 311	37 392.5	4 966.2	11 642.3	2 770.3
临　汾	8 782	44 965.7	5 972.0	27 944.7	4 447.6
呼和浩特	11 931	45 796.0	5 068.7	17 957.2	2 834.2
包　头	8 468	54 033.7	6 900.1	27 414.8	3 781.7
赤　峰	11 100	45 185.3	5 648.2	13 431.4	4 981.9
沈　阳	34 711	154 709.6	21 027.4	25 894.9	14 027.1
大　连	31 341	126 151.0	17 148.7	54 123.7	8 194.1
鞍　山	14 527	59 983.8	8 154.1	46 012.1	7 491.8
抚　顺	8 877	36 651.8	4 982.4	11 348.3	4 041.1
本　溪	4 314	31 352.3	4 262.0	15 759.8	2 046.8
.锦　州	11 282	38 002.7	5 166.0	19 643.4	3 791.0
长　春	20 797	90 631.5	11 753.9	32 006.0	6 970.3
吉　林	13 873	66 835.0	8 406.8	34 457.0	5 829.9
哈　尔　滨	33 636	154 961.2	20 814.6	86 227.4	14 093.0
齐齐哈尔	12 113	58 141.9	7 655.3	34 494.8	5 229.0
牡　丹　江	7 230	36 147.6	3 865.4	19 144.6	3 865.4
上　海	177 210	597 935.7	76 315.5	173 181.1	40 321.7
南　京	52 336	202 431.6	23 847.4	62 535.3	13 336.5
无　锡	42 500	124 243.0	15 992.3	19 316.7	2 329.7
徐　州	31 607	123 976.5	16 171.0	76 371.4	9 801.2
常　州	24 961	81 881.5	10 680.2	15 965.6	3 925.1
苏　州	66 435	195 297.1	25 473.5	18 393.5	9 967.8
南　通	29 877	109 654.1	14 302.7	37 013.5	9 133.1
连　云　港	16 668	61 176.9	7 979.6	48 333.0	6 807.8
扬　州	18 584	68 206.0	8 896.4	30 893.9	4 485.7
镇　江	14 346	52 653.8	6 867.9	28 108.9	3 419.0
杭　州	53 902	205 108.1	23 929.3	39 650.1	8 060.2

重点城市城镇生活污染排放及处理情况（二）（续表）

（2013）

城 市 名 称	城镇生活污水 排放量/万吨	城镇生活化学需氧 量产生量/吨	城镇生活氨氮 产生量/吨	城镇生活化学需氧 量排放量/吨	城镇生活氨氮 排放量/吨
宁　波	41 167	155 996.2	19 154.0	26 491.2	9 757.3
温　州	43 497	153 752.5	20 726.7	103 501.0	15 825.0
湖　州	11 647	59 077.1	6 272.7	16 722.4	3 940.9
绍　兴	21 396	76 849.2	10 024.9	38 170.3	4 767.8
合　肥	42 108	123 438.9	14 622.7	47 130.8	6 341.8
芜　湖	15 857	54 974.8	6 203.4	30 909.8	4 904.0
马 鞍 山	9 388	30 509.6	3 614.2	16 285.7	2 686.3
福　州	32 268	114 391.0	14 854.0	67 654.0	9 509.0
厦　门	22 936	81 092.8	10 530.0	23 893.0	5 634.0
泉　州	35 526	125 938.0	16 353.1	85 642.2	12 876.4
南　昌	33 466	88 366.2	9 759.2	37 879.0	6 060.4
九　江	13 864	77 985.9	6 065.6	42 555.0	5 508.0
济　南	29 788	109 589.2	14 668.1	30 316.7	4 981.9
青　岛	40 880	144 022.6	19 276.9	24 775.6	6 100.4
淄　博	19 825	63 175.3	9 524.6	14 991.1	2 934.3
枣　庄	10 328	45 480.8	6 087.4	18 149.1	3 605.8
烟　台	23 110	94 553.9	12 657.7	30 530.2	6 559.1
潍　坊	25 986	113 276.3	15 161.6	21 848.5	5 455.4
济　宁	26 853	94 089.4	12 593.5	34 391.2	6 480.2
泰　安	15 298	71 015.6	9 274.3	32 385.0	5 792.4
日　照	7 634	34 701.8	4 644.7	15 749.5	2 929.1
郑　州	48 646	129 385.2	16 604.4	26 947.3	8 477.0
开　封	12 680	37 732.2	4 865.8	16 165.8	2 464.8
洛　阳	21 755	71 306.4	9 151.0	16 800.3	4 222.9
平 顶 山	14 292	42 878.0	5 502.7	19 730.5	3 764.5
安　阳	12 205	43 077.3	5 528.3	18 725.0	3 487.1
焦　作	14 872	40 152.3	5 152.9	7 006.7	2 179.2
三 门 峡	7 194	23 980.5	3 537.1	8 489.2	2 452.7
武　汉	66 521	242 433.0	26 224.2	90 091.7	12 732.5
宜　昌	13 899	49 882.9	6 235.4	27 030.6	4 901.4
荆　州	16 060	64 240.0	8 216.5	53 837.0	7 434.2
长　沙	43 000	129 943.9	15 750.8	59 931.0	8 656.0
株　洲	16 194	55 668.0	7 009.1	35 907.5	5 108.1
湘　潭	11 387	37 573.2	4 554.6	18 021.8	2 927.8
岳　阳	18 151	67 051.1	8 003.3	52 768.5	6 534.9
常　德	16 132	61 441.1	7 447.4	49 038.0	5 719.7
张 家 界	3 868	15 954.8	1 933.9	13 080.0	1 558.1
广　州	135 179	309 795.8	39 026.2	106 077.0	17 442.0

重点城市城镇生活污染排放及处理情况（二）（续表）

（2013）

城 市 名 称	城镇生活污水 排放量/万吨	城镇生活化学需氧量 产生量/吨	城镇生活氨氮 产生量/吨	城镇生活化学需氧 量排放量/吨	城镇生活氨氮 排放量/吨
韶 关	10 053	33 669.2	4 474.9	28 999.0	3 854.0
深 圳	139 899	304 888.9	41 060.0	67 508.0	12 150.0
珠 海	17 916	42 845.6	4 966.8	15 395.0	3 058.0
汕 头	19 932	85 278.7	11 091.8	62 301.0	9 184.0
湛 江	18 411	64 632.8	8 299.6	38 946.0	5 450.0
南 宁	26 696	99 296.5	10 066.9	62 244.0	7 944.0
柳 州	21 064	59 166.0	7 504.0	32 363.4	5 340.8
桂 林	14 470	48 863.0	6 557.6	29 865.1	4 690.8
北 海	5 550	18 968.7	2 217.6	11 541.0	941.0
海 口	11 120	38 586.0	4 763.0	6 341.2	3 798.9
重 庆	108 937	435 221.2	55 682.7	218 600.5	36 210.8
成 都	99 860	259 706.9	31 540.8	102 595.1	13 144.2
自 贡	6 643	32 001.5	3 909.8	17 661.5	2 440.5
攀 枝 花	3 798	20 275.0	2 484.0	13 194.5	1 738.0
泸 州	9 709	46 681.4	5 743.9	39 864.8	4 922.1
德 阳	9 680	41 374.2	5 061.7	21 616.6	3 121.1
绵 阳	15 393	55 177.8	6 603.7	32 035.2	4 400.6
南 充	13 143	65 510.6	7 955.2	46 389.8	5 748.0
宜 宾	9 270	46 425.0	5 550.4	36 016.7	4 602.1
贵 阳	21 774	61 867.2	8 195.7	26 324.2	4 489.9
遵 义	11 552	59 147.5	7 393.4	49 995.1	5 824.5
昆 明	48 882	126 536.5	14 169.3	4 839.9	4 543.0
曲 靖	12 014	62 536.4	6 870.2	49 481.4	6 128.4
玉 溪	4 860	22 441.5	2 707.0	15 567.5	2 273.3
拉 萨	2 114	10 012.7	1 182.1	7 926.7	994.3
西 安	32 672	156 331.8	20 105.0	62 905.5	10 675.5
铜 川	2 261	13 085.3	1 604.9	8 810.7	1 081.9
宝 鸡	9 174	42 128.6	5 036.3	13 371.6	3 069.1
咸 阳	12 502	51 181.0	6 652.0	14 759.4	3 040.9
渭 南	7 351	36 495.1	4 861.6	20 441.8	3 188.6
延 安	5 438	23 852.9	2 937.0	15 844.9	2 196.3
兰 州	14 043	63 735.6	8 426.1	32 806.4	4 977.0
金 昌	1 650	6 714.6	887.7	1 248.5	400.5
西 宁	7 660	35 917.6	4 567.3	16 332.3	3 496.0
银 川	13 922	34 117.0	4 292.1	3 026.5	2 617.8
石 嘴 山	2 957	12 220.2	1 537.4	3 665.5	1 058.1
乌 鲁 木 齐	18 816	66 805.8	7 612.5	13 709.5	4 613.4
克 拉 玛 依	3 972	6 084.7	526.2	674.0	69.3

重点城市城镇生活污染排放及处理情况（三）
（2013）

单位：吨

城　市 名　称	城镇生活二氧化硫排放量	城镇生活氮氧化物排放量	城镇生活烟尘排放量
总　计	1 030 894	220 752	569 635
北　京	34 967	13 638	28 258
天　津	8 959	5 221	18 400
石 家 庄	9 564	2 802	6 635
唐　山	9 446	4 410	1 090
秦 皇 岛	7 887	2 367	5 641
邯　郸	14 835	2 906	16 090
保　定	16 336	1 553	9 609
太　原	33 396	6 738	26 727
大　同	9 270	1 700	7 339
阳　泉	1 501	1 162	4 162
长　治	4 884	4 538	20 376
临　汾	13 348	4 673	2 753
呼和浩特	4 257	665	3 763
包　头	14 352	3 073	14 068
赤　峰	19 897	2 120	8 240
沈　阳	14 389	5 154	15 276
大　连	16 366	2 403	12 015
鞍　山	8 880	1 492	6 528
抚　顺	4 224	600	4 000
本　溪	4 629	939	4 538
锦　州	5 116	929	5 160
长　春	7 344	1 545	7 919
吉　林	6 885	1 525	6 750
哈 尔 滨	50 012	22 985	80 792
齐齐哈尔	20 387	5 667	20 406
牡 丹 江	13 818	8 344	8 370
上　海	42 947	23 474	6 451
南　京	1 750	400	1 000
无　锡	512	300	387
徐　州	1 502	373	1 413
常　州	82	594	384
苏　州	786	405	347
南　通	2 121	240	978
连 云 港	6 274	812	3 026
扬　州	3 089	582	1 365
镇　江	3 156	566	2 143
杭　州	633	335	135

重点城市城镇生活污染排放及处理情况（三）（续表）

（2013）

单位：吨

城　市 名　称	城镇生活二氧化硫排放量	城镇生活氮氧化物排放量	城镇生活烟尘排放量
宁　波	2 131	545	3 406
温　州	669	423	467
湖　州	2 057	208	239
绍　兴	870	199	244
合　肥	2 710	130	3 188
芜　湖	1 958	340	1 568
马　鞍　山	1 920	258	900
福　州	1 279	169	547
厦　门	486	152	190
泉　州	3 199	559	1 771
南　昌	641	58	254
九　江	6 705	1 919	2 191
济　南	26 087	3 629	8 355
青　岛	27 497	2 683	9 954
淄　博	4 973	1 223	3 988
枣　庄	13 218	1 595	5 305
烟　台	8 053	540	5 608
潍　坊	12 480	1 316	6 905
济　宁	13 601	2 771	9 958
泰　安	19 695	1 240	8 739
日　照	13 153	1 206	5 300
郑　州	11 975	1 780	9 150
开　封	15 771	1 397	1 141
洛　阳	25 500	2 512	1 375
平　顶　山	477	384	400
安　阳	7 840	980	1 069
焦　作	5 065	1 192	5 065
三　门　峡	23 746	2 595	3 572
武　汉	5 720	1 416	1 001
宜　昌	5 825	910	3 766
荆　州	6 035	872	2 696
长　沙	2 366	153	2 946
株　洲	2 592	542	1 080
湘　潭	711	310	919
岳　阳	7 720	874	2 758
常　德	6 088	1 134	3 796
张　家　界	1 920	172	800
广　州	663	276	214

重点城市城镇生活污染排放及处理情况（三）（续表）

（2013）

单位：吨

城 市 名 称	城镇生活二氧化硫排放量	城镇生活氮氧化物排放量	城镇生活烟尘排放量
韶 关	2 238	226	619
深 圳	238	885	0
珠 海	21	44	4
汕 头	86	117	55
湛 江	2 580	403	263
南 宁	8 748	1 068	4 631
柳 州	5 140	409	1 537
桂 林	4 515	456	221
北 海	1 240	141	395
海 口	11	17	5
重 庆	53 261	4 487	4 401
成 都	4 891	2 109	661
自 贡	3 942	382	309
攀 枝 花	581	114	114
泸 州	7 214	773	1 353
德 阳	1 360	291	132
绵 阳	6 120	680	180
南 充	4 291	293	137
宜 宾	11 178	703	2 236
贵 阳	35 493	1 753	5 530
遵 义	48 708	2 084	2 688
昆 明	5 263	970	328
曲 靖	8 207	1 248	1 230
玉 溪	999	216	45
拉 萨	678	40	199
西 安	23 831	10 951	14 012
铜 川	3 128	617	1 202
宝 鸡	8 801	2 020	10 249
咸 阳	21 501	3 022	9 275
渭 南	4 381	525	1 939
延 安	4 699	683	4 832
兰 州	7 413	1 950	1 088
金 昌	3 824	590	1 613
西 宁	7 129	1 419	4 793
银 川	5 697	1 237	3 016
石 嘴 山	5 375	447	1 897
乌鲁木齐	6 691	1 425	4 920
克拉玛依	222	64	163

重点城市机动车污染排放情况
（2013）

单位：万吨

城 市 名 称	机动车污染物排放总量			
	总颗粒物	氮氧化物	一氧化碳	碳氢化合物
总 计	29.0	319.8	1 727.2	211.9
北 京	0.4	7.6	69.0	7.7
天 津	0.6	5.6	45.2	5.1
石 家 庄	0.8	7.0	23.9	3.1
唐 山	0.6	6.7	31.6	3.8
秦 皇 岛	0.5	3.7	16.4	2.1
邯 郸	1.0	6.7	21.7	3.5
保 定	0.1	7.4	46.7	5.4
太 原	0.3	3.0	16.8	2.0
大 同	0.2	2.8	14.8	1.8
阳 泉	…	2.5	22.8	2.5
长 治	0.2	1.6	6.6	0.8
临 汾	0.6	5.3	27.1	3.5
呼 和 浩 特	0.4	3.0	10.0	1.4
包 头	0.5	3.3	11.4	1.7
赤 峰	0.4	2.2	7.7	1.2
沈 阳	0.4	4.4	25.0	3.1
大 连	0.6	4.6	20.3	2.7
鞍 山	0.2	2.4	13.1	1.5
抚 顺	0.1	1.7	8.2	1.1
本 溪	0.1	1.0	4.2	0.5
锦 州	0.2	2.2	7.2	1.1
长 春	0.5	5.4	36.9	4.6
吉 林	0.4	4.0	21.6	2.8
哈 尔 滨	0.5	3.8	25.3	3.1
齐 齐 哈 尔	0.4	3.3	14.0	1.9
牡 丹 江	0.2	1.7	11.7	1.5
上 海	0.7	9.4	42.5	6.2
南 京	0.3	3.4	15.7	1.8
无 锡	0.2	3.1	14.7	1.6
徐 州	0.5	6.0	17.2	2.4
常 州	0.1	2.0	10.9	1.2
苏 州	0.5	6.7	51.7	6.0
南 通	0.1	1.9	14.2	1.6
连 云 港	0.1	1.5	6.1	0.7
扬 州	0.2	1.8	8.6	1.0
镇 江	0.1	1.1	5.5	0.6
杭 州	0.4	4.2	23.6	2.6

重点城市机动车污染排放情况（续表）

（2013）

单位：万吨

城 市 名 称	机动车污染物排放总量			
	总颗粒物	氮氧化物	一氧化碳	碳氢化合物
宁　波	0.2	2.8	17.4	2.0
温　州	0.1	1.7	14.0	1.6
湖　州	0.1	0.9	9.2	1.1
绍　兴	0.1	1.4	8.9	1.0
合　肥	0.3	2.5	11.4	1.4
芜　湖	0.1	1.2	7.5	0.8
马 鞍 山	0.1	0.6	2.5	0.3
福　州	0.2	2.2	9.8	1.2
厦　门	0.2	2.1	8.9	1.0
泉　州	0.1	1.8	14.3	1.6
南　昌	0.3	3.3	15.5	2.0
九　江	0.2	2.0	6.9	1.0
济　南	0.3	2.8	15.7	1.7
青　岛	0.4	4.1	20.7	2.4
淄　博	0.2	2.0	12.7	1.5
枣　庄	0.2	1.7	8.6	1.0
烟　台	0.3	3.3	21.5	2.6
潍　坊	0.6	5.9	31.5	4.0
济　宁	0.5	3.7	14.7	1.9
泰　安	0.1	1.6	9.1	1.1
日　照	0.1	1.1	5.0	0.6
郑　州	0.6	6.1	29.6	3.5
开　封	0.2	1.7	6.1	0.9
洛　阳	0.4	3.3	13.0	1.6
平 顶 山	0.2	1.9	8.8	1.1
安　阳	0.3	3.2	13.0	1.8
焦　作	0.3	3.6	15.8	1.9
三 门 峡	0.1	1.3	8.0	0.9
武　汉	0.5	5.0	22.8	2.7
宜　昌	0.2	2.0	9.8	1.3
荆　州	0.1	1.4	8.7	1.0
长　沙	0.2	3.1	16.9	1.9
株　洲	0.1	1.0	5.4	0.7
湘　潭	0.1	0.6	4.2	0.4
岳　阳	0.1	0.9	5.4	0.6
常　德	0.1	1.3	6.8	0.8
张 家 界	...	0.5	2.7	0.3
广　州	0.6	6.7	30.1	3.5

重点城市机动车污染排放情况（续表）

（2013）

单位：万吨

城 市 名 称	机动车污染物排放总量			
	总颗粒物	氮氧化物	一氧化碳	碳氢化合物
韶　关	0.1	0.6	4.0	0.5
深　圳	0.8	7.7	43.6	4.7
珠　海	0.3	1.9	11.2	1.3
汕　头	0.4	2.0	14.8	1.8
湛　江	0.1	1.2	6.7	0.8
南　宁	0.4	4.0	21.9	2.6
柳　州	0.1	1.2	5.9	0.7
桂　林	0.1	1.3	8.0	1.0
北　海	…	0.5	3.3	0.4
海　口	0.1	1.0	5.5	0.6
重　庆	0.7	11.0	62.0	8.0
成　都	0.3	4.6	28.3	3.1
自　贡	…	0.6	3.7	0.4
攀 枝 花	0.1	0.7	2.1	0.3
泸　州	…	1.0	8.9	0.9
德　阳	0.1	1.1	6.5	0.8
绵　阳	0.1	1.1	6.2	0.7
南　充	0.1	1.4	5.1	0.7
宜　宾	0.1	0.9	5.0	0.6
贵　阳	0.2	2.5	11.0	1.4
遵　义	0.1	1.3	8.1	1.0
昆　明	0.2	4.2	23.9	3.4
曲　靖	0.3	2.3	14.4	2.1
玉　溪	0.2	1.8	13.7	2.0
拉　萨	0.1	1.2	12.8	1.3
西　安	0.2	4.1	24.2	2.6
铜　川	…	0.6	3.6	0.5
宝　鸡	…	1.4	15.6	1.6
咸　阳	0.1	1.6	7.1	0.8
渭　南	0.2	2.4	8.0	1.1
延　安	0.2	1.4	6.6	1.2
兰　州	0.1	2.2	20.8	2.5
金　昌	…	0.4	4.2	0.5
西　宁	0.1	1.7	12.0	1.8
银　川	0.4	3.5	14.6	1.9
石 嘴 山	0.1	0.7	2.8	0.4
乌鲁木齐	0.4	5.2	20.6	2.8
克拉玛依	0.2	1.8	7.2	1.0

重点城市污水处理情况（一）

（2013）

城 市 名 称	污水处理厂数/ 座	污水处理厂 设计处理能力/ （万吨/日）	本年运行费用/ 万元	污水处理厂累计 完成投资/万元	新增固定资产/ 万元
总　计	2 804	10 831.9	2 627 183.1	23 223 856.7	1 114 712.8
北　京	138	434.2	116 032.3	971 057.4	11 139.1
天　津	60	272.2	86 783.0	702 224.1	1 749.4
石 家 庄	27	185.1	38 353.5	348 992.3	16 804.7
唐　山	25	118.0	23 204.7	270 105.7	2 136.2
秦 皇 岛	17	55.8	13 775.7	77 473.5	124.9
邯　郸	28	81.8	15 882.7	156 322.9	20 813.4
保　定	33	110.5	26 481.7	294 847.6	1 738.7
太　原	14	77.8	17 159.6	147 612.7	36 269.3
大　同	17	27.5	8 667.2	106 429.9	34 025.8
阳　泉	7	17.2	5 502.7	54 434.0	114.1
长　治	21	38.7	8 697.2	108 583.8	1 648.8
临　汾	18	30.9	4 922.2	74 665.1	295.5
呼和浩特	14	36.8	9 847.0	116 039.7	5 030.0
包　头	12	59.1	13 170.4	217 527.1	38 291.7
赤　峰	15	55.0	8 361.3	121 589.8	5 872.9
沈　阳	44	226.3	41 645.2	260 724.4	21 497.8
大　连	24	113.8	26 848.0	193 761.2	5 612.8
鞍　山	10	61.5	10 262.7	133 094.6	6 467.0
抚　顺	8	59.3	3 801.3	56 848.0	18.2
本　溪	6	42.0	3 460.4	47 637.1	11 044.0
锦　州	8	35.0	9 918.8	75 581.1	148.5
长　春	13	98.8	19 987.6	223 604.0	24 004.3
吉　林	8	65.0	11 707.3	108 476.0	30.0
哈 尔 滨	20	139.8	28 521.4	284 873.6	27 312.8
齐 齐 哈 尔	9	40.0	6 875.8	90 235.9	5 285.6
牡 丹 江	6	20.0	5 308.4	46 090.6	80.0
上　海	56	771.1	161 761.9	1 410 181.0	25 405.3
南　京	65	215.8	69 727.7	595 274.5	17 438.0
无　锡	77	231.0	77 945.1	540 704.0	30 382.6
徐　州	36	100.7	20 781.9	189 333.3	2 367.8
常　州	46	112.6	35 807.4	362 801.2	1 363.9
苏　州	128	357.5	114 955.2	1 104 024.3	125 823.9
南　通	89	117.9	32 990.9	318 616.3	10 063.2
连 云 港	25	38.9	9 594.7	127 770.0	10 706.7
扬　州	42	75.4	19 508.0	137 373.8	4 779.8
镇　江	35	49.3	14 012.0	166 804.9	17 346.7
杭　州	41	299.1	113 877.8	734 738.7	12 300.6

重点城市污水处理情况（一）（续表）

（2013）

城　市　名　称	污水处理厂数/座	污水处理厂设计处理能力/（万吨/日）	本年运行费用/万元	污水处理厂累计完成投资/万元	新增固定资产/万元
宁　波	36	170.4	40 868.3	397 742.4	65 308.9
温　州	39	91.3	34 720.7	263 179.5	17 770.0
湖　州	34	66.8	32 556.3	194 673.4	8 850.6
绍　兴	6	139.2	87 029.9	375 265.0	8 445.8
合　肥	17	113.3	17 425.2	255 052.6	26 452.5
芜　湖	10	52.3	5 994.3	119 407.8	1 169.4
马鞍山	11	35.0	7 653.3	91 238.5	8 113.5
福　州	20	107.0	14 102.3	150 041.0	6 250.4
厦　门	15	84.7	12 055.1	127 753.3	1 139.3
泉　州	73	111.8	20 173.9	205 260.6	22 006.8
南　昌	14	114.7	15 699.9	131 051.5	48.7
九　江	19	49.1	8 808.7	81 544.5	1 311.0
济　南	38	105.8	29 945.7	194 495.1	29 267.3
青　岛	34	172.3	69 537.9	513 562.6	35 664.0
淄　博	14	85.0	24 476.4	136 675.7	6 527.0
枣　庄	18	62.9	10 517.3	93 304.2	5 431.8
烟　台	37	87.2	20 488.2	205 958.6	16 970.0
潍　坊	35	162.6	48 647.6	288 719.9	44 337.7
济　宁	40	137.0	28 161.4	248 243.0	12 580.3
泰　安	18	50.5	7 839.6	83 081.1	7 665.1
日　照	12	26.0	7 224.4	78 528.1	1 327.4
郑　州	20	149.5	32 828.6	334 623.3	6 321.1
开　封	11	34.6	4 093.9	58 543.4	1 694.5
洛　阳	18	74.1	10 262.4	136 037.8	14 740.2
平顶山	8	39.0	7 443.0	80 801.0	160.0
安　阳	12	36.4	7 790.4	74 509.7	19.9
焦　作	19	55.9	12 130.3	109 762.8	8 894.8
三门峡	9	23.5	3 594.5	30 688.0	7 861.1
武　汉	22	225.0	41 460.7	340 499.7	941.3
宜　昌	32	58.2	8 301.3	106 376.2	7 288.0
荆　州	11	44.0	7 886.8	77 393.6	19.0
长　沙	24	155.6	33 093.5	369 254.9	978.1
株　洲	20	42.6	7 627.8	124 229.7	1 528.9
湘　潭	6	32.0	7 954.6	65 438.4	23 428.6
岳　阳	12	41.0	7 821.9	111 155.3	3 852.0
常　德	10	42.0	6 858.9	54 849.0	…
张家界	7	9.6	1 674.3	25 199.3	…
广　州	33	450.9	84 212.7	1 012 379.5	8 134.6

重点城市污水处理情况（一）（续表）

（2013）

城 市 名 称	污水处理厂数/ 座	污水处理厂 设计处理能力/ （万吨/日）	本年运行费用/ 万元	污水处理厂累计 完成投资/万元	新增固定资产/ 万元
韶 关	12	27.5	5 558.6	31 575.4	631.3
深 圳	42	470.7	65 663.5	492 178.4	14 039.0
珠 海	12	60.3	19 416.9	149 608.9	11 355.6
汕 头	20	59.3	14 873.1	244 139.6	2 454.5
湛 江	9	51.5	8 040.0	99 544.5	6 248.8
南 宁	11	92.4	23 125.7	205 736.0	4 779.4
柳 州	10	60.5	20 465.7	139 668.1	131.4
桂 林	23	46.0	11 160.5	87 106.2	5 542.3
北 海	2	15.0	3 304.0	50 939.0	21.0
海 口	10	56.2	9 726.7	93 433.8	819.3
重 庆	149	285.5	107 786.6	750 738.0	20 250.9
成 都	152	252.6	65 722.4	540 456.8	19 007.6
自 贡	5	14.5	4 374.0	30 484.0	122.0
攀 枝 花	10	15.1	2 932.2	34 549.1	7.7
泸 州	9	12.3	5 686.9	54 758.0	550.0
德 阳	10	28.7	7 334.6	56 143.4	176.0
绵 阳	11	39.1	10 801.5	91 411.2	3 492.0
南 充	12	28.8	8 638.6	60 120.7	2 538.5
宜 宾	15	14.3	4 518.6	40 312.4	218.1
贵 阳	15	66.9	9 553.2	162 398.0	1 291.9
遵 义	25	25.7	3 692.6	49 632.1	1 672.0
昆 明	20	128.5	36 048.4	223 666.2	156.4
曲 靖	10	27.0	5 091.7	63 426.8	1 657.4
玉 溪	10	17.8	3 106.0	48 485.7	10 541.5
拉 萨	1	5.0	2 000.0	12 210.0	…
西 安	26	147.4	26 747.4	304 862.8	2 811.2
铜 川	12	8.2	1 958.2	14 280.0	211.6
宝 鸡	13	31.1	6 841.0	77 034.0	2 407.1
咸 阳	15	38.4	8 772.3	98 253.4	7 603.4
渭 南	12	24.5	5 795.3	85 210.0	9 102.3
延 安	22	12.9	4 043.2	87 727.8	4 608.6
兰 州	7	65.4	10 899.2	86 613.3	1 354.5
金 昌	5	14.5	3 611.0	29 223.0	76.0
西 宁	9	28.4	6 045.8	73 530.6	2 314.0
银 川	8	41.5	8 190.0	73 634.2	0.2
石 嘴 山	3	11.5	3 831.0	36 162.4	193.1
乌鲁木齐	8	83.0	15 424.2	78 868.6	28 293.0
克拉玛依	3	14.8	1 300.0	42 760.0	…

重点城市污水处理情况（二）

（2013）

单位：万吨

城 市 名 称	污水实际 处理量	污水再生 利用量	工业用水量	市政用水量	景观用水量
总 计	3 035 223.0	108 792.5	68 626.3	7 822.0	32 344.1
北 京	130 928.9	10 410.2	5 034.2	1 529.0	3 847.0
天 津	72 118.3	2 327.4	1 537.6	400.0	389.7
石 家 庄	51 725.9	3 343.2	1 919.4	277.0	1 146.8
唐 山	28 625.1	7 351.1	5 617.4	549.5	1 184.1
秦 皇 岛	16 862.4	365.0	…	365.0	…
邯 郸	21 122.9	3 951.1	3 507.4	294.2	149.5
保 定	33 040.1	936.3	260.6	530.8	144.9
太 原	18 740.8	2 749.4	2 657.2	35.1	57.1
大 同	6 088.8	1 523.7	1 523.7	…	…
阳 泉	5 125.7	213.5	213.5	…	…
长 治	8 886.7	2 161.5	2 151.0	10.5	…
临 汾	5 777.0	431.4	431.4	…	…
呼 和 浩 特	10 344.0	772.4	579.2	…	193.2
包 头	13 637.5	5 631.3	5 224.4	406.9	…
赤 峰	10 689.1	372.8	372.8	…	…
沈 阳	53 650.5	3 208.4	3 208.4	…	…
大 连	34 805.3	1 184.2	451.7	0.5	732.0
鞍 山	8 283.4	…	…	…	…
抚 顺	12 003.8	349.7	349.7	…	…
本 溪	10 102.8	…	…	…	…
锦 州	10 437.2	387.9	387.9	…	…
长 春	21 850.4	990.7	698.7	…	292.0
吉 林	17 309.7	7.4	7.4	…	…
哈 尔 滨	29 335.9	197.3	197.3	…	…
齐 齐 哈 尔	6 897.1	…	…	…	…
牡 丹 江	6 643.3	…	…	…	…
上 海	208 182.7	164.9	16.8	148.1	…
南 京	71 031.9	381.9	373.9	…	8.0
无 锡	59 857.4	2 121.3	212.8	108.5	1 800.0
徐 州	28 832.8	1 691.7	1 627.5	8.0	56.2
常 州	32 233.8	3 118.1	2 286.6	230.6	600.9
苏 州	87 891.9	2 429.4	2 417.9	11.5	…
南 通	35 640.0	5.5	5.5	…	…
连 云 港	9 709.7	9.5	…	…	9.5
扬 州	21 542.8	908.3	904.4	…	3.9
镇 江	14 515.6	400.4	9.5	44.2	346.7
杭 州	85 799.1	7 582.1	7 299.3	266.3	16.6

重点城市污水处理情况（二）（续表）

（2013）

单位：万吨

城　市 名　称	污水实际 处理量	污水再生 利用量	工业用水量	市政用水量	景观用水量
宁　波	45 791.0	3 328.3	2 802.6	147.0	378.7
温　州	29 015.0	30.4	…	30.0	0.4
湖　州	17 034.9	2 534.6	2 195.3	339.3	…
绍　兴	40 342.0	579.0	579.0	…	…
合　肥	41 919.1	540.0	540.0	…	…
芜　湖	15 137.1	…	…	…	…
马 鞍 山	10 298.8	…	…	…	…
福　州	27 438.7	…	…	…	…
厦　门	24 273.8	195.4	…	87.9	107.5
泉　州	28 689.1	2 075.6	…	…	2 075.6
南　昌	32 209.7	…	…	…	…
九　江	11 279.0	525.2	313.9	10.7	200.6
济　南	30 538.3	1 244.7	1 104.7	35.0	105.0
青　岛	48 709.3	418.2	123.5	130.2	164.5
淄　博	27 287.7	2 484.2	106.3	…	2 377.9
枣　庄	19 508.5	…	…	…	…
烟　台	26 187.8	182.5	182.5	…	…
潍　坊	45 854.9	…	…	…	…
济　宁	36 586.3	2 796.1	2 796.1	…	…
泰　安	15 797.6	526.0	526.0	…	…
日　照	8 469.4	212.0	210.8	1.2	…
郑　州	49 660.9	1 825.0	…	…	1 825.0
开　封	10 642.4	…	…	…	…
洛　阳	20 926.4	1 635.1	1 635.1	…	…
平 顶 山	13 061.8	365.0	365.0	…	…
安　阳	9 390.6	921.6	638.6	283.0	…
焦　作	16 584.1	219.0	…	73.0	146.0
三 门 峡	6 160.2	…	…	…	…
武　汉	67 708.1	86.7	…	43.6	43.2
宜　昌	13 968.5	0.5	…	0.4	0.1
荆　州	11 747.1	…	…	…	…
长　沙	42 964.1	30.1	…	30.1	…
株　洲	13 470.8	…	…	…	…
湘　潭	10 946.1	18.1	2.1	3.5	12.5
岳　阳	11 674.6	…	…	…	…
常　德	11 020.3	…	…	…	…
张 家 界	2 600.0	…	…	…	…
广　州	135 443.0	9 118.4	776.8	72.6	8 269.0

重点城市污水处理情况（二）（续表）
（2013）

单位：万吨

城 市 名 称	污水实际 处理量	污水再生 利用量	工业用水量	市政用水量	景观用水量
韶 关	7 743.4	…	…	…	…
深 圳	140 580.0	3 986.9	17.3	82.9	3 886.6
珠 海	18 753.2	32.1	32.1	…	…
汕 头	18 628.2	152.9	…	82.0	70.9
湛 江	15 022.9	…	…	…	…
南 宁	25 155.0	2.0	1.3	0.7	…
柳 州	18 401.9	…	…	…	…
桂 林	12 169.6	28.0	…	…	28.0
北 海	4 979.0	29.2	29.2	…	…
海 口	17 768.1	…	…	…	…
重 庆	92 973.2	658.1	449.4	135.4	73.3
成 都	82 777.6	33.3	…	25.1	8.2
自 贡	6 098.8	21.9	20.8	…	1.1
攀 枝 花	2 906.1	14.2	…	7.3	6.9
泸 州	3 509.5	6.1	…	…	6.1
德 阳	8 700.9	8.6	3.5	0.5	4.6
绵 阳	10 397.9	…	…	…	…
南 充	8 369.2	87.6	70.7	10.0	6.9
宜 宾	4 113.9	2.0	2.0	…	…
贵 阳	20 789.5	593.4	…	…	593.4
遵 义	7 237.4	…	…	…	…
昆 明	43 268.3	…	…	…	…
曲 靖	6 767.9	144.4	144.4	…	…
玉 溪	3 738.6	…	…	…	…
拉 萨	1 800.0	…	…	…	…
西 安	33 227.1	1 248.3	1 041.8	93.9	112.6
铜 川	1 452.3	5.0	5.0	…	…
宝 鸡	9 356.4	…	…	…	…
咸 阳	11 302.3	735.6	…	…	735.6
渭 南	6 507.6	…	…	…	…
延 安	2 784.0	…	…	…	…
兰 州	13 294.9	253.7	225.7	…	28.0
金 昌	1 794.0	200.0	200.0	…	…
西 宁	7 071.1	…	…	…	…
银 川	13 577.1	…	…	…	…
石 嘴 山	2 538.5	64.5	…	64.5	…
乌鲁木齐	16 457.1	583.1	…	485.1	98.0
克拉玛依	2 603.6	331.4	…	331.4	…

重点城市污水处理情况（三）

（2013）

单位：万吨

城市名称	污泥产生量	污泥处置量	土地利用量	填埋处置量	建筑材料利用量	焚烧处置量	污泥倾倒丢弃量
总　计	1 931.9	1 931.9	367.2	820.3	242.4	502.0	…
北　京	124.0	124.0	60.0	30.2	14.9	18.8	…
天　津	39.1	39.1	2.6	34.1	1.6	0.6	…
石　家　庄	42.4	42.4	9.7	31.9	…	0.7	…
唐　山	23.8	23.8	14.4	7.9	…	1.6	…
秦　皇　岛	9.7	9.7	5.1	4.6	…	…	…
邯　郸	11.4	11.4	…	10.6	0.2	0.6	…
保　定	22.3	22.3	0.5	21.5	0.3	0.1	…
太　原	15.9	15.9	14.6	1.3	…	…	…
大　同	3.2	3.2	…	3.1	…	…	…
阳　泉	1.9	1.9	0.4	1.4	…	…	…
长　治	4.7	4.7	…	4.4	…	0.3	…
临　汾	2.5	2.5	1.4	0.7	…	0.5	…
呼和浩特	3.4	3.4	0.4	1.0	0.1	1.9	…
包　头	1.8	1.8	…	1.8	…	…	…
赤　峰	6.1	6.1	…	6.1	…	…	…
沈　阳	32.8	32.8	5.8	27.1	…	…	…
大　连	13.7	13.7	0.6	7.0	6.1	…	…
鞍　山	6.1	6.1	…	6.1	…	…	…
抚　顺	2.6	2.6	…	2.6	…	…	…
本　溪	3.6	3.6	…	3.6	0.1	…	…
锦　州	5.2	5.2	…	5.0	…	0.1	…
长　春	11.6	11.6	…	11.6	…	…	…
吉　林	7.4	7.4	0.2	4.1	…	3.0	…
哈　尔　滨	16.9	16.9	1.9	15.0	…	…	…
齐齐哈尔	1.6	1.6	…	1.6	…	…	…
牡　丹　江	1.1	1.1	…	1.1	…	…	…
上　海	126.1	126.1	0.4	102.7	11.0	12.0	…
南　京	19.1	19.1	0.7	6.8	2.0	9.6	…
无　锡	63.8	63.8	3.4	10.2	10.7	39.5	…
徐　州	15.7	15.7	2.2	2.1	0.9	10.5	…
常　州	35.1	35.1	…	…	7.7	27.4	…
苏　州	76.8	76.8	13.4	12.4	0.4	50.7	…
南　通	26.9	26.9	0.4	0.1	7.2	19.2	…
连　云　港	2.5	2.5	…	0.1	0.9	1.5	…
扬　州	10.2	10.2	…	3.5	0.5	6.2	…
镇　江	6.6	6.6	…	0.3	0.7	5.6	…
杭　州	175.1	175.1	24.9	1.7	68.4	80.1	…

重点城市污水处理情况（三）（续表）

（2013）

单位：万吨

城 市 名 称	污泥产生量	污泥处置量	土地利用量	填埋处置量	建筑材料利用量	焚烧处置量	污泥倾倒丢弃量
宁　波	31.7	31.7	8.4	3.4	1.4	18.4	…
温　州	20.8	20.8	1.5	9.7	3.1	6.4	…
湖　州	10.1	10.1	…	3.3	2.5	4.2	…
绍　兴	88.6	88.6	…	…	1.8	86.8	…
合　肥	23.3	23.3	8.9	1.1	…	13.3	…
芜　湖	6.8	6.8	…	1.1	…	5.7	…
马 鞍 山	4.4	4.4	3.2	1.2	…	…	…
福　州	11.4	11.4	3.2	8.2	…	…	…
厦　门	12.4	12.4	2.7	7.6	2.2	…	…
泉　州	13.1	13.1	0.5	1.6	2.9	8.0	…
南　昌	10.1	10.1	…	0.2	6.0	3.8	…
九　江	2.8	2.8	…	1.7	0.2	0.8	…
济　南	17.4	17.4	16.0	0.5	…	1.0	…
青　岛	41.6	41.6	17.8	5.2	6.3	12.3	…
淄　博	29.2	29.2	…	22.0	2.3	4.9	…
枣　庄	5.8	5.8	0.2	…	…	5.6	…
烟　台	19.2	19.2	1.5	15.9	…	1.9	…
潍　坊	29.2	29.2	4.9	14.9	1.3	8.1	…
济　宁	13.5	13.5	4.9	2.9	1.8	3.9	…
泰　安	6.8	6.8	1.4	0.7	1.0	3.7	…
日　照	3.5	3.5	2.8	…	0.6	0.1	…
郑　州	45.5	45.5	17.8	27.6	…	…	…
开　封	5.8	5.8	…	5.8	…	…	…
洛　阳	13.7	13.7	3.5	10.2	…	…	…
平 顶 山	6.4	6.4	…	6.4	…	…	…
安　阳	5.9	5.9	0.4	3.3	…	2.2	…
焦　作	9.1	9.1	2.1	4.4	2.5	…	…
三 门 峡	2.7	2.7	0.1	2.6	…	…	…
武　汉	24.6	24.6	0.1	22.3	2.1	…	…
宜　昌	2.7	2.7	…	2.0	…	0.7	…
荆　州	4.5	4.5	1.3	2.3	0.9	…	…
长　沙	5.5	5.5	…	5.5	…	…	…
株　洲	6.2	6.2	…	2.7	3.5	…	…
湘　潭	3.4	3.4	…	0.1	3.1	0.2	…
岳　阳	4.6	4.6	0.1	4.5	…	…	…
常　德	3.5	3.5	…	2.6	0.2	0.6	…
张 家 界	1.1	1.1	…	1.1	…	…	…
广　州	69.3	69.3	41.7	2.1	19.2	6.3	…

重点城市污水处理情况（三）（续表）
（2013）

单位：万吨

城 市 名 称	污泥产生量	污泥处置量	土地利用量	填埋处置量	建筑材料利用量	焚烧处置量	污泥倾倒丢弃量
韶 关	0.9	0.9	0.1	0.9	…	…	…
深 圳	57.3	57.3	3.4	54.0	…	…	…
珠 海	8.0	8.0	…	1.7	1.8	4.5	…
汕 头	9.6	9.6	…	9.6	…	…	…
湛 江	6.4	6.4	6.2	0.2	…	…	…
南 宁	9.0	9.0	8.9	…	…	…	…
柳 州	4.8	4.8	0.3	0.3	4.2	…	…
桂 林	1.4	1.4	0.9	0.4	0.1	…	…
北 海	2.0	2.0	…	2.0	…	…	…
海 口	4.5	4.5	4.5	…	…	…	…
重 庆	40.2	40.2	2.8	21.9	13.9	1.6	…
成 都	31.9	31.9	14.5	2.6	10.3	4.4	…
自 贡	2.3	2.3	…	2.3	…	…	…
攀 枝 花	1.0	1.0	…	1.0	…	…	…
泸 州	2.1	2.1	…	2.1	…	…	…
德 阳	3.9	3.9	0.8	2.5	0.1	0.5	…
绵 阳	5.0	5.0	…	0.3	4.5	0.1	…
南 充	4.8	4.8	…	4.8	…	…	…
宜 宾	2.2	2.2	…	2.1	0.1	…	…
贵 阳	9.2	9.2	…	9.1	0.1	…	…
遵 义	3.0	3.0	…	3.0	…	…	…
昆 明	24.0	24.0	0.1	23.8	…	0.1	…
曲 靖	2.7	2.7	2.0	0.6	…	…	…
玉 溪	1.3	1.3	1.1	0.3	…	…	…
拉 萨	36.0	36.0	…	36.0	…	…	…
西 安	23.7	23.7	3.1	15.2	5.4	…	…
铜 川	0.8	0.8	…	0.8	…	…	…
宝 鸡	4.7	4.7	0.2	4.4	…	…	…
咸 阳	7.4	7.4	2.0	2.3	3.0	…	…
渭 南	4.0	4.0	0.1	4.0	…	…	…
延 安	1.9	1.9	…	1.9	…	…	…
兰 州	10.8	10.8	…	10.8	…	…	…
金 昌	1.2	1.2	0.7	0.4	…	0.1	…
西 宁	4.5	4.5	…	4.5	…	…	…
银 川	7.5	7.5	6.9	0.4	…	0.2	…
石 嘴 山	0.9	0.9	…	…	…	0.9	…
乌鲁木齐	1.8	1.8	0.3	1.5	…	…	…
克拉玛依	0.8	0.8	…	0.8	…	…	…

重点城市污水处理情况（四）

（2013）

单位：吨

城 市名 称	污染物去除量						
	化学需氧量	氨氮	油类	总氮	总磷	挥发酚	氰化物
总 计	7 946 888.5	707 262.5	36 830.8	590 608.5	84 407.0	620.4	708.5
北 京	427 243.9	58 232.9	7 718.7	57 816.4	5 231.0	5.4	0.9
天 津	258 587.0	25 909.8	1 723.8	26 591.0	2 298.8	6.7	0.1
石 家 庄	156 800.2	19 900.9	68.1	5 010.8	617.4	…	…
唐 山	84 219.1	7 028.3	274.1	8 498.2	1 242.1	…	…
秦 皇 岛	49 181.5	4 142.7	…	1 553.9	209.9	…	…
邯 郸	57 140.3	6 038.8	39.3	1 080.4	327.4	…	…
保 定	91 705.6	6 832.5	37.7	4 181.3	699.4	…	…
太 原	59 677.3	7 490.2	826.3	7 431.3	1 256.7	3.3	0.3
大 同	18 590.5	2 434.2	…	2 014.7	166.7	…	…
阳 泉	13 282.7	1 399.8	462.4	1 851.3	231.5	0.6	…
长 治	27 071.1	2 422.1	171.4	2 790.4	322.5	0.2	0.1
临 汾	17 712.9	1 547.3	…	1 259.4	108.1		
呼 和 浩 特	27 980.8	2 261.1	…	…	…		
包 头	30 488.4	3 523.7	537.1	8 035.7	1 078.3	…	11.6
赤 峰	53 474.6	3 160.3	…	239.5	43.0		
沈 阳	128 204.0	7 317.8	86.4	1 991.3	474.1	…	…
大 连	79 562.4	9 476.3	115.5	7 385.5	930.1	…	…
鞍 山	14 782.8	1 141.3	…	1 324.6	139.5	…	…
抚 顺	23 431.1	1 425.6	…	1 425.6	…	…	…
本 溪	18 369.6	2 103.0	…	2 732.8	66.4	…	…
锦 州	20 446.5	1 764.9	65.8	1 961.3	314.7	…	…
长 春	58 625.5	4 783.6	56.8	1 605.3	197.2	…	…
吉 林	47 835.3	2 576.9	1 038.2	2 650.6	707.1	31.2	0.2
哈 尔 滨	68 733.8	6 721.6	229.3	7 405.9	1 250.1	…	…
齐 齐 哈 尔	21 014.7	2 426.4	…	1 894.5	241.3	…	…
牡 丹 江	17 607.2	979.3	2.2	865.6	214.8	…	…
上 海	603 536.7	44 347.7	4 607.1	40 528.0	7 242.6	209.8	8.1
南 京	134 083.1	10 568.4	453.6	7 274.1	1 140.4	53.4	…
无 锡	174 640.7	13 995.5	1 368.2	13 960.9	1 390.3	…	0.1
徐 州	47 584.3	6 110.0	…	4 272.3	655.6	…	…
常 州	89 323.1	6 828.9	134.4	4 788.6	976.3	30.9	23.3
苏 州	284 280.6	20 887.5	395.2	15 762.6	3 077.1	5.5	…
南 通	106 015.3	8 440.2	407.5	1 146.9	523.3	…	…
连 云 港	21 142.5	1 670.3	…	1 036.4	195.5	1.0	…
扬 州	52 446.9	4 784.6	…	117.3	446.7	…	…
镇 江	27 868.2	3 368.8	271.6	2 343.1	285.6	…	…
杭 州	416 213.6	19 548.5	15.7	5 689.7	1 743.5	0.1	…

重点城市污水处理情况（四）（续表）

（2013）

单位：吨

城 市 名 称	污染物去除量						
	化学需氧量	氨氮	油类	总氮	总磷	挥发酚	氰化物
宁 波	131 220.7	9 743.8	927.2	7 199.2	1 375.9	…	151.1
温 州	74 465.0	6 034.3	149.6	1 355.5	331.9	…	426.7
湖 州	42 354.7	2 335.5	166.8	1 421.0	239.1	…	…
绍 兴	292 499.1	17 435.1	2 097.6	17 208.9	1 577.9	…	…
合 肥	68 014.7	7 772.0	355.4	6 771.9	1 432.0	33.1	…
芜 湖	23 236.1	1 702.7	15.2	626.1	272.3	…	…
马 鞍 山	15 158.9	970.9	…	1 702.3	178.7	…	…
福 州	53 477.0	5 341.8	368.0	6 481.6	542.5	…	…
厦 门	63 175.1	6 662.8	183.8	5 036.0	713.4	11.6	0.7
泉 州	131 112.1	4 718.8	550.0	3 763.4	364.5	1.5	21.7
南 昌	49 820.2	3 774.9	…	2 074.2	689.2	…	…
九 江	14 855.0	1 189.7	23.6	910.5	45.1	…	…
济 南	78 845.4	10 433.2	…	8 276.4	1 162.5	…	…
青 岛	200 485.9	16 101.6	67.0	16 984.7	2 664.8	…	41.5
淄 博	85 716.9	7 968.7	114.3	431.9	975.2	…	…
枣 庄	31 213.0	2 937.6	138.3	1 482.2	200.1	…	…
烟 台	75 462.1	7 796.7	17.8	7 578.0	970.2	…	…
潍 坊	166 525.5	12 721.5	324.5	12 673.7	938.9	7.9	…
济 宁	66 544.1	6 237.9	…	5 174.8	665.9	…	…
泰 安	39 456.4	3 420.2	…	1 381.9	265.5	…	…
日 照	22 938.4	1 889.8	…	2 129.5	178.7	…	0.4
郑 州	124 327.7	14 885.0	1 335.2	16 614.6	2 322.3	51.4	1.7
开 封	23 610.1	2 716.8	…	413.3	200.0	…	…
洛 阳	51 215.6	5 722.1	24.8	8 212.7	1 014.2	0.1	…
平 顶 山	29 242.7	2 074.6	…	3 408.9	400.1	…	…
安 阳	23 318.7	1 894.3	…	1 908.5	563.4	…	…
焦 作	63 535.0	3 487.8	…	4 683.2	656.2	…	…
三 门 峡	15 198.8	1 071.7	…	2 292.6	310.1	…	…
武 汉	94 622.7	10 262.5	71.7	6 824.2	1 349.5	…	…
宜 昌	24 252.3	1 773.4	13.3	1 639.5	167.5	…	…
荆 州	17 946.8	1 158.3	…	762.2	121.2	…	…
长 沙	69 975.1	7 313.0	0.5	6 048.4	842.3	…	…
株 洲	21 206.8	1 812.8	…	1 319.8	175.7	…	…
湘 潭	22 416.1	1 924.0	36.2	1 711.8	165.8	…	…
岳 阳	18 841.0	2 028.4	…	1 877.9	103.5	…	…
常 德	14 427.3	1 397.2	493.1	1 308.1	202.7	23.2	8.3
张 家 界	2 873.7	374.7	…	…	0.4	…	…
广 州	258 429.9	22 209.6	494.5	18 268.0	3 790.0	…	…

重点城市污水处理情况（四）（续表）

（2013）

单位：吨

城 市 名 称	污染物去除量						
	化学需氧量	氨氮	油类	总氮	总磷	挥发酚	氰化物
韶　关	6 167.6	558.0	79.2	780.6	66.1	…	…
深　圳	242 040.6	29 432.0	84.7	21 850.6	5 184.2	…	…
珠　海	31 411.2	2 796.6	448.4	5 326.5	799.9	3.3	…
汕　头	33 537.5	3 393.2	140.7	2 825.3	276.2	0.4	…
湛　江	28 686.9	2 633.8	…	562.7	409.6	…	…
南　宁	41 985.3	5 219.8	0.5	3 583.8	613.2	…	…
柳　州	27 184.7	2 606.0	110.6	1 393.3	191.3	…	…
桂　林	20 808.1	1 773.3	…	1 684.2	439.5	…	…
北　海	9 264.2	1 073.9	…	796.7	216.7	…	…
海　口	33 502.9	961.6	0.4	1 797.7	632.5	…	…
重　庆	231 104.3	19 712.7	962.1	26 197.6	2 868.7	0.4	…
成　都	157 111.8	18 396.6	728.7	26 523.9	1 757.9	22.3	11.4
自　贡	14 363.7	1 471.6	…	926.6	129.2	…	…
攀 枝 花	6 533.9	673.8	…	162.8	12.6	…	…
泸　州	6 787.2	817.9	5.4	686.6	73.8	…	…
德　阳	19 760.3	1 940.8	854.7	1 455.4	77.7	…	…
绵　阳	23 142.6	2 203.1	2.7	1 325.9	144.7	…	…
南　充	19 142.7	2 218.7	450.8	442.0	49.5	…	…
宜　宾	10 408.4	948.3	35.0	554.7	108.3	…	…
贵　阳	37 274.5	4 186.7	24.4	2 104.1	205.7	0.5	…
遵　义	15 684.8	2 028.8	…	342.8	42.1	…	…
昆　明	112 619.2	9 837.1	1 101.3	11 538.7	2 187.3	…	…
曲　靖	10 756.2	1 345.2	124.9	668.0	127.9	7.5	…
玉　溪	7 031.4	605.8	3.5	453.2	46.3	…	…
拉　萨	810.0	72.0	198.0	81.0	12.6		
西　安	94 047.5	9 429.6	103.9	9 331.8	1 187.0	…	…
铜　川	4 274.6	523.0	…	586.9	50.4	…	…
宝　鸡	25 553.4	2 397.3	1.0	3 376.7	294.3	…	…
咸　阳	32 838.1	4 004.4	1 400.5	5 086.5	203.8	…	…
渭　南	17 716.1	3 127.0	172.2	3 844.9	171.6	…	…
延　安	8 092.5	744.1	…	708.6	6.2	…	…
兰　州	46 090.6	4 337.8	370.8	4 772.0	676.0	83.6	0.4
金　昌	5 778.4	513.8	114.8	555.4	17.4	0.4	…
西　宁	19 585.3	1 071.3	…	…	2.2	…	…
银　川	41 503.7	3 367.3	58.9	1 628.9	799.7	…	…
石 嘴 山	7 037.6	597.6	…	…	9.1	25.2	…
乌鲁木齐	57 424.7	6 654.3	0.2	1 202.0	5.1	…	…
克拉玛依	8 905.6	699.8	177.5	975.2	96.9	…	…

重点城市生活垃圾处理情况（一）

（2013）

城　市 名　称	生活垃圾处理厂数/ 座	实际处理量/ 万吨	本年运行费用/ 万元	生活垃圾处理厂累 计完成投资/万元	新增固定资产/ 万元
总　计	816	11 031.3	572 921.9	5 812 300.7	205 282.6
北　京	19	655.6	81 752.7	355 471.0	18 718.1
天　津	10	215.9	16 386.3	232 900.4	2.4
石 家 庄	14	56.2	1 919.2	44 213.2	320.0
唐　山	16	65.5	2 370.9	45 194.8	79.0
秦 皇 岛	4	15.1	1 517.4	8 250.5	2.0
邯　郸	16	127.8	3 106.8	69 203.6	4 800.0
保　定	7	66.6	1 363.0	24 739.2	3.0
太　原	6	171.4	13 346.2	72 241.0	387.9
大　同	5	13.8	485.0	7 563.0	…
阳　泉	2	18.1	677.0	8 374.0	…
长　治	9	33.0	2 173.5	27 167.3	1.0
临　汾	13	57.8	1 351.3	19 375.0	50.0
呼 和 浩 特	5	63.8	3 366.9	60 664.7	3 030.2
包　头	3	71.8	3 250.0	28 637.0	1 637.0
赤　峰	14	65.9	925.0	28 231.2	80.0
沈　阳	3	241.8	2 333.0	28 879.0	350.0
大　连	2	104.3	1 684.0	40 745.0	…
鞍　山	1	55.0	1 112.6	7 300.0	…
抚　顺	1	50.0	305.0	305.0	…
本　溪	2	38.4	550.0	14 751.5	…
锦　州	2	20.7	1 092.0	13 900.0	…
长　春	6	117.8	4 114.6	51 606.0	…
吉　林	6	54.1	2 640.0	23 428.2	5 739.0
哈 尔 滨	3	111.7	5 215.0	24 113.3	450.9
齐 齐 哈 尔	2	26.0	470.0	10 070.0	…
牡 丹 江	6	61.5	399.0	22 165.0	730.0
上　海	25	683.0	68 257.5	486 570.6	3 845.5
南　京	6	211.5	2 408.4	29 699.8	40.0
无　锡	5	128.3	16 587.0	102 260.9	303.4
徐　州	3	28.4	1 093.0	11 551.6	109.8
常　州	6	119.4	5 245.0	64 548.0	1 110.0
苏　州	7	183.7	12 953.8	179 676.6	425.8
南　通	5	114.8	5 847.0	83 802.0	1 200.0
连 云 港	7	57.6	4 952.8	37 650.0	381.0
扬　州	6	88.0	4 141.0	78 552.0	1 021.8
镇　江	4	50.5	871.9	17 610.0	663.0
杭　州	12	243.1	16 526.8	137 215.3	818.0

重点城市生活垃圾处理情况（一）（续表）

（2013）

城市名称	生活垃圾处理厂数/座	实际处理量/万吨	本年运行费用/万元	生活垃圾处理厂累计完成投资/万元	新增固定资产/万元
宁 波	9	260.4	23 171.6	246 105.0	32 790.3
温 州	20	180.0	18 371.5	114 171.4	3 022.7
湖 州	0	…	…	…	…
绍 兴	9	101.3	4 840.8	65 852.4	3 695.8
合 肥	25	101.1	3 959.0	62 601.6	44 657.5
芜 湖	20	49.2	5 465.5	39 831.2	6 093.7
马 鞍 山	18	36.7	735.8	14 893.0	1 080.0
福 州	4	52.3	1 901.5	54 108.0	6 254.9
厦 门	1	89.3	5 176.4	140 300.0	300.0
泉 州	27	21.3	1 014.1	22 805.6	146.0
南 昌	1	83.9	3 491.0	11 000.0	689.8
九 江	12	25.3	572.0	14 192.2	162.0
济 南	5	70.2	4 560.5	31 276.5	454.3
青 岛	6	97.8	4 088.5	47 357.5	4 703.5
淄 博	1	8.9	267.3	3 600.0	100.0
枣 庄	1	12.8	400.0	5 000.0	…
烟 台	8	104.5	5 594.8	48 242.0	40.0
潍 坊	8	138.7	3 592.0	67 864.5	1 050.0
济 宁	4	78.9	4 335.0	57 479.6	326.8
泰 安	4	23.3	505.0	11 740.0	19.0
日 照	4	43.3	1 365.0	25 720.0	61.1
郑 州	5	120.2	7 243.0	33 898.0	28.0
开 封	3	12.6	414.0	7 943.0	…
洛 阳	10	78.1	3 050.0	31 101.0	120.0
平 顶 山	7	53.6	1 316.0	30 171.0	1 217.0
安 阳	3	48.7	832.2	14 166.9	129.0
焦 作	7	49.8	1 910.0	29 695.7	575.1
三 门 峡	6	40.8	713.0	16 476.0	582.0
武 汉	5	107.3	2 918.0	90 815.2	…
宜 昌	20	66.2	1 706.7	53 774.8	3 652.2
荆 州	10	31.7	1 558.9	46 784.0	120.0
长 沙	4	211.2	14 372.5	44 338.0	1.8
株 洲	6	54.0	1 977.6	22 052.6	190.0
湘 潭	7	48.3	2 218.2	32 740.7	3 070.7
岳 阳	10	74.1	2 175.2	41 465.1	940.0
常 德	6	31.3	1 281.5	27 470.0	6 500.0
张 家 界	1	16.0	161.5	3 800.0	…
广 州	7	530.6	12 839.8	152 242.1	10 779.9

重点城市生活垃圾处理情况（一）（续表）

（2013）

城 市 名 称	生活垃圾处理厂数/座	实际处理量/万吨	本年运行费用/万元	生活垃圾处理厂累计完成投资/万元	新增固定资产/万元
韶 关	9	49.2	1 691.2	15 940.6	1 199.0
深 圳	16	896.1	53 377.2	344 663.5	10 665.4
珠 海	2	65.7	6 370.1	60 441.0	30.6
汕 头	4	74.6	1 720.1	11 729.5	73.0
湛 江	4	40.2	951.5	14 373.0	8.0
南 宁	4	101.3	2 705.0	38 013.4	990.4
柳 州	1	40.2	964.3	16 651.1	0.4
桂 林	7	50.6	1 432.0	11 570.0	476.0
北 海	1	21.7	1 340.0	24 780.0	…
海 口	0	…	…	…	…
重 庆	45	539.4	25 258.7	386 642.2	2 756.7
成 都	11	153.1	4 240.1	182 480.3	226.1
自 贡	2	27.8	449.0	8 248.7	50.0
攀 枝 花	4	21.8	984.0	9 218.1	384.0
泸 州	4	36.3	1 046.0	11 758.0	124.1
德 阳	5	36.0	2 357.1	19 283.4	843.0
绵 阳	8	55.1	666.4	26 168.0	694.4
南 充	6	62.8	3 079.0	21 759.0	600.0
宜 宾	9	52.6	2 934.6	33 009.4	820.0
贵 阳	4	103.1	1 673.1	30 541.5	1 521.7
遵 义	11	78.7	1 981.0	34 854.0	1 400.0
昆 明	6	21.1	861.8	16 210.0	342.0
曲 靖	2	8.4	335.0	7 120.0	…
玉 溪	11	36.1	2 062.0	14 267.2	508.9
拉 萨	1	18.4	151.0	5 187.8	…
西 安	3	248.0	2 305.0	25 030.4	…
铜 川	4	18.2	219.0	5 643.0	150.0
宝 鸡	8	36.4	723.0	43 619.0	51.7
咸 阳	8	47.0	813.4	18 890.0	2.0
渭 南	5	26.9	623.0	11 337.0	550.0
延 安	11	36.9	907.0	23 282.9	15.0
兰 州	10	47.1	740.8	3 037.6	…
金 昌	2	13.0	265.0	2 206.4	…
西 宁	9	85.8	1 007.5	16 914.1	80.0
银 川	3	48.3	572.6	14 480.0	425.0
石 嘴 山	3	14.3	203.4	5 627.0	12.3
乌 鲁 木 齐	3	160.3	2 973.7	24 726.0	431.4
克 拉 玛 依	3	17.7	52.0	10 897.4	…

重点城市生活垃圾处理情况（二）

（2013）

单位：吨

城 市 名 称	渗滤液中污染物排放量					
	化学需氧量	氨氮	油类	总磷	挥发酚	氰化物
总　　计	80 247.1	8 117.3	58.5	126.3	6.8	0.3
北　　京	7 835.1	680.9	2.1	7.3	0.3	…
天　　津	473.2	38.5	…	0.1	…	…
石 家 庄	239.4	23.5	0.3	0.3	0.1	…
唐　　山	1 045.4	122.1	1.1	2.2	0.2	…
秦 皇 岛	2.2	1.1	…	…	…	…
邯　　郸	3 467.0	136.2	2.8	5.2	0.6	…
保　　定	1 049.9	103.6	1.0	1.5	0.2	…
太　　原	169.1	18.0	0.1	0.2	…	…
大　　同	142.9	12.8	0.1	0.1	…	…
阳　　泉	240.0	24.0	0.4	0.3	…	…
长　　治	334.6	17.3	0.4	0.5	0.1	…
临　　汾	285.3	24.3	2.0	2.2	0.1	…
呼和浩特	355.6	66.7	0.3	0.6	0.1	…
包　　头	241.8	16.1	0.1	0.2	…	…
赤　　峰	350.9	25.0	0.6	0.5	…	…
沈　　阳	37.2	2.0	0.7	2.9	…	…
大　　连	2.7	0.3	…	…	…	…
鞍　　山	1 379.1	473.4	0.3	0.3	0.1	…
抚　　顺	1 000.0	100.0	1.5	1.6	…	…
本　　溪	157.8	14.5	…	…	…	…
锦　　州	0.6	0.1	…	…	…	…
长　　春	5 067.4	572.8	3.5	25.1	0.2	…
吉　　林	944.1	94.4	1.2	1.7	0.2	…
哈 尔 滨	20.7	3.5	0.2	0.2	0.1	…
齐齐哈尔	274.0	31.0	…	0.4	…	…
牡 丹 江	795.5	89.8	0.3	0.4	0.1	…
上　　海	5 804.0	293.5	2.8	5.5	1.0	…
南　　京	598.7	88.5	0.5	0.9	0.3	…
无　　锡	14.3	2.6	0.1	0.1	…	…
徐　　州	591.6	7.5	…	…	…	…
常　　州	5.2	0.2	…	…	…	…
苏　　州	158.3	44.0	2.8	0.1	…	…
南　　通	251.6	28.5	0.2	0.9	…	…
连 云 港	1 008.0	86.0	0.4	1.5	…	…
扬　　州	883.2	128.8	…	0.3	…	…
镇　　江	155.6	41.1	…	…	…	…
杭　　州	163.8	17.5	0.7	2.1	…	…

重点城市生活垃圾处理情况（二）（续表）

（2013）

单位：吨

城 市 名 称	渗滤液中污染物排放量					
	化学需氧量	氨氮	油类	总磷	挥发酚	氰化物
宁 波	1 065.2	87.7	…	…	…	…
温 州	2 173.3	170.0	1.6	3.1	0.4	…
湖 州	…	…	…	…	…	…
绍 兴	820.3	68.7	0.6	1.9	…	…
合 肥	968.5	99.7	0.5	0.7	…	…
芜 湖	205.6	17.3	0.1	0.3	…	…
马 鞍 山	161.8	13.5	0.1	0.3	…	…
福 州	754.9	58.5	0.8	1.4	0.1	…
厦 门	89.0	13.4	0.1	…	…	…
泉 州	203.5	17.1	2.1	0.7	0.1	…
南 昌	22.3	3.1	0.1	0.1	…	…
九 江	1 006.1	89.8	0.6	1.3	0.1	…
济 南	259.1	27.4	0.4	0.6	0.1	…
青 岛	283.4	30.3	0.1	0.1	…	…
淄 博	0.2	…	…	…	…	…
枣 庄	0.6	…	…	…	…	…
烟 台	43.6	4.9	0.4	0.4	0.1	…
潍 坊	738.3	81.6	0.5	0.5	…	…
济 宁	217.2	38.4	0.4	0.5	…	…
泰 安	77.3	4.4	0.1	0.2	…	…
日 照	250.9	24.7	0.4	0.5	0.1	…
郑 州	981.2	156.3	1.2	3.0	0.2	…
开 封	303.7	40.6	0.3	0.4	0.1	…
洛 阳	709.6	73.7	0.4	0.3	0.1	…
平 顶 山	169.2	19.0	0.1	…	…	…
安 阳	7.8	0.9	…	…	…	…
焦 作	197.7	23.5	0.1	0.2	…	…
三 门 峡	51.7	5.9	…	…	…	…
武 汉	3 624.1	250.1	…	…	…	…
宜 昌	536.2	113.0	0.5	0.8	…	…
荆 州	388.3	83.2	0.4	0.6	0.1	…
长 沙	491.0	83.1	0.2	0.5	…	…
株 洲	524.8	43.5	0.4	0.9	0.1	…
湘 潭	663.2	60.8	0.9	1.9	0.1	…
岳 阳	2 654.9	205.5	1.2	2.9	0.2	…
常 德	1 045.0	78.9	0.4	1.0	0.1	…
张 家 界	3.8	0.9	…	…	…	…
广 州	368.2	25.7	…	…	…	…

重点城市生活垃圾处理情况（二）（续表）

（2013）

单位：吨

城 市 名 称	渗滤液中污染物排放量					
	化学需氧量	氨氮	油类	总磷	挥发酚	氰化物
韶 关	1 174.8	94.6	0.2	1.9	…	…
深 圳	1 337.1	217.7	0.1	0.7	…	…
珠 海	275.3	18.4	…	…	…	…
汕 头	943.5	93.4	…	…	…	…
湛 江	866.5	99.6	0.1	0.2	…	…
南 宁	541.3	46.6	0.8	4.5	…	…
柳 州	27.2	2.7	0.1	0.1	…	…
桂 林	702.4	58.9	1.0	0.7	…	…
北 海	9.4	0.9	…	…	…	…
海 口	…	…	…	…	…	…
重 庆	627.0	159.3	0.9	0.9	…	…
成 都	1 247.1	129.9	1.8	0.9	0.1	…
自 贡	357.5	44.7	0.4	0.3	…	…
攀 枝 花	142.6	16.3	…	…	…	…
泸 州	77.1	9.3	…	…	…	…
德 阳	45.7	1.1	0.1	…	…	…
绵 阳	83.6	17.2	0.4	0.3	…	…
南 充	766.3	123.3	0.5	1.3	…	…
宜 宾	161.9	18.3	0.1	0.1	…	…
贵 阳	2 591.7	298.0	…	0.6	…	…
遵 义	956.5	121.0	0.3	0.3	…	…
昆 明	1 935.8	213.0	4.2	10.9	0.6	…
曲 靖	119.1	19.2	…	…	…	…
玉 溪	977.6	105.2	0.1	0.4	…	…
拉 萨	154.2	10.3	0.1	0.2	…	…
西 安	2 037.0	301.0	…	…	…	…
铜 川	2.5	0.8	…	…	…	…
宝 鸡	247.9	39.7	0.1	0.1	…	…
咸 阳	40.8	3.0	1.5	…	…	…
渭 南	…	…	…	…	…	…
延 安	…	…	…	0.1	…	…
兰 州	430.3	25.0	0.2	0.2	…	…
金 昌	46.7	4.7	0.1	…	…	…
西 宁	941.8	79.6	0.4	1.4	…	…
银 川	201.0	19.4	0.2	0.4	0.1	…
石 嘴 山	13.1	1.1	…	…	…	…
乌鲁木齐	2 483.3	174.8	2.8	9.9	0.2	…
克拉玛依	74.7	4.7	1.0	0.1	…	…

重点城市生活垃圾处理情况（三）

（2013）

单位：千克

城　市 名　称	渗滤液中污染物排放量					
	汞	镉	六价铬	总铬	铅	砷
总　计	55.5	279.7	36.4	467.9	853.1	257.8
北　京	0.6	16.3	1.8	15.9	147.4	3.3
天　津	…	…	…	0.1	0.4	0.2
石　家　庄	0.2	0.9	0.1	1.9	3.9	1.0
唐　山	0.4	3.0	0.5	5.9	12.2	4.7
秦　皇　岛	…	…	…	…	…	…
邯　郸	1.8	14.2	…	27.5	69.3	17.6
保　定	0.3	3.1	…	5.2	12.7	4.2
太　原	0.1	0.3	…	0.5	1.5	0.5
大　同	…	0.2	…	0.5	1.2	0.3
阳　泉	0.2	1.2	…	2.4	4.5	0.9
长　治	…	…	…	…	…	…
临　汾	0.1	0.6	0.1	1.7	2.3	1.1
呼和浩特	0.2	…	…	…	8.7	…
包　头	…	…	…	…	…	…
赤　峰	…	…	…	…	…	…
沈　阳	0.9	0.1	0.3	0.3	0.8	0.9
大　连	…	0.5	0.5	0.5	0.9	0.5
鞍　山	2.9	2.9	…	1.5	7.7	0.4
抚　顺	1.0	5.0	…	10.0	19.0	4.0
本　溪	…	…	…	0.2	0.1	0.2
锦　州	…	…	…	…	…	…
长　春	…	…	…	4.1	13.9	1.6
吉　林	1.0	6.7	…	14.1	27.3	4.6
哈　尔　滨	0.3	1.1	…	2.1	5.7	2.1
齐齐哈尔	0.1	0.5	…	1.0	2.9	1.0
牡　丹　江	1.2	1.6	1.5	4.2	5.7	1.5
上　海	2.4	11.5	0.1	44.6	67.9	25.4
南　京	0.6	2.6	2.1	5.2	15.6	5.1
无　锡	…	0.1	…	…	0.1	…
徐　州	…	…	…	…	…	…
常　州	0.1	0.4	…	0.8	2.5	0.8
苏　州	…	0.2	…	4.5	12.1	0.5
南　通	0.1	0.4	…	1.2	2.4	0.7
连　云　港	10.8	3.6	0.9	11.5	2.6	1.4
扬　州	…	…	…	…	…	…
镇　江	1.0	5.0	1.2	10.0	30.0	10.0
杭　州	0.3	9.3	0.6	29.7	66.5	3.2

重点城市生活垃圾处理情况（三）（续表）

（2013）　　　　　　　　　　　　　　　　　　　　　　　　　　　　　单位：千克

城　市 名　称	渗滤液中污染物排放量					
	汞	镉	六价铬	总铬	铅	砷
宁　波	…	…	…	…	…	…
温　州	1.2	6.1	…	16.6	24.4	7.6
湖　州	…	…	…	…	…	…
绍　兴	0.4	2.2	…	6.5	10.9	2.9
合　肥	0.3	2.2	…	6.2	6.9	1.9
芜　湖	0.2	0.5	0.7	1.8	1.5	0.7
马 鞍 山	0.4	0.5	0.1	2.8	2.9	2.9
福　州	…	…	…	…	…	…
厦　门	…	…	…	…	…	…
泉　州	0.2	78.5	…	6.0	10.8	2.8
南　昌	…	0.1	…	1.8	2.1	1.8
九　江	0.3	2.0	1.3	8.3	7.6	5.2
济　南	0.3	1.5	…	3.1	6.6	1.5
青　岛	…	0.1	…	…	0.4	0.1
淄　博	…	…	…	…	0.1	…
枣　庄	…	…	…	…	…	…
烟　台	0.2	1.4	1.3	2.5	3.4	3.3
潍　坊	0.2	1.4	1.7	3.4	8.3	1.0
济　宁	0.5	3.2	0.1	6.5	11.1	2.1
泰　安	…	…	…	…	…	…
日　照	0.2	1.5	1.5	2.5	5.5	2.0
郑　州	0.2	6.2	…	7.9	20.7	6.0
开　封	0.1	0.2	0.3	0.6	1.7	0.5
洛　阳	0.2	1.4	…	1.6	5.1	1.5
平 顶 山	0.1	0.1	…	…	0.1	…
安　阳	…	…	…	…	…	…
焦　作	1.0	0.9	0.2	0.7	2.7	0.5
三 门 峡	…	0.2	…	0.3	0.6	0.2
武　汉	…	…	…	…	…	…
宜　昌	1.1	1.0	…	1.6	5.0	1.5
荆　州	0.4	0.4	…	1.3	2.1	0.8
长　沙	0.1	0.8	0.4	1.7	1.9	1.5
株　洲	…	2.6	1.0	5.0	3.5	4.5
湘　潭	0.5	4.7	0.3	1.4	11.4	2.9
岳　阳	0.9	8.1	2.4	10.4	15.6	3.1
常　德	…	0.9	2.4	5.0	0.4	7.8
张 家 界	…	0.2	…	0.4	1.0	0.4
广　州	…	0.2	0.3	1.1	1.4	…

重点城市生活垃圾处理情况（三）（续表）
（2013）

单位：千克

城 市 名 称	渗滤液中污染物排放量					
	汞	镉	六价铬	总铬	铅	砷
韶 关	0.2	0.8	0.6	2.6	3.3	2.0
深 圳	12.2	2.6	0.8	23.9	10.3	9.4
珠 海	…	0.1	…	2.0	0.6	0.4
汕 头	0.3	0.1	…	0.2	0.6	0.7
湛 江	0.1	0.1	…	2.7	1.3	0.7
南 宁	0.2	2.5	7.6	9.3	8.9	3.4
柳 州	…	0.1	…	1.4	0.7	0.4
桂 林	0.5	2.5	…	1.9	5.3	0.8
北 海	…	0.2	0.1	0.1	0.2	…
海 口	…	…	…	…	…	…
重 庆	0.6	2.3	0.7	4.8	14.3	4.1
成 都	0.2	1.2	…	9.2	7.7	1.7
自 贡	0.1	0.9	1.6	4.0	8.3	4.6
攀 枝 花	…	…	…	…	…	…
泸 州	…	…	…	0.5	0.3	0.1
德 阳	…	…	…	2.1	0.4	0.7
绵 阳	0.4	1.1	…	3.1	5.5	1.3
南 充	0.1	0.7	0.2	4.2	8.9	2.7
宜 宾	…	0.2	0.2	0.8	3.1	0.5
贵 阳	…	0.1	0.2	1.2	0.5	0.8
遵 义	0.2	0.8	…	2.1	4.2	1.1
昆 明	0.8	3.3	0.1	10.3	19.0	5.3
曲 靖	…	0.1	…	0.1	0.2	…
玉 溪	0.2	0.4	0.6	1.3	1.6	0.4
拉 萨	0.1	0.5	…	1.0	2.6	0.6
西 安	…	…	…	…	…	…
铜 川	…	…	…	…	…	…
宝 鸡	…	…	…	0.2	0.1	0.1
咸 阳	…	…	…	…	…	…
渭 南	…	…	…	…	…	…
延 安	…	…	…	…	…	…
兰 州	0.1	0.6	…	1.0	2.4	0.9
金 昌	…	0.2	…	0.5	1.2	0.3
西 宁	0.3	0.2	…	0.5	1.0	0.3
银 川	0.2	1.6	…	2.8	5.7	1.7
石 嘴 山	…	…	…	…	…	…
乌鲁木齐	3.0	36.3	…	60.0	12.2	47.9
克拉玛依	…	…	0.1	0.4	…	0.2

重点城市危险（医疗）废物集中处置情况（一）

（2013）

城市名称	危险废物集中处置厂数/个	医疗废物集中处置厂数/个	本年运行费用/万元	危险（医疗）废物集中处置厂累计完成投资/万元	新增固定资产/万元
总 计	517	90	432 020.1	2 075 892.7	123 561.4
北 京	2	2	27 413.9	99 577.7	2 837.3
天 津	11	0	14 574.6	35 943.3	364.5
石 家 庄	4	1	1 099.5	7 891.9	50.0
唐 山	2	3	864.5	2 161.0	1 208.0
秦 皇 岛	6	0	1 375.0	6 330.0	49.0
邯 郸	0	1	235.2	1 805.0	0.2
保 定	1	2	1 210.0	6 492.7	1 799.7
太 原	4	1	733.8	3 164.7	417.7
大 同	1	1	264.0	1 622.0	92.0
阳 泉	0	0	…	…	…
长 治	0	0	…	…	…
临 汾	0	1	357.0	2 006.0	…
呼 和 浩 特	0	1	925.0	1 114.3	91.8
包 头	0	4	524.5	21 908.0	…
赤 峰	1	1	331.0	1 228.1	…
沈 阳	3	1	5 921.1	30 316.0	427.0
大 连	12	1	24 344.7	42 006.3	1 398.0
鞍 山	0	1	231.0	1 395.0	…
抚 顺	0	0	…	…	…
本 溪	2	0	295.0	1 865.0	…
锦 州	1	1	450.0	1 810.0	…
长 春	1	1	3 267.0	5 290.0	…
吉 林	1	1	302.0	3 571.0	…
哈 尔 滨	5	1	2 703.7	38 878.0	478.0
齐 齐 哈 尔	0	1	184.0	2 069.5	…
牡 丹 江	0	1	71.0	300.0	…
上 海	35	2	71 461.5	196 890.2	9 663.4
南 京	15	1	10 491.3	54 927.0	4 474.4
无 锡	37	0	18 291.1	80 846.1	4 692.2
徐 州	2	0	700.0	20 700.0	300.0
常 州	37	0	12 109.0	39 386.0	255.0
苏 州	29	0	24 857.8	125 059.3	3 410.1
南 通	16	0	4 774.3	27 913.0	…
连 云 港	2	0	1 620.0	10 300.0	300.0
扬 州	8	1	970.5	7 054.7	111.6
镇 江	3	0	2 635.0	11 370.0	…
杭 州	12	1	21 415.7	52 737.9	3 165.6

重点城市危险（医疗）废物集中处置情况（一）（续表）

（2013）

城 市 名 称	危险废物集中 处置厂数/个	医疗废物集中 处置厂数/个	本年运行费用/ 万元	危险（医疗）废物集中处置厂累计完成 投资/万元	新增固定资产/ 万元
宁　波	16	1	8 930. 4	51 003. 7	5 642. 7
温　州	8	1	3 294. 0	14 396. 0	6 196. 0
湖　州	2	0	2 100. 0	7 500. 0	…
绍　兴	8	0	6 245. 0	21 868. 0	1 738. 0
合　肥	1	1	3 975. 0	12 109. 0	164. 0
芜　湖	1	1	737. 5	2 704. 0	…
马 鞍 山	0	0	…	…	…
福　州	6	0	2 159. 4	17 710. 0	1 190. 0
厦　门	3	0	2 540. 2	6 782. 1	300. 0
泉　州	1	1	1 305. 6	9 520. 0	100. 0
南　昌	4	0	1 840. 0	6 272. 1	92. 1
九　江	6	0	400. 0	126 445. 0	…
济　南	1	1	4 301. 0	27 312. 0	3 528. 0
青　岛	10	0	6 773. 9	27 233. 7	368. 7
淄　博	18	1	2 634. 3	14 873. 4	254. 4
枣　庄	1	0	1 250. 0	3 400. 0	600. 0
烟　台	3	1	3 778. 0	25 248. 0	4 080. 0
潍　坊	9	1	3 091. 4	10 754. 0	330. 0
济　宁	0	1	1 000. 0	1 800. 0	…
泰　安	1	1	467. 0	3 792. 0	863. 0
日　照	0	1	170. 0	830. 0	…
郑　州	0	1	2 280. 0	3 500. 0	260. 0
开　封	0	1	733. 0	1 486. 0	…
洛　阳	0	1	1 200. 0	1 850. 0	…
平 顶 山	0	1	410. 0	1 780. 0	…
安　阳	0	1	181. 0	1 700. 0	…
焦　作	0	1	360. 0	980. 0	…
三 门 峡	0	1	100. 0	1 000. 0	…
武　汉	9	1	7 592. 2	29 201. 3	1 563. 6
宜　昌	3	0	1 165. 0	10 546. 0	27. 0
荆　州	1	1	350. 0	2 306. 2	10. 0
长　沙	0	0	…	…	…
株　洲	1	0	530. 0	1 931. 0	35. 0
湘　潭	0	0	…	…	…
岳　阳	16	1	400. 0	16 545. 0	…
常　德	1	0	534. 0	2 008. 0	58. 0
张 家 界	0	1	278. 0	1 191. 0	…
广　州	24	1	8 850. 5	81 283. 5	759. 0

重点城市危险（医疗）废物集中处置情况（一）（续表）

（2013）

城　市名　称	危险废物集中处置厂数/个	医疗废物集中处置厂数/个	本年运行费用/万元	危险（医疗）废物集中处置厂累计完成投资/万元	新增固定资产/万元
韶　关	9	1	592.0	8 279.8	22.7
深　圳	7	1	20 056.4	132 316.0	1 514.2
珠　海	7	0	5 767.0	25 311.8	3 307.0
汕　头	1	1	1 311.7	1 817.2	36.6
湛　江	2	1	1 491.3	7 400.0	…
南　宁	4	0	2 373.8	20 758.4	274.3
柳　州	6	1	1 674.0	4 713.0	1 161.0
桂　林	0	1	290.0	2 002.0	…
北　海	0	1	250.0	1 349.0	…
海　口	0	1	1 131.8	643.0	143.0
重　庆	9	4	11 991.9	59 644.1	1 349.3
成　都	7	1	6 306.8	16 009.3	922.4
自　贡	0	1	…	3 748.0	…
攀 枝 花	1	1	311.0	4 066.0	30.0
泸　州	0	2	408.5	2 015.0	1 069.3
德　阳	1	2	132.8	10 923.0	1 600.0
绵　阳	7	1	4 644.0	15 580.7	54.7
南　充	0	0	…	…	…
宜　宾	0	1	792.9	1 505.2	36.6
贵　阳	0	1	1 999.0	22 400.0	750.0
遵　义	4	1	2 231.0	5 866.5	415.0
昆　明	0	1	900.0	12 220.0	120.0
曲　靖	0	0	…	…	…
玉　溪	0	0	…	…	…
拉　萨	0	0	…	…	…
西　安	3	1	6 691.3	13 950.5	5.0
铜　川	0	0	…	…	…
宝　鸡	0	2	400.0	5 731.4	…
咸　阳	4	1	3 080.7	38 174.8	2 850.5
渭　南	3	1	6 892.0	48 677.0	40 027.2
延　安	16	1	5 437.0	15 907.6	351.0
兰　州	3	0	3 150.0	26 696.2	1 683.4
金　昌	0	1	90.0	698.0	…
西　宁	6	1	963.3	36 340.6	…
银　川	2	0	643.3	12 695.2	58.3
石 嘴 山	0	0	…	6 471.0	2 000.0
乌鲁木齐	4	1	503.7	8 945.9	20.0
克拉玛依	4	1	553.0	8 246.0	15.0

重点城市危险（医疗）废物集中处置情况（二）

（2013）

城　市名　称	危险废物设计处置能力/（吨/日）	危险废物实际处置量/吨	工业危险废物处置量	医疗废物处置量	其他危险废物处置量	危险废物综合利用量/吨
总　计	55 230	2 061 854	1 532 181	417 106	112 568	3 471 531
北　京	855	94 685	79 631	15 054	0	58 880
天　津	1 120	34 582	14 591	19 991	0	165 176
石 家 庄	68	8 122	8 062	60	0	704
唐　山	36	3 107	100	3 007	0	255
秦 皇 岛	399	15 676	13 801	1 743	132	7 091
邯　郸	10	1 400	0	1 400	0	0
保　定	173	8 112	5 013	3 099	0	0
太　原	95	8 041	86	7 955	0	256
大　同	12	1 379	0	1 379	0	632
阳　泉	0	0	0	0	0	0
长　治	0	0	0	0	0	0
临　汾	8	785	0	785	0	0
呼和浩特	18	2 056	0	2 056	0	0
包　头	167	1 124	0	1 124	0	0
赤　峰	5	740	0	740	0	0
沈　阳	155	22 855	17 300	4 918	637	0
大　连	1 367	18 630	14 035	4 595	0	89 264
鞍　山	8	992	0	992	0	0
抚　顺	0	0	0	0	0	0
本　溪	15	1 665	1 665	0	0	0
锦　州	212	900	0	900	0	29 829
长　春	58	16 439	11 374	5 065	0	627
吉　林	155	13 800	13 800	0	0	0
哈 尔 滨	197	7 675	35	7 640	0	694
齐齐哈尔	8	716	0	716	0	0
牡 丹 江	5	779	0	779	0	0
上　海	2 187	199 778	135 132	30 182	34 464	173 160
南　京	404	28 897	21 926	6 971	0	48 111
无　锡	5 904	31 720	27 082	4 638	0	581 680
徐　州	10	2 083	1 842	241	0	0
常　州	2 495	12 099	10 542	1 558	0	186 816
苏　州	3 067	140 462	109 144	5 361	25 957	255 597
南　通	306	17 947	14 859	3 088	0	24 422
连 云 港	49	7 603	6 096	1 507	0	0
扬　州	241	7 200	5 500	1 700	0	5 012
镇　江	112	32 427	30 604	1 823	0	22 665
杭　州	2 048	190 878	178 212	12 667	0	223 462

重点城市危险（医疗）废物集中处置情况（二）（续表）

（2013）

城 市 名 称	危险废物设计 处置能力/ （吨/日）	危险废物实际 处置量/吨	工业危险废物 处置量	医疗废物 处置量	其他危险废物 处置量	危险废物综合 利用量/吨
宁 波	1 496	115 100	103 670	9 866	1 563	100 719
温 州	306	13 839	11 776	2 063	0	46 888
湖 州	80	9 229	3 919	5 176	134	0
绍 兴	201	15 740	11 340	4 400	0	18 691
合 肥	84	17 893	13 039	4 853	0	0
芜 湖	32	1 915	0	1 915	0	0
马 鞍 山	0	0	0	0	0	0
福 州	329	20 433	14 780	5 653	0	1 114
厦 门	154	13 248	9 970	3 278	0	3 629
泉 州	208	7 060	4 260	2 800	0	0
南 昌	457	4 350	634	3 716	0	97 119
九 江	377	1 564	1 564	0	0	0
济 南	124	14 942	2 689	5 561	6 693	2 160
青 岛	798	18 398	13 907	4 490	0	14 791
淄 博	779	4 480	323	4 156	0	66 672
枣 庄	15	3 271	810	2 461	0	0
烟 台	217	28 619	23 683	4 936	0	4 173
潍 坊	549	19 970	16 637	3 334	0	7 007
济 宁	12	3 276	0	3 276	0	0
泰 安	8	2 360	0	2 360	0	450
日 照	5	799	0	799	0	0
郑 州	30	9 982	0	9 982	0	0
开 封	0	1 468	0	1 468	0	0
洛 阳	24	5 544	0	5 544	0	0
平 顶 山	4	1 366	0	1 366	0	0
安 阳	5	1 220	0	1 220	0	0
焦 作	5	1 650	0	1 650	0	0
三 门 峡	5	1 000	0	1 000	0	0
武 汉	223	42 344	28 687	13 657	0	21 139
宜 昌	3 043	4 355	2 935	1 420	0	0
荆 州	18	765	0	765	0	0
长 沙	0	0	0	0	0	0
株 洲	8	2 400	600	1 800	0	0
湘 潭	0	0	0	0	0	0
岳 阳	272	1 600	0	1 600	0	31 710
常 德	16	2 462	0	2 462	0	0
张 家 界	3	925	0	925	0	0
广 州	3 728	130 763	109 713	17 515	3 535	122 613

重点城市危险（医疗）废物集中处置情况（二）（续表）

（2013）

城 市 名 称	危险废物设计 处置能力/ （吨/日）	危险废物实际 处置量/吨	工业危险废物 处置量	医疗废物 处置量	其他危险废物 处置量	危险废物综合 利用量/吨
韶　关	307	3 189	0	3 189	0	57 310
深　圳	1 525	141 467	131 257	9 944	266	164 953
珠　海	467	5 823	5 823	0	0	35 990
汕　头	30	2 855	0	2 855	0	3 048
湛　江	51	4 438	0	4 438	0	1 794
南　宁	110	10 300	3 302	6 998	0	3 510
柳　州	87	10 877	8 096	2 781	0	1 550
桂　林	10	2 488	0	2 488	0	0
北　海	5	1 200	0	1 200	0	0
海　口	5	2 810	0	2 810	0	0
重　庆	2 343	48 797	35 741	10 647	2 409	331 825
成　都	257	19 163	3 653	15 511	0	14 826
自　贡	3	548	0	548	0	0
攀 枝 花	6	1 089	0	1 089	0	79
泸　州	17	4 007	0	4 007	0	0
德　阳	65	1 127	0	1 127	0	526
绵　阳	768	1 349	125	1 224	0	81 260
南　充	0	0	0	0	0	0
宜　宾	5	1 559	0	1 559	0	0
贵　阳	125	5 454	176	4 971	307	0
遵　义	143	6 664	3 997	2 657	10	1 435
昆　明	30	10 210	0	10 210	0	0
曲　靖	0	0	0	0	0	0
玉　溪	0	0	0	0	0	0
拉　萨	0	0	0	0	0	0
西　安	3 364	91 721	71 471	18 250	2 000	3 046
铜　川	0	0	0	0	0	0
宝　鸡	10	2 772	0	2 772	0	0
咸　阳	720	22 324	20 800	1 524	0	10 313
渭　南	934	2 025	0	2 025	0	260 000
延　安	6 100	41 081	40 523	558	0	33 353
兰　州	396	109 312	106 980	2 309	23	0
金　昌	3	438	0	438	0	0
西　宁	1 258	36 927	0	2 547	34 380	13 731
银　川	51	2 273	806	1 410	58	1 000
石 嘴 山	83	0	0	0	0	1 664
乌鲁木齐	152	6 259	2 600	3 659	0	4 836
克拉玛依	220	1 624	1 461	163	0	32 272

重点城市危险（医疗）废物集中处置情况（三）

（2013）

单位：吨

城市名称	焚烧废气中污染物排放量		
	二氧化硫	氮氧化物	烟尘
总　计	424.4	1 162.4	603.9
北　京	0.2	8.4	0.8
天　津	61.8	35.8	13.4
石 家 庄	8.9	5.4	7.8
唐　山	0.8	4.3	26.4
秦 皇 岛	2.4	10.3	0.9
邯　郸	0.5	1.3	0.3
保　定	0.8	2.4	2.5
太　原	0.3	9.9	0.9
大　同	0.1	1.7	0.1
阳　泉	…	…	…
长　治	…	…	…
临　汾	0.7	1.7	0.7
呼和浩特	…	2.0	0.2
包　头	0.3	1.1	11.1
赤　峰	0.1	0.9	0.1
沈　阳	1.9	7.0	2.9
大　连	3.8	13.2	36.7
鞍　山	…	…	…
抚　顺	…	…	…
本　溪	1.0	1.7	0.5
锦　州	3.0	1.3	10.0
长　春	0.5	3.8	1.9
吉　林	…	…	…
哈 尔 滨	1.6	6.7	0.9
齐齐哈尔	0.2	0.8	0.7
牡 丹 江	0.2	0.7	1.7
上　海	33.7	416.1	81.7
南　京	19.7	20.3	8.7
无　锡	20.6	31.6	30.0
徐　州	0.4	1.5	0.7
常　州	4.3	7.2	19.2
苏　州	21.4	131.9	97.9
南　通	13.2	33.1	9.6
连 云 港	3.4	7.7	2.2
扬　州	0.4	…	0.3
镇　江	…	…	…
杭　州	15.5	7.9	2.9

重点城市危险（医疗）废物集中处置情况（三）（续表）

（2013）

单位：吨

城 市 名 称	焚烧废气中污染物排放量		
	二氧化硫	氮氧化物	烟尘
宁 波	12.6	24.0	6.7
温 州	24.3	13.3	9.4
湖 州	19.4	5.1	5.1
绍 兴	13.9	7.9	8.8
合 肥	5.3	15.6	2.2
芜 湖	0.9	1.3	22.7
马 鞍 山	...	...	...
福 州	1.4	12.0	3.2
厦 门	3.0	11.0	0.9
泉 州	3.5	10.2	1.9
南 昌	1.5	4.4	0.5
九 江	1.4	8.9	11.1
济 南	33.3	35.8	5.5
青 岛	1.1	1.1	1.1
淄 博	0.9	3.0	0.3
枣 庄	1.0	3.9	0.6
烟 台	1.6	11.1	1.4
潍 坊	2.3	2.5	4.2
济 宁	1.0	2.9	1.9
泰 安	...	...	...
日 照	1.0	1.0	0.1
郑 州	2.4	5.2	0.4
开 封	...	...	...
洛 阳	5.3	4.1	2.7
平 顶 山	...	...	...
安 阳	...	...	...
焦 作	...	...	...
三 门 峡	...	...	...
武 汉	7.5	11.2	4.9
宜 昌	...	...	...
荆 州	3.0	0.3	0.4
长 沙	...	...	...
株 洲	...	1.2	0.1
湘 潭	...	...	...
岳 阳	0.5	1.5	0.8
常 德	...	...	...
张 家 界	0.3	0.6	0.1
广 州	6.6	60.9	6.2

重点城市危险（医疗）废物集中处置情况（三）（续表）

（2013）

单位：吨

城 市 名 称	焚烧废气中污染物排放量		
	二氧化硫	氮氧化物	烟尘
韶 关	1.2	1.1	1.5
深 圳	3.8	9.4	1.6
珠 海	2.4	3.2	1.1
汕 头	2.0	2.4	0.2
湛 江	4.0	7.2	3.4
南 宁	5.1	6.4	0.9
柳 州	2.1	10.7	5.1
桂 林	0.7	2.0	24.1
北 海	…	…	…
海 口	…	…	…
重 庆	9.5	19.4	9.1
成 都	3.2	12.7	0.7
自 贡	…	…	…
攀 枝 花	0.1	1.3	0.1
泸 州	0.9	1.6	1.0
德 阳	…	0.1	0.6
绵 阳	0.3	0.4	0.4
南 充	…	…	…
宜 宾	…	…	…
贵 阳	…	6.3	1.5
遵 义	2.3	2.6	0.3
昆 明	0.6	9.1	0.9
曲 靖	…	…	…
玉 溪	…	…	…
拉 萨	…	…	…
西 安	5.7	16.4	2.0
铜 川	…	…	…
宝 鸡	…	…	…
咸 阳	0.5	1.2	11.0
渭 南	1.0	1.8	28.5
延 安	…	…	…
兰 州	0.3	2.0	1.8
金 昌	…	…	…
西 宁	0.8	2.4	0.3
银 川	0.6	2.4	1.9
石 嘴 山	…	…	…
乌鲁木齐	0.7	2.6	26.9
克拉玛依	0.1	0.1	2.1

10

各工业行业环境统计

ANNUAL STATISTIC REPORT ON ENVIRONMENT IN CHINA
2013

按行业分重点调查工业废水排放及处理情况（一）

（2013）

行业名称	汇总工业企业数/个	工业废水排放量/万吨	直接排入环境的	排入污水处理厂的	工业废水处理量/万吨	废水治理设施数/套	废水治理设施处理能力/（万吨/日）	废水治理设施运行费用/万元
行业总计	147 657	1 914 249.7	1 439 260.0	474 989.7	4 924 811.2	80 298	25 641.6	6 286 649.2
煤炭开采和洗选业	6 462	142 867.5	135 009.2	7 858.3	177 379.1	4 536	1 300.6	167 694.0
石油和天然气开采业	249	11 029.5	9 841.1	1 188.4	105 076.6	629	528.1	307 312.0
黑色金属矿采选业	4 457	24 893.8	24 866.8	27.0	255 933.8	1 980	1 557.2	149 411.2
有色金属矿采选业	3 301	52 906.8	52 792.3	114.5	186 496.0	2 276	718.0	130 444.1
非金属矿采选业	983	6 089.7	5 866.1	223.5	11 152.6	471	77.2	19 152.6
开采辅助活动	58	650.2	588.9	61.3	785.1	26	4.6	2 568.0
其他采矿业	39	257.5	257.5	…	313.2	16	2.1	468.5
农副食品加工业	11 442	139 060.7	116 822.6	22 238.1	116 727.0	5 827	836.1	168 489.1
食品制造业	3 826	60 374.7	39 928.0	20 446.7	52 208.3	2 539	336.9	109 360.4
酒、饮料和精制茶制造业	4 049	72 674.9	53 953.2	18 721.6	66 049.5	2 143	414.6	115 797.9
烟草制品业	156	2 361.6	1 416.3	945.3	3 112.5	148	20.8	8 562.2
纺织业	8 719	214 573.6	97 211.4	117 362.2	202 896.4	6 091	1 161.7	451 798.1
纺织服装、服饰业	1 599	17 023.3	10 400.4	6 622.9	17 128.7	922	108.4	24 466.5
皮革、毛皮、羽毛及其制品和制鞋业	2 133	24 465.0	17 167.4	7 297.5	20 884.6	1 219	169.1	52 441.8
木材加工和木、竹、藤、棕、草制品业	2 644	4 861.8	4 385.5	476.3	2 992.1	614	23.6	8 262.1
家具制造业	502	827.0	479.6	347.4	526.2	192	2.5	1 973.9
造纸和纸制品业	4 856	285 451.9	242 348.7	43 103.3	400 696.3	4 006	2 409.5	578 579.9
印刷和记录媒介复制业	773	1 509.3	734.0	775.3	1 102.0	244	14.4	6 350.9
文教、工美、体育和娱乐用品制造业	670	1 955.6	1 132.6	823.0	1 573.2	410	10.4	5 227.1
石油加工、炼焦和核燃料加工业	1 280	76 903.1	66 302.5	10 600.6	76 047.4	1 328	378.9	429 454.1
化学原料和化学制品制造业	12 217	265 572.7	201 970.6	63 602.1	409 677.4	9 479	2 163.6	930 536.6
医药制造业	3 332	53 959.2	31 590.0	22 369.1	47 395.7	2 735	215.7	178 213.7
化学纤维制造业	496	37 905.8	24 954.6	12 951.1	35 018.5	438	166.4	82 126.0
橡胶和塑料制品业	2 687	12 623.9	7 032.2	5 591.7	16 544.5	1 012	82.2	24 677.7
非金属矿物制品业	32 532	29 032.6	24 374.0	4 658.6	82 625.9	4 965	534.2	100 085.3
黑色金属冶炼和压延加工业	4 202	94 761.9	82 825.6	11 936.3	2 186 335.4	4 101	9 815.8	1 111 983.5
有色金属冶炼和压延加工业	4 166	27 600.5	24 144.7	3 455.8	98 189.4	2 860	525.9	172 529.0
金属制品业	8 994	34 752.1	22 869.6	11 882.5	56 601.6	5 884	444.9	229 771.2
通用设备制造业	3 236	9 637.6	5 374.8	4 262.9	8 588.5	1 356	45.6	29 044.8
专用设备制造业	1 640	8 396.0	4 177.6	4 218.4	6 311.3	768	33.6	15 132.1
汽车制造业	1 966	16 861.7	7 169.6	9 692.2	15 303.3	1 470	76.3	66 050.1
铁路、船舶、航空航天和其他运输设备制造业	984	13 187.7	8 142.6	5 045.1	9 358.3	941	57.5	22 220.9
电气机械和器材制造业	1 948	10 471.2	4 716.1	5 755.1	9 301.6	1 315	60.5	47 438.8
计算机、通信和其他电子设备制造业	2 665	49 926.7	18 702.2	31 224.6	53 437.2	2 729	295.0	246 443.5
仪器仪表制造业	380	1 983.7	1 008.4	975.3	1 586.3	276	8.1	4 664.0
其他制造业	1 296	5 483.4	3 017.4	2 466.0	4 739.5	558	50.6	31 218.5
废弃资源综合利用业	494	2 248.6	1 617.9	630.7	2 457.4	204	21.1	10 213.6
金属制品、机械和设备修理业	195	1 263.5	717.3	546.2	1 318.2	142	6.8	3 154.8
电力、热力生产和供应业	5 830	95 934.2	81 527.3	14 406.9	178 858.6	3 304	950.5	236 036.0
燃气生产和供应业	50	999.4	921.6	77.8	1 258.2	33	5.0	6 380.5
水的生产和供应业	5	14.2	12.6	1.6	5.5	7	0.5	45.0
其他行业	144	895.8	889.2	6.6	818.4	104	6.7	869.4

按行业分重点调查工业废水排放及处理情况（二）

（2013）

单位：吨

行业名称	工业废水中污染物产生量				
	化学需氧量	氨氮	石油类	挥发酚	氰化物
行业总计	20 901 661.5	1 321 141.5	261 789.7	69 996.5	5 772.0
煤炭开采和洗选业	350 268.9	10 655.1	6 932.6	253.3	10.2
石油和天然气开采业	409 034.0	69 606.0	52 870.3	123.3	…
黑色金属矿采选业	181 178.0	817.8	386.8	5.0	10.0
有色金属矿采选业	187 538.5	6 085.4	147.9	0.1	23.5
非金属矿采选业	53 434.8	685.6	131.0	0.1	…
开采辅助活动	1 983.4	89.8	1 936.4	8.4	…
其他采矿业	321.4	9.8	0.3	…	…
农副食品加工业	2 178 699.2	56 940.3	794.3	0.4	8.8
食品制造业	1 051 649.6	75 471.7	487.5	4.0	…
酒、饮料和精制茶制造业	2 527 164.0	34 045.3	60.7	0.1	0.2
烟草制品业	9 994.1	482.7	17.3	…	…
纺织业	1 779 960.0	55 745.5	331.9	19.4	5.0
纺织服装、服饰业	89 876.2	3 149.4	29.7	0.2	…
皮革、毛皮、羽毛及其制品和制鞋业	350 626.9	18 322.6	3 916.6	4.5	…
木材加工和木、竹、藤、棕、草制品业	61 152.1	1 115.4	8.2	2.3	…
家具制造业	2 905.5	349.1	39.2	…	4.5
造纸和纸制品业	5 079 414.2	52 306.1	551.4	1 259.0	0.1
印刷和记录媒介复制业	5 800.8	321.0	52.7	…	…
文教、工美、体育和娱乐用品制造业	6 651.1	482.7	50.8	…	4.6
石油加工、炼焦和核燃料加工业	729 655.5	159 879.5	90 711.8	43 088.5	998.7
化学原料和化学制品制造业	2 330 184.1	572 521.0	10 721.1	5 292.0	369.2
医药制造业	725 171.3	31 372.2	1 483.2	32.3	1.3
化学纤维制造业	528 792.4	6 862.7	1 248.5	19.7	2.0
橡胶和塑料制品业	54 276.2	3 106.9	335.7	1.3	0.3
非金属矿物制品业	131 679.9	5 970.9	1 307.1	14.0	0.1
黑色金属冶炼和压延加工业	1 027 269.4	26 738.2	64 971.9	18 285.7	2 084.7
有色金属冶炼和压延加工业	152 396.7	76 545.0	3 703.3	1 240.5	311.9
金属制品业	111 864.5	7 531.1	5 506.8	2.3	1 697.3
通用设备制造业	37 638.4	1 657.8	1 495.0	0.1	14.9
专用设备制造业	20 186.0	1 425.8	976.1	…	7.3
汽车制造业	99 783.4	4 326.4	5 297.0	3.6	12.9
铁路、船舶、航空航天和其他运输设备制造业	41 913.2	2 924.0	2 362.3	0.2	4.9
电气机械和器材制造业	44 650.3	1 947.1	397.1	3.1	6.6
计算机、通信和其他电子设备制造业	180 497.2	11 945.5	570.5	…	83.3
仪器仪表制造业	5 374.6	283.6	33.6	…	6.9
其他制造业	42 716.9	995.5	148.5	54.1	94.9
废弃资源综合利用业	11 230.4	793.4	17.2	1.1	…
金属制品、机械和设备修理业	4 419.7	217.2	260.0	0.2	0.5
电力、热力生产和供应业	241 060.4	15 282.5	1 036.6	1.2	0.1
燃气生产和供应业	17 850.7	1 234.4	460.5	276.3	7.4
水的生产和供应业	13.6	0.8	0	…	…
其他行业	35 384.0	898.9	0	…	…

按行业分重点调查工业废水排放及处理情况（三）

（2013）

单位：吨

行业名称	工业废水中污染物产生量					
	汞	镉	六价铬	总铬	铅	砷
行业总计	20.936	2 243.096	3 250.031	6 990.343	2 962.439	11 228.723
煤炭开采和洗选业	0.006	0.002	0.070	0.202	0.109	0.063
石油和天然气开采业	…	…	…	0.317	…	…
黑色金属矿采选业	0.002	0.696	0.110	6.481	3.999	11.693
有色金属矿采选业	3.726	24.199	0.945	1.817	177.533	167.077
非金属矿采选业	0.024	0.334	0.403	0.410	1.187	1.893
开采辅助活动	…	…	…	0.001	…	…
其他采矿业	…	0.002	…	…	0.090	0.158
农副食品加工业	…	…	…	0.004	…	…
食品制造业	…	0.002	…	0.004	0.011	…
酒、饮料和精制茶制造业	…	…	…	0.003	…	…
烟草制品业	…	…	…	…	…	…
纺织业	…	…	1.691	2.566	0.086	0.014
纺织服装、服饰业	…	…	0.069	0.109	…	…
皮革、毛皮、羽毛及其制品和制鞋业	0.001	0.871	146.235	775.010	0.013	0.027
木材加工和木、竹、藤、棕、草制品业	…	…	…	…	…	…
家具制造业	…	…	0.083	2.634	…	…
造纸和纸制品业	…	…	0.020	0.063	0.088	0.012
印刷和记录媒介复制业	…	…	1.720	3.134	0.007	…
文教、工美、体育和娱乐用品制造业	…	811	6.970	8.371	0.005	…
石油加工、炼焦和核燃料加工业	0.011	0.005	0.242	0.289	0.001	0.725
化学原料和化学制品制造业	11.840	71.081	135.525	217.390	192.763	1 500.540
医药制造业	…	…	…	0.289	25	15.041
化学纤维制造业	…	…	0.054	0.054	…	…
橡胶和塑料制品业	…	…	3.789	9.923	…	…
非金属矿物制品业	0.016	0.167	0.073	0.139	0.207	0.513
黑色金属冶炼和压延加工业	0.074	12.698	63.609	78.958	23.577	39.935
有色金属冶炼和压延加工业	4.981	2 056.617	33.378	80.036	2 383.593	9 338.260
金属制品业	0.053	71.603	2 576.546	5 384.584	104.368	137.978
通用设备制造业	0.002	0.006	55.920	87.735	0.469	0.004
专用设备制造业	0.002	0.013	40.195	44.232	0.257	0.045
汽车制造业	…	0.078	79.327	124.278	0.088	0.016
铁路、船舶、航空航天和其他运输设备制造业	…	0.095	47.530	60.487	0.010	0.014
电气机械和器材制造业	0.016	3.995	12.674	17.675	64.232	1.574
计算机、通信和其他电子设备制造业	0.096	0.406	7.004	23.270	6.244	0.759
仪器仪表制造业	…	0.004	18.438	25.748	0.051	0.001
其他制造业	…	…	8.537	17.811	2.143	0.013
废弃资源综合利用业	…	0.146	0.039	0.771	1.326	0.077
金属制品、机械和设备修理业	…	0.058	7.988	13.887	0.010	…
电力、热力生产和供应业	0.084	0.016	0.094	0.908	0.060	12.292
燃气生产和供应业	…	…	…	…	…	…
水的生产和供应业	…	…	0.750	0.750	…	…
其他行业	…	…	…	…	…	…

按行业分重点调查工业废水排放及处理情况（四）

（2013）

单位：吨

行业名称	工业废水中污染物排放量				
	化学需氧量	氨氮	石油类	挥发酚	氰化物
行业总计	2 852 943.9	224 769.1	17 389.2	1 259.1	162.0
煤炭开采和洗选业	123 678.1	4 107.5	2 263.7	0.4	…
石油和天然气开采业	12 561.3	855.1	946.0	5.0	…
黑色金属矿采选业	10 574.1	389.0	166.9	…	…
有色金属矿采选业	44 610.3	2 243.4	76.6	0.1	3.2
非金属矿采选业	6 369.1	276.4	12.6	0.1	…
开采辅助活动	677.0	49.6	26.4	0.4	…
其他采矿业	132.3	4.3	…	…	…
农副食品加工业	471 292.2	19 041.7	156.0	0.2	1.3
食品制造业	111 038.3	9 569.4	80.6	0.2	…
酒、饮料和精制茶制造业	200 550.7	9 475.6	29.2	…	…
烟草制品业	2 345.4	149.2	7.4	…	…
纺织业	254 179.6	17 919.0	60.6	3.3	…
纺织服装、服饰业	17 464.5	1 412.7	6.0	…	…
皮革、毛皮、羽毛及其制品和制鞋业	54 823.1	4 385.8	339.8	1.9	…
木材加工和木、竹、藤、棕、草制品业	15 129.5	383.2	5.7	0.1	…
家具制造业	698.4	66.3	2.0	…	…
造纸和纸制品业	533 014.2	17 779.1	164.6	54.3	…
印刷和记录媒介复制业	2 600.2	119.9	27.3	…	…
文教、工美、体育和娱乐用品制造业	1 789.8	151.2	20.8	…	0.2
石油加工、炼焦和核燃料加工业	73 659.4	13 889.4	2 126.8	1 031.6	40.3
化学原料和化学制品制造业	321 985.0	76 419.5	2 638.5	79.6	44.0
医药制造业	97 237.9	7 458.6	329.0	7.8	0.2
化学纤维制造业	157 201.5	3 093.2	167.8	1.0	0.2
橡胶和塑料制品业	14 805.7	1 143.7	136.9	0.2	0.1
非金属矿物制品业	33 227.1	1 730.8	249.7	1.2	…
黑色金属冶炼和压延加工业	68 256.4	5 711.4	2 702.1	55.0	33.3
有色金属冶炼和压延加工业	28 270.5	12 837.0	482.3	6.8	2.3
金属制品业	34 660.6	2 652.9	1 136.9	1.2	31.1
通用设备制造业	10 726.8	624.9	513.5	0.1	0.5
专用设备制造业	7 037.5	639.6	330.0	…	…
汽车制造业	16 679.8	1 232.6	649.9	3.6	0.3
铁路、船舶、航空航天和其他运输设备制造业	16 948.4	1 422.4	943.8	0.2	0.2
电气机械和器材制造业	8 355.5	753.2	139.5	0.1	0.2
计算机、通信和其他电子设备制造业	35 041.3	3 139.2	179.7	…	3.2
仪器仪表制造业	1 475.0	98.7	11.8	…	0.2
其他制造业	6 261.8	355.2	49.4	1.4	0.7
废弃资源综合利用业	3 295.9	177.4	8.6	0.9	…
金属制品、机械和设备修理业	1 183.2	67.0	31.2	0.2	…
电力、热力生产和供应业	33 752.2	2 192.3	146.4	0.8	0.1
燃气生产和供应业	1 183.5	173.0	23.1	1.2	0.3
水的生产和供应业	8.5	0.1	…	…	…
其他行业	18 162.0	579.1	…	…	…

按行业分重点调查工业废水排放及处理情况（五）

（2013）　　　　　　　　　　　　　　　　　　　　　　　　　　单位：吨

行业名称	工业废水中污染物排放量					
	汞	镉	六价铬	总铬	铅	砷
行业总计	0.783	17.875	58.139	161.863	74.084	111.607
煤炭开采和洗选业	0.005	0.002	0.069	0.201	0.093	0.062
石油和天然气开采业	…	…	…	0.173	…	…
黑色金属矿采选业	…	0.069	0.087	0.087	0.407	0.090
有色金属矿采选业	0.254	2.244	0.466	0.700	25.277	31.463
非金属矿采选业	0.007	0.228	0.036	0.038	0.348	1.602
开采辅助活动	…	…	…	0.001	…	…
其他采矿业	…	0.001	…	…	0.031	0.071
农副食品加工业	…	…	…	0.003	…	…
食品制造业	…	…	…	0.003	…	…
酒、饮料和精制茶制造业	…	…	…	0.003	…	…
烟草制品业	…	…	…	…	…	…
纺织业	…	…	0.986	1.025	0.086	0.008
纺织服装、服饰业	…	…	0.014	0.039	…	…
皮革、毛皮、羽毛及其制品和制鞋业	…	0.751	2.037	66.741	0.010	0.026
木材加工和木、竹、藤、棕、草制品业	…	…	…	…	…	…
家具制造业	…	…	0.001	0.008	…	…
造纸和纸制品业	…	…	…	0.013	…	0.002
印刷和记录媒介复制业	…	…	0.053	0.077	0.002	…
文教、工美、体育和娱乐用品制造业	…	0.001	0.103	0.139	0.003	…
石油加工、炼焦和核燃料加工业	0.011	0.005	…	0.047	0.001	0.023
化学原料和化学制品制造业	0.219	1.659	0.372	6.275	8.431	38.380
医药制造业	…	…	…	0.005	…	0.011
化学纤维制造业	…	…	0.054	0.054	…	…
橡胶和塑料制品业	…	…	0.060	0.229	…	…
非金属矿物制品业	0.014	0.002	0.059	0.066	0.173	0.222
黑色金属冶炼和压延加工业	0.017	0.193	2.664	3.798	1.506	4.794
有色金属冶炼和压延加工业	0.226	12.428	2.698	5.141	32.366	25.322
金属制品业	0.003	0.175	36.601	60.786	1.210	0.238
通用设备制造业	0.001	0.002	0.951	1.510	0.087	0.004
专用设备制造业	0.001	0.013	0.581	0.966	0.094	0.042
汽车制造业	…	0.004	8.315	9.024	0.013	0.016
铁路、船舶、航空航天和其他运输设备制造业	…	0.008	0.351	0.714	0.008	0.013
电气机械和器材制造业	0.006	0.048	0.090	0.228	2.483	0.076
计算机、通信和其他电子设备制造业	0.011	0.028	0.450	2.328	1.152	0.181
仪器仪表制造业	…	0.004	0.179	0.357	0.051	0.001
其他制造业	…	…	0.066	0.203	0.142	…
废弃资源综合利用业	…	0.004	0.002	0.005	0.095	0.002
金属制品、机械和设备修理业	…	0.002	0.015	0.059	0.003	…
电力、热力生产和供应业	0.006	0.002	0.027	0.066	0.009	8.959
燃气生产和供应业	…	…	…	…	…	…
水的生产和供应业	…	…	0.750	0.750	…	…
其他行业	…	…	…	…	…	…

按行业分重点调查工业废气排放及处理情况（一）

（2013）

行业名称	工业废气排放总量（标态）/亿米³	废气治理设施数/套	废气治理设施处理能力（标态）/（万米³/小时）	废气治理设施运行费用/万元
行业总计	669 360.9	234 316	1 435 110.3	14 977 779.0
煤炭开采和洗选业	2 363.1	6 019	7 441.4	43 399.7
石油和天然气开采业	1 114.3	257	492.0	23 203.4
黑色金属矿采选业	3 127.4	1 329	2 367.4	21 194.9
有色金属矿采选业	332.4	710	972.8	7 446.4
非金属矿采选业	770.0	559	1 357.9	12 913.1
开采辅助活动	70.0	38	46.9	1 057.8
其他采矿业	31.1	39	68.8	1 045.4
农副食品加工业	5 068.6	7 273	18 100.1	75 475.0
食品制造业	2 498.4	3 571	7 387.5	42 258.4
酒、饮料和精制茶制造业	2 031.8	3 035	6 190.0	39 471.8
烟草制品业	553.7	1 041	2 552.8	11 750.4
纺织业	2 875.1	8 754	12 964.6	96 373.0
纺织服装、服饰业	195.6	1 224	1 075.6	7 692.3
皮革、毛皮、羽毛及其制品和制鞋业	327.4	1 507	1 310.1	7 300.1
木材加工和木、竹、藤、棕、草制品业	2 811.4	3 083	8 418.7	26 201.7
家具制造业	620.4	687	3 237.2	2 596.5
造纸和纸制品业	6 720.6	4 961	17 665.6	163 132.3
印刷和记录媒介复制业	245.8	384	953.0	4 468.1
文教、工美、体育和娱乐用品制造业	201.7	602	1 043.7	2 730.3
石油加工、炼焦和核燃料加工业	21 344.6	3 720	30 907.9	569 080.2
化学原料和化学制品制造业	31 536.2	21 473	65 391.6	771 324.2
医药制造业	1 741.4	3 988	4 951.9	49 982.5
化学纤维制造业	2 234.4	1 097	3 535.3	37 421.7
橡胶和塑料制品业	3 762.0	4 707	9 526.8	62 056.7
非金属矿物制品业	120 337.2	66 461	227 319.9	1 198 821.6
黑色金属冶炼和压延加工业	173 002.5	17 017	352 674.7	3 240 295.2
有色金属冶炼和压延加工业	32 635.6	9 587	58 246.6	852 753.9
金属制品业	5 478.7	8 894	15 000.5	99 716.9
通用设备制造业	1 258.6	3 003	5 080.3	18 663.1
专用设备制造业	1 271.8	2 448	6 957.8	22 112.2
汽车制造业	4 896.1	3 482	11 323.5	76 016.5
铁路、船舶、航空航天和其他运输设备制造业	1 585.0	4 990	8 127.9	33 343.6
电气机械和器材制造业	2 423.5	4 989	6 805.3	45 290.3
计算机、通信和其他电子设备制造业	6 437.9	7 050	14 829.8	97 913.1
仪器仪表制造业	128.9	365	640.1	2 272.3
其他制造业	693.2	1 136	2 340.6	7 640.4
废弃资源综合利用业	340.0	508	1 028.9	5 757.3
金属制品、机械和设备修理业	215.5	253	962.3	2 389.5
电力、热力生产和供应业	225 446.6	23 852	513 707.0	7 190 117.8
燃气生产和供应业	596.1	159	2 009.5	4 526.5
水的生产和供应业	0.1	0	…	…
其他行业	36.1	64	96.0	572.9

按行业分重点调查工业废气排放及处理情况（二）

（2013）

单位：万吨

行业名称	二氧化硫产生量	二氧化硫排放量	氮氧化物产生量	氮氧化物排放量	烟（粉）尘产生量	烟（粉）尘排放量
行业总计	6 071.3	1 689.2	1 813.6	1 464.9	73 583.2	1 022.5
煤炭开采和洗选业	16.6	12.6	4.6	4.6	100.9	38.2
石油和天然气开采业	2.8	2.1	2.4	2.4	5.4	0.8
黑色金属矿采选业	2.8	2.3	0.7	0.7	45.6	11.5
有色金属矿采选业	2.1	1.4	0.4	0.4	9.2	1.5
非金属矿采选业	4.7	3.6	1.2	1.2	33.0	4.7
开采辅助活动	1.7	0.3	0.1	0.1	0.4	0.1
其他采矿业	0.1	…	…	…	7.2	0.2
农副食品加工业	30.7	23.6	9.3	8.9	207.0	19.5
食品制造业	20.0	14.9	5.2	5.1	105.2	5.5
酒、饮料和精制茶制造业	16.9	13.1	4.0	3.9	92.7	6.8
烟草制品业	1.8	1.1	0.4	0.4	4.6	0.5
纺织业	30.8	25.5	7.3	7.3	123.0	9.0
纺织服装、服饰业	1.8	1.7	0.5	0.5	5.3	0.8
皮革、毛皮、羽毛及其制品和制鞋业	2.8	2.6	0.6	0.6	5.4	1.1
木材加工和木、竹、藤、棕、草制品业	4.5	4.2	1.5	1.4	131.8	14.4
家具制造业	0.4	0.3	0.1	0.1	1.4	0.4
造纸和纸制品业	69.1	44.9	19.6	19.3	472.1	14.9
印刷和记录媒介复制业	0.5	0.4	0.1	0.1	2.0	0.2
文教、工美、体育和娱乐用品制造业	0.2	0.2	0.1	0.1	0.6	0.2
石油加工、炼焦和核燃料加工业	247.7	79.3	40.0	38.5	779.5	40.8
化学原料和化学制品制造业	383.2	128.2	59.7	54.7	1 593.6	60.0
医药制造业	13.2	10.6	3.0	2.9	54.2	4.2
化学纤维制造业	15.4	8.6	5.1	5.0	122.1	2.1
橡胶和塑料制品业	11.7	8.5	3.0	2.8	44.1	3.7
非金属矿物制品业	230.1	196.0	301.0	271.6	22 620.4	258.8
黑色金属冶炼和压延加工业	342.8	235.1	107.1	99.7	6 839.0	193.5
有色金属冶炼和压延加工业	1 168.3	122.3	28.4	26.4	1 497.6	36.0
金属制品业	85.9	8.1	2.3	2.3	45.3	8.0
通用设备制造业	2.6	2.1	0.8	0.8	12.2	3.4
专用设备制造业	2.5	1.6	0.9	0.9	16.6	1.8
汽车制造业	1.6	1.2	0.7	0.7	13.7	2.2
铁路、船舶、航空航天和其他运输设备制造业	2.4	1.5	1.0	1.0	16.6	2.2
电气机械和器材制造业	1.9	1.1	0.5	0.4	5.0	0.7
计算机、通信和其他电子设备制造业	0.8	0.7	0.6	0.5	2.5	0.5
仪器仪表制造业	0.1	0.1	…	…	0.4	0.1
其他制造业	6.2	6.1	1.2	1.2	9.6	2.3
废弃资源综合利用业	0.8	0.5	0.2	0.1	13.5	0.4
金属制品、机械和设备修理业	0.1	0.1	…	…	7.0	0.7
电力、热力生产和供应业	3 340.7	720.6	1 198.8	896.9	38 508.3	270.3
燃气生产和供应业	2.9	1.6	1.1	1.1	28.5	0.6
水的生产和供应业	…	…	…	…	…	…
其他行业	0.3	0.3	0.1	0.1	1.0	0.2

按行业分重点调查一般工业固体废物产生及处置利用情况

行业名称	一般工业固体废物产生量	一般工业固体废物综合利用量	一般工业固体废物处置量	一般工业固体废物贮存量	一般工业固体废物倾倒丢弃量	一般工业固体废弃物综合利用率/%
行业总计	313 015	195 561	80 355	40 883	114	62.3
煤炭开采和洗选业	39 233	28 719	9 040	2 788	12	72.4
石油和天然气开采业	135	105	29	…	…	77.9
黑色金属矿采选业	67 851	17 031	41 527	9 417	31	25.1
有色金属矿采选业	37 977	13 964	11 376	13 396	22	36.7
非金属矿采选业	3 158	2 001	612	553	22	63.0
开采辅助活动	117	26	90	0	0	22.6
其他采矿业	57	51	3	2	0	90.1
农副食品加工业	2 106	1 976	127	3	4	93.8
食品制造业	546	525	23	1	…	96.1
酒、饮料和精制茶制造业	962	883	79	…	…	91.8
烟草制品业	213	107	71	78	…	45.9
纺织业	687	584	102	1	…	85.1
纺织服装、服饰业	31	24	7	…	…	78.5
皮革、毛皮、羽毛及其制品和制鞋业	57	48	10	…	…	82.9
木材加工和木、竹、藤、棕、草制品业	236	229	6	…	…	97.1
家具制造业	14	13	1	…	…	94.6
造纸和纸制品业	2 055	1 734	317	7	1	84.4
印刷和记录媒介复制业	20	15	6	…	…	72.1
文教、工美、体育和娱乐用品制造业	8	5	3	…	…	61.7
石油加工、炼焦和核燃料加工业	3 398	2 219	1 099	97	0	65.3
化学原料和化学制品制造业	27 908	17 917	4 420	6 027	8	63.9
医药制造业	281	250	28	3	…	89.0
化学纤维制造业	346	312	32	3	…	90.0
橡胶和塑料制品业	184	166	18	…	…	90.3
非金属矿物制品业	7 073	6 885	236	57	1	97.3
黑色金属冶炼和压延加工业	44 076	39 902	2 588	1 852	9	90.5
有色金属冶炼和压延加工业	11 181	5 698	3 998	1 594	1	50.7
金属制品业	910	784	64	63	…	86.1
通用设备制造业	156	118	38	…	…	75.4
专用设备制造业	202	164	35	3	…	81.2
汽车制造业	321	280	41	…	…	87.3
铁路、船舶、航空航天和其他运输设备制造业	232	198	34	…	…	85.4
电气机械和器材制造业	80	69	10	…	…	86.7
计算机、通信和其他电子设备制造业	97	60	37	…	…	62.0
仪器仪表制造业	5	4	1	…	…	86.1
其他制造业	47	39	8	…	…	82.8
废弃资源综合利用业	242	209	26	8	…	86.2
金属制品、机械和设备修理业	16	13	3	…	…	81.9
电力、热力生产和供应业	60 714	52 153	4 208	4 926	3	85.9
燃气生产和供应业	70	68	2	0	…	96.8
水的生产和供应业	…	…	0	0	0	100.0
其他行业	11	11	…	0	0	99.8

按行业分重点调查危险废物产生及处置利用情况

（2013）

单位：万吨

行业名称	危险废物产生量	危险废物综合利用量	危险废物处置量	危险废物贮存量	危险废物倾倒丢弃量/吨	危险废物处置利用率/%
行业总计	3 157	1 700	701	811	0	74.8
煤炭开采和洗选业	…	…	…	…	0	90.5
石油和天然气开采业	39	7	32	1	0	98.2
黑色金属矿采选业	…	…	…	…	0	93.5
有色金属矿采选业	83	67	2	14	0	83.2
非金属矿采选业	652	51	9	593	0	9.1
开采辅助活动	…	…	…	0	0	100.0
其他采矿业	…	…	0	…	0	100.0
农副食品加工业	1	…	1	…	0	99.7
食品制造业	2	2	…	…	0	100.0
酒、饮料和精制茶制造业	…	…	…	…	0	97.4
烟草制品业	…	…	…	…	0	97.0
纺织业	2	1	1	…	0	95.5
纺织服装、服饰业	…	…	…	…	0	98.2
皮革、毛皮、羽毛及其制品和制鞋业	3	1	2	…	0	97.0
木材加工和木、竹、藤、棕、草制品业	…	…	…	…	0	98.4
家具制造业	…	…	…	…	0	92.1
造纸和纸制品业	308	288	20	…	0	100.0
印刷和记录媒介复制业	1	…	1	…	0	99.5
文教、工美、体育和娱乐用品制造业	1	…	1	…	0	99.2
石油加工、炼焦和核燃料加工业	148	107	38	3	0	97.9
化学原料和化学制品制造业	681	448	232	20	0	97.1
医药制造业	51	32	18	1	0	98.3
化学纤维制造业	38	33	5	…	0	99.8
橡胶和塑料制品业	5	1	4	…	0	95.7
非金属矿物制品业	24	11	13	…	0	99.7
黑色金属冶炼和压延加工业	139	134	21	2	0	99.0
有色金属冶炼和压延加工业	564	363	61	156	0	73.1
金属制品业	62	22	38	3	0	95.6
通用设备制造业	12	3	9	…	0	97.8
专用设备制造业	4	…	3	1	0	83.6
汽车制造业	30	4	26	…	0	98.9
铁路、船舶、航空航天和其他运输设备制造业	10	1	9	…	0	99.1
电气机械和器材制造业	28	11	17	1	0	98.1
计算机、通信和其他电子设备制造业	161	92	69	1	0	99.6
仪器仪表制造业	1	…	1	…	0	99.1
其他制造业	3	1	1	…	0	84.7
废弃资源综合利用业	9	4	4	1	0	85.9
金属制品、机械和设备修理业	5	1	4	…	0	99.4
电力、热力生产和供应业	88	15	59	14	0	84.0
燃气生产和供应业	1	…	…	…	0	99.9
水的生产和供应业	…	0	0	…	0	…
其他行业	0	0	0	0	0	…

按行业分重点调查工业汇总情况（一）

（2013）

行业名称	汇总工业企业数/个	工业总产值（现价）/万元	工业煤炭消耗量/万吨	燃料煤	工业用水总量/万吨	取水量	重复用水量
行业总计	147 657	375 719.7	377 737.7	285 937.4	36 880 690.7	4 631 233.2	32 249 457.6
煤炭开采和洗选业	6 462	12 689.2	25 704.6	1 753.8	240 053.0	93 757.0	146 296.1
石油和天然气开采业	249	7 584.3	78.6	78.6	143 701.0	32 981.3	110 719.7
黑色金属矿采选业	4 457	2 479.4	145.4	128.2	416 583.2	70 081.0	346 502.2
有色金属矿采选业	3 301	2 003.2	101.2	99.1	253 542.5	66 468.5	187 074.0
非金属矿采选业	983	1 058.3	289.0	280.7	32 662.0	11 496.4	21 165.6
开采辅助活动	58	340.9	12.8	12.8	2 302.5	646.4	1 656.0
其他采矿业	39	16.9	23.9	3.7	502.4	333.1	169.4
农副食品加工业	11 442	13 926.1	1 971.4	1 956.3	374 101.3	172 627.6	201 473.7
食品制造业	3 826	6 092.3	1 240.0	1 236.9	132 746.0	71 415.2	61 330.8
酒、饮料和精制茶制造业	4 049	6 558.4	1 084.7	1 078.8	232 545.9	99 211.2	133 334.7
烟草制品业	156	7 638.9	64.8	63.9	15 730.6	3 614.5	12 116.2
纺织业	8 719	12 239.6	2 141.3	2 094.9	359 065.2	267 518.8	91 546.4
纺织服装、服饰业	1 599	1 189.9	113.1	112.2	24 233.9	19 995.4	4 238.5
皮革、毛皮、羽毛及其制品和制鞋业	2 133	1 599.1	172.6	172.4	33 163.5	29 113.1	4 050.4
木材加工和木、竹、藤、棕、草制品业	2 644	1 209.8	283.0	245.7	11 871.0	6 858.4	5 012.5
家具制造业	502	605.5	15.9	15.8	1 366.8	1 221.7	145.1
造纸和纸制品业	4 856	7 051.5	4 723.7	4 705.1	1 211 333.1	344 578.2	866 754.8
印刷和记录媒介复制业	773	978.0	29.3	28.9	4 222.3	2 020.3	2 202.0
文教、工美、体育和娱乐用品制造业	670	1 152.9	12.5	12.5	3 111.0	2 514.0	597.0
石油加工、炼焦和核燃料加工业	1 280	36 186.5	36 432.9	7 597.1	2 383 566.6	148 013.9	2 235 552.7
化学原料和化学制品制造业	12 217	31 592.5	20 993.3	11 626.4	5 959 269.1	442 176.6	5 517 092.5
医药制造业	3 332	8 327.2	811.4	805.5	301 540.1	69 996.9	231 543.2
化学纤维制造业	496	3 621.2	1 074.5	1 042.4	402 123.0	57 700.1	344 422.9
橡胶和塑料制品业	2 687	6 881.6	713.9	710.6	67 700.3	17 973.3	49 727.1
非金属矿物制品业	32 532	15 479.1	29 865.5	22 703.5	404 846.7	76 656.4	328 190.2
黑色金属冶炼和压延加工业	4 202	44 052.1	31 606.8	11 220.3	7 532 899.6	319 841.9	7 213 057.7
有色金属冶炼和压延加工业	4 166	18 999.0	6 094.6	5 343.3	889 425.2	73 496.3	815 928.9
金属制品业	8 994	10 396.3	437.2	416.4	318 478.5	60 548.3	257 930.2
通用设备制造业	3 236	9 504.4	153.0	149.3	31 083.2	12 977.8	18 105.4
专用设备制造业	1 640	6 347.2	163.7	144.4	44 620.7	10 989.0	33 631.7
汽车制造业	1 966	26 293.9	102.3	99.7	122 648.6	23 111.6	99 537.0
铁路、船舶、航空航天和其他运输设备制造业	984	8 254.2	203.2	198.6	34 659.5	17 537.5	17 122.0
电气机械和器材制造业	1 948	11 253.6	69.2	67.9	51 225.3	14 655.5	36 569.8
计算机、通信和其他电子设备制造业	2 665	28 572.7	40.9	40.1	251 077.8	62 382.8	188 695.0
仪器仪表制造业	380	1 452.5	5.9	5.9	6 264.2	2 607.2	3 657.0
其他制造业	1 296	1 710.6	308.7	212.3	20 407.4	6 845.0	13 562.4
废弃资源综合利用业	494	464.1	34.4	33.0	5 526.4	3 057.8	2 468.7
金属制品、机械和设备修理业	195	932.2	7.7	7.7	2 042.5	1 750.5	292.0
电力、热力生产和供应业	5 830	18 565.0	209 417.3	209 250.5	14 496 195.0	1 908 140.9	12 588 054.1
燃气生产和供应业	50	276.2	980.1	168.7	61 016.3	3 337.7	57 678.8
水的生产和供应业	5	0.3	0.1	0.1	5.7	5.7	…
其他行业	144	143.0	13.5	13.5	1 231.7	978.6	253.1

按行业分重点调查工业汇总情况（二）

（2013）

行业名称	工业炉窑数/台	工业锅炉数/台	35蒸吨及以上	20（含）~35蒸吨	10（含）~20蒸吨	10蒸吨以下
行业总计	94 149	102 466	14 044	7 447	13 816	67 159
煤炭开采和洗选业	120	7 096	57	204	845	5 990
石油和天然气开采业	3 376	2 619	26	452	160	1 981
黑色金属矿采选业	150	539	14	13	86	426
有色金属矿采选业	81	523	3	15	58	447
非金属矿采选业	350	154	24	9	9	112
开采辅助活动	63	228	6	14	15	193
其他采矿业	11	12	0	0	4	8
农副食品加工业	415	8 686	591	470	623	7 002
食品制造业	160	4 216	170	218	555	3 273
酒、饮料和精制茶制造业	220	4 067	123	325	748	2 871
烟草制品业	15	355	35	65	150	105
纺织业	174	8 483	157	276	1 596	6 454
纺织服装、服饰业	10	1 520	4	16	112	1 388
皮革、毛皮、羽毛及其制品和制鞋业	35	1 629	3	4	70	1 552
木材加工和木、竹、藤、棕、草制品业	153	2 727	26	70	259	2 372
家具制造业	61	263	1	8	23	231
造纸和纸制品业	371	5 013	534	213	773	3 493
印刷和记录媒介复制业	6	400	0	2	26	372
文教、工美、体育和娱乐用品制造业	39	222	0	1	7	214
石油加工、炼焦和核燃料加工业	3 374	1 832	592	292	378	570
化学原料和化学制品制造业	5 727	10 678	1 346	614	1 319	7 399
医药制造业	75	3 474	102	109	426	2 837
化学纤维制造业	60	813	117	92	207	397
橡胶和塑料制品业	76	2 678	71	139	363	2 105
非金属矿物制品业	39 677	3 666	92	181	396	2 997
黑色金属冶炼和压延加工业	11 345	1 565	555	173	146	691
有色金属冶炼和压延加工业	15 372	1 515	206	89	211	1 009
金属制品业	4 356	2 155	9	47	115	1 984
通用设备制造业	2 307	911	9	35	68	799
专用设备制造业	1 644	727	23	41	152	511
汽车制造业	1 351	1 105	10	82	144	869
铁路、船舶、航空航天和其他运输设备制造业	972	845	66	157	133	489
电气机械和器材制造业	367	775	2	20	73	680
计算机、通信和其他电子设备制造业	447	954	10	19	107	818
仪器仪表制造业	31	162	40	0	11	111
其他制造业	728	548	4	23	52	469
废弃资源综合利用业	176	127	0	6	6	115
金属制品、机械和设备修理业	25	134	1	1	20	112
电力、热力生产和供应业	110	18 898	8 988	2 939	3 350	3 621
燃气生产和供应业	47	89	27	13	16	33
水的生产和供应业	0	2	0	0	0	2
其他行业	72	61	0	0	4	57

各地区独立火电厂污染排放及处理情况（一）

（2013）

地 区 名 称	汇总工业 企业数/ 个	机组数/ 台	废气治理 设施数/ 套	脱硫 设施数	脱硝 设施数	除尘 设施数
全 国	1 853	4 825	9 990	3 547	1 076	5 140
北 京	10	31	60	12	28	20
天 津	18	49	132	50	16	66
河 北	107	246	560	216	63	275
山 西	120	298	551	206	72	269
内蒙古	103	299	557	203	37	314
辽 宁	72	227	455	141	27	253
吉 林	40	113	210	51	14	135
黑龙江	81	239	444	94	18	322
上 海	24	65	110	41	19	50
江 苏	176	442	1 091	420	122	514
浙 江	151	429	1 036	407	131	473
安 徽	53	121	261	88	39	124
福 建	35	87	187	71	34	81
江 西	19	44	97	30	14	49
山 东	272	731	1 580	598	80	867
河 南	94	206	462	189	54	217
湖 北	42	101	206	84	25	96
湖 南	31	67	119	39	22	58
广 东	84	223	436	149	93	178
广 西	16	32	81	29	18	34
海 南	8	23	28	11	4	12
重 庆	30	54	136	49	8	75
四 川	45	78	142	47	16	78
贵 州	23	73	153	56	22	75
云 南	16	41	102	41	16	45
西 藏	2	21	0	0	0	0
陕 西	53	145	245	56	31	137
甘 肃	22	64	137	41	21	74
青 海	7	17	13	3	0	9
宁 夏	25	61	137	48	24	65
新 疆	74	198	262	77	8	175

各地区独立火电厂污染排放及处理情况（二）

（2013）

地 区 名 称	燃料煤消耗量/ 万吨			燃料煤平均含硫量/ %	燃料油消耗量/ 万吨	燃料油平均含硫量/ %
		发电消耗量	供热消耗量			
全 国	199 695.2	182 717.8	16 977.4	0.96	604.0	0.73
北 京	882.9	591.5	291.4	0.59	0	0
天 津	2 914.9	2 503.4	411.5	0.80	…	0.06
河 北	11 004.7	9 985.1	1 019.6	1.03	0.8	0.43
山 西	12 929.1	12 224.1	705.0	1.18	1.1	0.66
内蒙古	19 442.5	17 830.5	1 612.0	0.76	1.0	0.66
辽 宁	8 653.4	7 005.2	1 648.2	0.68	1.3	0.22
吉 林	4 532.0	3 463.7	1 068.3	0.42	0.2	0.43
黑龙江	5 591.8	4 181.7	1 410.1	0.34	1.0	0.23
上 海	3 305.3	3 184.7	120.6	0.62	13.8	0.91
江 苏	17 546.9	15 700.9	1 846.0	0.72	498.6	0.80
浙 江	10 058.2	8 401.3	1 656.9	0.66	4.1	0.99
安 徽	8 274.8	8 012.9	261.9	0.54	0.1	0.62
福 建	4 914.5	4 730.1	184.4	0.66	0.8	0.29
江 西	2 992.2	2 992.2	0	1.10	0.3	0.69
山 东	16 395.0	14 078.3	2 316.7	1.12	0.5	0.99
河 南	12 926.2	12 271.9	654.4	0.91	1.8	0.50
湖 北	4 293.8	4 120.9	172.9	1.27	0.6	0.60
湖 南	3 490.2	3 411.2	79.0	1.14	0.5	0.35
广 东	11 953.0	11 710.2	242.8	0.72	54.7	0.22
广 西	2 702.8	2 681.2	21.6	1.76	0.2	0.12
海 南	679.1	679.1	0	0.90	…	0.11
重 庆	2 196.2	2 142.6	53.7	2.68	0.3	0.24
四 川	3 025.7	2 974.0	51.7	2.04	0.3	0.40
贵 州	5 905.3	5 896.6	8.8	2.46	1.1	1.88
云 南	2 598.3	2 542.2	56.0	1.64	0.3	0.74
西 藏	0	0	0	0	11.1	0.04
陕 西	6 146.0	6 008.8	137.1	1.29	0.4	1.40
甘 肃	3 309.1	3 016.0	293.0	0.73	0.1	0.32
青 海	404.5	404.5	0	0.75	…	0.37
宁 夏	5 349.1	5 188.3	160.8	1.13	0.1	0.52
新 疆	5 277.7	4 784.6	493.1	1.37	8.7	0.56

各地区独立火电厂污染排放及处理情况（三）

（2013）

单位：吨

地 区 名 称	工业废气排放量/ 亿米³	工业二氧化硫 产生量	工业二氧化硫 排放量	工业氮氧化物 产生量	工业氮氧化物 排放量	工业烟（粉） 尘产生量	工业烟（粉） 尘排放量
全 国	195 646.0	32 188 060.0	6 341 322.0	11 620 825.6	8 618 096.6	362 883 863.2	1 838 659.4
北 京	1 478.0	84 230.7	9 943.2	54 509.9	37 925.6	1 814 919.7	4 004.7
天 津	2 219.9	405 334.0	60 126.6	199 316.7	151 200.1	4 572 489.5	4 377.0
河 北	13 749.9	1 949 870.7	261 636.9	709 980.6	550 092.6	19 679 568.2	76 759.8
山 西	16 325.8	2 607 989.2	539 408.1	769 889.2	522 990.6	21 275 210.1	159 365.7
内蒙古	16 418.1	2 544 996.1	595 030.9	902 731.5	813 808.1	32 739 442.4	229 756.7
辽 宁	6 708.6	956 417.4	297 304.5	426 989.2	353 738.9	14 381 128.9	96 427.9
吉 林	3 394.3	330 615.1	114 239.9	242 189.7	222 687.3	7 460 587.5	60 970.8
黑龙江	4 162.5	318 687.2	155 007.7	308 349.7	291 377.9	11 065 722.4	190 217.7
上 海	3 324.9	359 400.5	59 402.5	235 327.6	172 908.2	4 371 853.8	20 023.4
江 苏	14 129.3	2 161 929.5	347 332.8	921 650.5	572 965.8	26 324 789.6	115 005.0
浙 江	7 860.9	1 129 197.2	216 241.9	522 485.9	360 660.6	10 801 627.8	45 947.5
安 徽	6 495.2	768 556.0	101 337.8	442 789.6	295 020.1	19 764 955.2	38 135.9
福 建	4 589.1	582 885.9	69 177.2	298 463.6	152 308.1	5 237 021.9	19 497.4
江 西	2 699.3	599 008.3	96 337.2	182 773.6	151 868.1	4 947 711.2	19 986.0
山 东	14 555.9	3 060 159.7	603 757.4	841 428.0	651 789.2	33 768 594.2	99 086.8
河 南	13 083.2	1 902 638.8	354 960.9	752 137.8	632 207.3	30 153 903.6	127 650.5
湖 北	4 176.7	911 644.6	183 445.5	264 982.3	193 992.1	9 144 281.4	37 154.4
湖 南	3 007.7	749 455.9	118 783.4	363 931.4	211 832.4	8 913 312.1	23 362.1
广 东	10 461.1	1 411 340.6	225 438.2	499 164.1	331 073.3	11 287 157.0	36 926.2
广 西	4 570.2	844 880.7	99 129.6	159 476.8	130 206.6	4 474 272.4	12 341.7
海 南	3 177.3	105 931.5	13 087.7	39 642.3	31 138.5	835 420.1	1 604.4
重 庆	1 779.6	961 304.1	207 458.4	162 397.1	130 955.9	8 062 305.2	57 972.5
四 川	2 054.5	1 070 896.9	244 229.6	181 261.9	153 032.2	8 539 993.3	38 618.7
贵 州	16 292.8	2 217 819.0	513 530.0	428 416.6	357 644.0	15 194 154.8	109 950.0
云 南	1 614.6	740 903.0	138 797.0	137 591.7	125 314.3	7 322 341.3	13 405.9
西 藏	12.4	472.9	472.9	736.9	736.9	28.1	28.1
陕 西	5 247.6	1 282 268.0	269 635.8	632 166.1	335 519.6	14 000 752.8	65 671.2
甘 肃	2 967.3	393 423.1	84 671.4	220 346.5	159 189.5	5 144 841.7	25 144.4
青 海	337.2	51 661.0	23 826.9	20 453.5	20 453.5	775 521.2	8 154.8
宁 夏	4 333.8	1 145 277.1	143 220.7	411 559.0	230 238.4	13 663 831.3	51 815.4
新 疆	4 418.4	538 865.3	194 349.4	287 686.5	273 221.1	7 166 124.7	49 296.9

各地区水泥行业污染排放及处理情况（一）

（2013）

地区名称	汇总工业企业数/个	水泥窑数/台	新型干法窑	废气治理设施数/套	脱硝设施数 设施数	除尘设施数	煤炭消耗量/万吨
全　国	3 679	2 846	1 634	48 899	538	46 592	21 951.9
北　京	8	10	10	541	4	537	113.2
天　津	10	9	3	216	2	211	27.3
河　北	194	117	78	2 546	36	2 420	836.7
山　西	155	101	53	2 609	15	2 558	440.5
内蒙古	94	74	61	1 697	0	1 696	728.4
辽　宁	116	67	44	1 396	14	1 336	454.4
吉　林	48	38	24	589	9	580	750.2
黑龙江	70	48	30	863	3	840	251.5
上　海	6	8	2	97	0	97	8.2
江　苏	225	151	69	1 880	13	1 831	989.0
浙　江	160	101	71	2 901	25	2 667	830.6
安　徽	122	117	92	1 420	38	1 311	1 853.9
福　建	101	85	44	1 907	23	1 876	678.2
江　西	141	118	59	1 923	7	1 685	715.5
山　东	252	168	84	2 729	11	2 712	1 225.6
河　南	254	120	96	3 123	11	3 003	985.9
湖　北	138	110	63	2 255	15	2 076	861.0
湖　南	228	175	66	1 948	43	1 862	1 307.5
广　东	196	167	59	1 866	30	1 729	1 286.2
广　西	171	145	55	2 368	22	2 340	1 160.7
海　南	11	10	10	322	4	318	193.3
重　庆	76	92	51	1 615	21	1 591	684.6
四　川	229	166	97	2 812	50	2 666	1 277.8
贵　州	116	112	61	1 237	61	1 140	803.0
云　南	184	170	114	2 785	22	2 507	1 089.6
西　藏	10	18	4	211	0	145	39.6
陕　西	116	90	61	643	35	606	655.4
甘　肃	89	86	45	1 176	15	1 156	574.0
青　海	18	19	15	621	4	555	207.2
宁　夏	28	34	28	191	2	189	221.0
新　疆	113	120	85	2 412	3	2 352	701.8

各地区水泥行业污染排放及处理情况（二）

（2013）

单位：吨

地 区 名 称	工业废气 排放量/ 万米³	工业二氧化硫 产生量	工业二氧化硫 排放量	工业氮氧化物 产生量	工业氮氧化物 排放量	工业烟（粉） 尘产生量	工业烟（粉） 尘排放量
全 国	568 346 309.4	318 957.1	306 370.7	2 225 716.5	1 969 036.4	179 710 267.5	649 417.7
北 京	3 930 915.2	964.7	964.7	11 241.9	9 636.8	1 660 750.4	2 862.2
天 津	1 022 237.6	238.2	238.2	2 880.9	2 846.0	370 626.7	425.8
河 北	37 848 214.1	8 147.9	8 087.8	105 967.4	86 278.6	6 739 331.3	41 458.2
山 西	10 784 936.4	8 057.7	7 272.8	54 719.9	50 516.8	2 419 593.3	18 476.0
内蒙古	20 403 922.1	10 022.1	7 714.9	53 300.8	53 300.8	4 732 443.0	48 882.7
辽 宁	14 822 384.9	3 249.7	3 249.7	52 504.5	48 359.0	5 050 931.1	14 333.9
吉 林	6 506 274.1	2 629.5	2 629.5	48 867.6	42 906.8	2 298 874.5	6 753.0
黑龙江	6 389 133.3	2 372.8	2 372.8	25 561.1	25 202.4	2 504 612.8	20 396.0
上 海	389 707.9	210.4	210.4	733.3	733.3	31 365.4	327.5
江 苏	16 862 558.7	8 114.1	7 067.1	79 242.1	75 377.4	3 789 273.0	10 349.8
浙 江	22 666 262.9	8 985.3	8 985.3	101 768.0	86 317.3	8 977 910.7	29 279.2
安 徽	52 754 842.7	19 082.9	19 082.9	193 489.8	173 781.9	14 244 940.0	28 467.3
福 建	22 438 832.7	8 353.8	8 353.8	80 677.5	69 449.6	9 042 453.5	25 057.0
江 西	14 904 633.2	10 993.1	10 993.1	72 801.9	67 511.9	5 560 002.7	27 108.3
山 东	30 759 103.1	13 532.0	13 130.6	135 742.2	132 972.7	13 526 216.4	33 076.4
河 南	28 502 948.8	18 509.9	13 981.9	115 882.7	115 882.7	10 214 343.3	24 008.7
湖 北	19 859 147.0	5 889.3	5 889.3	95 137.4	91 416.0	9 125 318.7	18 799.8
湖 南	24 780 451.3	22 299.5	21 657.9	98 631.9	79 477.3	6 658 453.8	47 239.3
广 东	28 411 882.6	15 211.2	14 486.8	116 368.2	93 646.0	6 621 648.9	28 516.0
广 西	25 671 834.6	9 767.3	9 584.6	115 409.3	112 470.4	5 089 200.5	18 298.4
海 南	5 085 105.2	3 512.5	3 512.5	22 452.5	21 016.0	523 836.1	2 851.9
重 庆	23 553 135.8	39 385.4	39 385.4	72 093.9	53 567.2	9 485 992.1	26 601.8
四 川	37 606 852.7	21 921.8	21 912.4	125 374.2	104 888.5	9 361 708.7	22 260.0
贵 州	23 414 129.3	15 616.0	15 084.2	96 041.7	66 260.2	11 515 309.9	28 026.0
云 南	29 984 204.4	14 073.8	14 073.5	94 075.5	92 339.0	8 730 790.8	21 344.6
西 藏	383 659.4	291.8	269.6	1 560.0	1 560.0	324 592.2	386.9
陕 西	14 294 742.3	13 214.7	13 036.2	96 876.2	62 326.5	4 448 270.1	24 088.3
甘 肃	13 467 164.3	20 196.6	19 286.9	48 782.1	43 668.2	5 983 604.9	14 171.7
青 海	4 689 224.0	3 081.6	3 081.6	15 555.2	14 256.4	1 907 074.0	16 841.3
宁 夏	6 316 433.2	2 271.9	2 271.9	21 311.6	20 775.5	2 455 097.5	9 753.9
新 疆	19 841 435.7	8 759.5	8 503.0	70 665.1	70 295.4	6 315 701.3	38 975.1

各地区钢铁行业污染排放及处理情况（一）

（2013）

地区名称	汇总工业企业数/个	烧结机数/台	有脱硫设施的	有脱硝设施的	有除尘设施的	球团设备数/套	有脱硫设施的	有脱硝设施的	有除尘设施的
全 国	740	1 258	447	41	837	598	52	16	365
北 京	0	0	0	0	0	0	0	0	0
天 津	8	15	4	0	13	3	0	0	3
河 北	170	337	182	8	259	195	9	8	148
山 西	91	138	26	2	111	17	0	0	16
内蒙古	21	36	8	0	23	16	2	0	10
辽 宁	38	60	14	0	47	38	3	3	22
吉 林	8	13	2	0	12	13	0	0	7
黑龙江	6	12	2	0	6	5	0	0	3
上 海	2	7	6	6	6	0	0	0	0
江 苏	77	114	30	0	44	46	1	0	9
浙 江	5	11	7	1	11	3	0	0	2
安 徽	22	17	8	0	15	29	12	2	20
福 建	13	25	10	0	18	5	1	0	2
江 西	12	16	5	2	15	13	0	0	9
山 东	42	85	45	6	67	23	2	1	16
河 南	22	41	21	11	34	24	6	1	15
湖 北	21	22	9	1	19	16	1	0	8
湖 南	27	32	8	0	14	6	1	0	4
广 东	4	9	6	0	6	0	0	0	0
广 西	8	13	7	0	8	3	0	0	3
海 南	0	0	0	0	0	0	0	0	0
重 庆	1	3	3	0	3	1	0	0	1
四 川	30	24	16	1	20	25	12	0	21
贵 州	9	8	2	0	7	21	0	0	21
云 南	47	60	16	2	26	15	1	0	9
西 藏	0	0	0	0	0	0	0	0	0
陕 西	6	13	3	0	11	2	0	0	1
甘 肃	7	13	2	0	8	9	0	0	3
青 海	3	1	1	0	1	3	0	0	2
宁 夏	4	9	0	0	4	4	0	0	0
新 疆	36	124	4	1	29	63	1	1	10

各地区钢铁行业污染排放及处理情况（二）

（2013）

地区名称	铁精矿消耗量/亿吨	铁精矿平均含硫率/%	炼焦煤消耗量/万吨	炼焦煤平均含硫率/%	高炉喷煤量/万吨	高炉喷煤平均含硫率/%	焦炭消耗量/万吨	焦炭平均含硫率/%
全　国	66.4	0.10	16 388	0.70	24 729	0.48	22 920	0.64
北　京	…	…	0	…	0	…	0	…
天　津	0.2	0.04	0	…	179	0.44	464	0.77
河　北	2.6	0.10	1 696	0.81	16 850	0.47	5 320	0.66
山　西	0.6	0.10	866	0.90	694	0.56	1 370	0.71
内蒙古	0.3	0.30	489	0.55	214	0.57	499	0.81
辽　宁	7.9	0.05	2 509	0.70	902	0.47	2 177	0.67
吉　林	0.2	0.15	350	0.52	136	0.51	373	0.49
黑龙江	0.1	0.13	59	0.39	90	0.35	294	0.32
上　海	0.2	0.02	729	0.78	255	0.43	554	0.64
江　苏	0.7	0.06	975	0.64	906	0.50	1 485	0.56
浙　江	0.1	0.20	397	0.61	169	0.48	325	0.49
安　徽	0.3	0.10	713	0.57	286	0.16	795	0.23
福　建	0.2	0.24	105	0.61	212	0.58	410	0.68
江　西	0.3	0.20	543	0.83	277	0.56	838	0.72
山　东	1.0	0.06	1·292	1.10	998	0.47	2 270	0.76
河　南	0.3	0.18	467	0.61	347	0.48	830	0.55
湖　北	0.3	0.10	1 329	0.85	381	0.47	305	0.69
湖　南	0.2	0.14	499	0.77	272	0.68	322	0.45
广　东	0.2	0.11	251	0.71	136	0.47	381	0.66
广　西	0.2	0.09	697	0.70	246	0.63	762	0.55
海　南	…	…	0	…	0	…	0	…
重　庆	0.1	0.12	409	1.07	87	0.51	281	0.79
四　川	49.6	0.32	724	0.61	289	0.68	890	0.69
贵　州	0.1	0.10	176	0.76	79	1.42	194	0.73
云　南	0.2	0.21	227	0.58	283	0.65	659	0.48
西　藏	…	…	0	…	0	…	0	…
陕　西	0.1	0.19	41	0.70	115	0.41	188	0.65
甘　肃	0.1	0.21	393	0.78	119	0.31	392	0.69
青　海	…	0.27	0	…	22	0.34	55	0.73
宁　夏	…	0.23	21	1.00	15	0.25	54	0.70
新　疆	0.2	0.36	433	0.73	170	0.80	435	1.06

各地区钢铁行业污染排放及处理情况（三）

（2013）

单位：吨

地 区 名 称	工业废气 排放量/ 万米³	工业二氧化硫 产生量	工业二氧化硫 排放量	工业氮氧化物 产生量	工业氮氧化物 排放量	工业烟（粉） 尘产生量	工业烟（粉） 尘排放量
全 国	516 434 965	2 751 588	1 993 286	610 068	554 628	22 839 709	619 440
北 京	0	0	0	0	0	0	0
天 津	9 762 906	80 410	23 022	61 257	12 157	364 564	6 096
河 北	145 882 584	572 644	413 978	161 146	161 142	6 915 319	234 711
山 西	28 151 863	135 834	120 479	34 456	32 057	1 793 373	63 627
内蒙古	11 763 478	139 903	119 203	14 809	14 809	621 722	24 147
辽 宁	40 678 575	105 684	86 530	43 578	43 578	2 027 190	43 272
吉 林	9 401 124	55 856	46 156	9 116	9 116	476 590	9 200
黑龙江	5 667 461	27 081	22 254	4 194	4 194	211 469	4 987
上 海	11 371 260	16 312	16 312	10 673	10 673	512 446	7 681
江 苏	36 459 013	142 499	92 490	44 909	44 901	1 283 704	24 353
浙 江	12 380 857	42 867	20 323	7 615	7 121	130 462	6 562
安 徽	17 988 012	62 460	49 431	16 658	16 658	450 186	10 196
福 建	7 980 036	92 097	46 705	8 872	8 872	626 375	13 549
江 西	14 800 742	111 922	107 968	16 662	16 662	539 924	9 737
山 东	45 472 580	174 716	108 910	43 597	43 597	1 911 031	37 106
河 南	20 814 249	132 542	106 244	14 488	14 488	546 438	12 568
湖 北	15 106 789	96 179	73 442	17 453	17 453	425 718	9 356
湖 南	7 633 344	91 990	67 092	13 919	13 919	410 117	6 937
广 东	4 478 883	50 558	28 754	6 947	6 947	160 202	2 351
广 西	11 951 071	64 715	24 412	10 893	10 893	493 398	8 996
海 南	0	0	0	0	0	0	0
重 庆	2 686 266	25 394	15 984	7 493	7 493	241 828	4 024
四 川	17 795 771	210 545	120 138	21 011	17 662	865 239	21 909
贵 州	2 494 163	16 696	16 696	615	615	83 008	5 564
云 南	9 581 544	114 075	95 906	11 442	11 355	475 185	15 053
西 藏	0	0	0	0	0	0	0
陕 西	9 493 767	44 678	41 625	6 046	6 046	275 067	4 466
甘 肃	7 759 922	44 501	44 501	7 918	7 918	346 353	7 160
青 海	1 111 635	11 796	8 936	1 133	1 133	62 256	934
宁 夏	419 658	7 305	7 305	1 048	1 048	92 914	4 972
新 疆	7 347 412	80 328	68 489	12 124	12 124	497 632	19 927

各地区制浆及造纸行业污染排放及处理情况（一）

（2013）

地区名称	汇总工业企业数/个	造纸生产线/条	化学浆生产线/条	化机浆生产线/条	废纸浆生产线/条	取水量/万吨	造纸生产线	化学浆生产线	化机浆生产线	废纸浆生产线
全 国	4 856	3 537	548	468	1 819	344 578	177 824	46 027	4 304	57 216
北 京	16	5	0	0	1	130	105	0	0	0
天 津	58	20	1	1	4	1 997	357	0	0	964
河 北	350	300	12	8	148	24 566	12 247	331	7	7 715
山 西	49	40	4	2	23	5 219	2 444	552	0	820
内蒙古	30	8	2	0	5	523	154	213	0	19
辽 宁	76	45	9	10	20	6 160	3 489	0	337	1 100
吉 林	45	41	7	6	14	6 031	3 927	1 204	18	616
黑龙江	47	41	15	10	19	3 465	2 440	316	40	104
上 海	60	31	0	0	1	1 362	1 145	0	0	183
江 苏	199	155	49	50	79	16 417	10 817	200	1	1 464
浙 江	635	522	116	112	154	39 592	35 285	10	0	1 822
安 徽	153	95	4	8	55	8 850	4 142	194	29	2 025
福 建	318	256	14	11	169	14 118	4 668	1 732	136	3 934
江 西	163	134	8	4	78	14 793	4 743	672	174	2 084
山 东	259	187	59	48	95	35 334	16 654	9 956	1 020	3 873
河 南	253	176	30	18	114	24 842	12 924	4 518	726	8 064
湖 北	88	67	14	12	39	10 642	5 793	390	153	2 394
湖 南	365	340	47	36	179	28 833	11 680	3 725	121	6 300
广 东	764	437	37	62	243	34 363	21 122	1 959	77	3 980
广 西	203	152	26	8	74	22 268	8 505	9 834	1 077	2 060
海 南	5	4	1	0	1	4 297	935	3 344	0	18
重 庆	76	59	5	2	37	7 140	2 345	854	0	2 997
四 川	293	203	31	23	130	12 658	6 066	1 990	21	1 948
贵 州	41	20	5	3	17	778	464	0	0	46
云 南	111	71	12	6	23	5 120	2 043	1 802	17	271
西 藏	1	1	0	0	1	32	9	0	0	0
陕 西	50	34	8	7	28	5 675	1 055	144	349	1 706
甘 肃	14	12	6	3	9	1 901	921	353	0	333
青 海	0	0	0	0	0	0	0	0	0	0
宁 夏	19	11	7	1	1	5 268	901	1 433	0	1
新 疆	115	70	19	17	58	2 203	444	300	0	375

各地区制浆及造纸行业污染排放及处理情况（二）

（2013）

地区名称	废水产生量/万吨	造纸生产线	化学浆生产线	化机浆生产线	废纸浆生产线	化学需氧量产生量/吨	造纸生产线	化学浆生产线	化机浆生产线	废纸浆生产线
全国	378 441	254 602	46 339	4 847	72 653	4 645 453	2 743 171	926 514	167 895	807 874
北京	83	83	0	0	0	323	276	0	0	47
天津	1 158	416	0	0	741	22 814	4 989	0	0	17 826
河北	27 013	16 397	548	5	10 063	231 154	139 679	4 888	108	86 479
山西	4 567	2 835	647	0	1 085	22 842	13 467	1 341	0	8 034
内蒙古	379	148	213	0	18	34 575	3 782	30 515	0	278
辽宁	4 344	2 837	407	83	1 017	35 389	15 592	7 633	5 927	6 236
吉林	5 640	3 621	1 236	29	754	95 033	47 310	23 608	12 112	12 003
黑龙江	2 995	2 632	255	40	68	24 132	18 001	5 245	289	598
上海	1 334	1 030	0	0	304	10 051	7 277	0	0	2 774
江苏	11 838	9 607	145	1	2 084	187 071	171 913	784	3	14 371
浙江	78 215	76 133	10	0	2 072	1 003 922	978 686	240	1	24 994
安徽	7 152	4 834	194	7	2 117	96 814	67 663	3 208	90	25 853
福建	30 629	17 690	2 625	126	10 188	225 802	115 321	16 633	5 965	87 884
江西	10 063	6 382	1 148	167	2 366	123 281	61 365	1 171	19 147	41 598
山东	35 406	17 996	11 254	1 442	4 714	573 874	227 387	187 476	38 616	120 395
河南	24 672	12 018	2 953	1 376	8 324	393 664	145 891	119 277	35 138	93 358
湖北	10 247	6 542	433	254	3 018	110 881	59 022	9 177	11 899	30 783
湖南	22 330	12 244	4 012	120	5 955	228 921	116 552	55 956	7 400	49 014
广东	34 211	26 062	1 672	18	6 459	390 668	303 388	18 872	1 104	67 304
广西	23 019	10 666	9 584	767	2 001	306 754	87 303	179 188	22 519	17 744
海南	2 911	625	2 283	0	3	54 499	8 932	45 539	0	28
重庆	7 599	3 277	853	0	3 469	123 919	21 159	67 570	0	35 190
四川	13 687	8 222	2 119	51	3 294	100 651	50 322	31 379	215	18 734
贵州	498	398	43	0	57	4 510	2 983	132	103	1 292
云南	10 140	8 093	1 715	14	318	35 183	12 206	18 264	864	3 849
西藏	36	36	0	0	0	4	4	0	0	0
陕西	3 591	1 599	123	349	1 520	70 519	27 411	5 455	6 394	31 259
甘肃	1 351	810	300	0	242	20 699	9 983	5 977	0	4 739
青海	0	0	0	0	0	0	0	0	0	0
宁夏	2 128	768	1 349	0	11	77 440	16 364	60 985	0	90
新疆	1 208	602	216	0	390	40 063	8 942	26 002	0	5 120

11

流域及入海陆源废水排放统计

ANNUAL STATISTIC REPORT ON ENVIRONMENT IN CHINA
2013

十大水系接纳工业废水及处理情况（一）

（2013）

流域	地区名称	工业废水排放量/万吨	直接排入环境的	排入污水处理厂的	工业废水处理量/万吨	废水治理设施数/套	废水治理设施治理能力/（万吨/日）	废水治理设施运行费用/万元
总计		**2 098 394**	**1 585 279**	**513 115**	**4 924 811**	**80 298**	**25 642**	**6 286 649.2**
辽河	内蒙古	10 982	7 596	3 386	11 570	224	54	10 885.6
	辽　宁	78 286	61 962	16 323	213 534	2 290	1 181	228 872.6
	吉　林	14 927	14 343	584	27 748	252	107	18 984.8
	合　计	104 194	83 901	20 293	252 851	2 766	1 342	258 743.0
海河	北　京	9 486	1 959	7 527	10 498	518	60	36 310.9
	天　津	18 692	7 015	11 677	29 161	1 088	151	171 400.4
	河　北	108 527	73 361	35 166	766 057	4 565	4 024	475 898.8
	山　西	17 656	15 912	1 745	36 795	1 088	220	60 809.7
	内蒙古	454	454	0	483	20	4	1 650.0
	山　东	26 862	18 784	8 078	34 950	583	320	67 676.7
	河　南	29 968	27 680	2 288	58 352	842	321	68 226.4
	合　计	211 645	145 165	66 480	936 295	8 704	5 100	881 972.9
淮河	江　苏	70 446	54 556	15 889	116 484	2 359	636	195 416.7
	安　徽	31 416	27 026	4 391	37 002	890	288	65 158.9
	山　东	145 084	89 580	55 504	253 777	4 265	1 224	424 004.4
	河　南	65 198	53 102	12 096	61 496	1 239	400	80 481.1
	合　计	312 143	224 263	87 880	468 759	8 753	2 549	765 061.1
松花江	内蒙古	6 947	4 309	2 638	10 690	164	47	10 432.6
	吉　林	27 725	21 568	6 157	34 059	408	159	49 492.1
	黑龙江	47 796	41 219	6 578	94 946	1 187	691	289 257.8
	合　计	82 468	67 096	15 372	139 695	1 759	898	349 182.5
珠江	福　建	2 685	2 460	225	3 785	216	40	21 134.4
	江　西	338	338	0	302	20	2	748.9
	湖　南	2 780	2 780	0	2 346	93	12	2 105.5
	广　东	170 463	133 568	36 895	293 037	9 918	1 506	562 850.4
	广　西	89 000	85 411	3 589	304 387	2 285	1 066	193 823.1
	海　南	6 744	5 775	969	6 533	309	53	33 453.5
	贵　州	5 946	5 941	5	9 156	406	157	12 128.0
	云　南	6 073	5 841	232	102 918	962	447	59 796.4
	合　计	284 028	242 114	41 915	722 464	14 209	3 284	886 040.2

十大水系接纳工业废水及处理情况（一）（续表）

（2013）

流域	地区名称		工业废水排放量/万吨	直接排入环境的	排入污水处理厂的	工业废水处理量/万吨	废水治理设施数/套	废水治理设施治理能力/（万吨/日）	废水治理设施运行费用/万元
长江	上	海	45 426	18 071	27 355	65 019	1 743	319	170 673.6
	江	苏	150 113	88 316	61 797	279 415	5 093	1 291	537 751.6
	浙	江	36 750	9 040	27 710	81 173	1 894	434	137 836.0
	安	徽	38 514	33 937	4 576	165 316	1 467	658	150 629.0
	福	建	0	0	0	0	0	0	0
	江	西	67 893	65 886	2 007	167 120	2 523	1 070	186 702.6
	河	南	7 593	7 593	0	13 590	239	55	17 124.0
	湖	北	84 993	71 869	13 124	181 540	2 084	1 029	168 277.0
	湖	南	89 531	86 561	2 970	261 054	2 943	1 142	186 443.3
	广	东	0	0	0	0	0	0	0
	广	西	509	467	42	524	55	2	265.1
	重	庆	33 451	32 206	1 245	34 451	1 669	243	61 117.8
	四	川	64 829	56 532	8 297	171 881	3 856	1 053	199 649.6
	贵	州	16 952	16 848	104	102 139	1 183	459	47 971.4
	云	南	7 142	6 779	363	26 598	765	146	34 512.6
	西	藏	0	0	0	0	0	0	0
	陕	西	4 791	4 654	137	12 712	438	97	14 486.9
	甘	肃	1 774	1 774	0	2 569	73	15	4 528.2
	青	海	0	0	0	0	0	0	0
	合	计	650 260	500 532	149 727	1 565 100	26 025	8 013	1 917 968.7
黄河	山	西	30 138	26 292	3 846	89 216	1 949	583	147 173.6
	内蒙古		14 345	4 755	9 590	82 198	632	399	82 424.7
	山	东	9 233	7 556	1 677	57 529	341	313	49 729.2
	河	南	28 030	23 754	4 276	39 614	986	251	66 051.8
	四	川	35	35	0	21	10	0	30.9
	陕	西	30 080	22 493	7 587	47 216	1 408	317	91 990.3
	甘	肃	9 536	6 891	2 645	19 008	326	103	19 786.0
	青	海	3 286	3 286	0	22 828	104	67	11 099.1
	宁	夏	15 708	14 274	1 434	23 902	367	175	35 309.2
	合	计	140 390	109 335	31 055	381 531	6 123	2 209	503 594.8

十大水系接纳工业废水及处理情况（一）（续表）

（2013）

流域	地名	区称	工业废水排放量/万吨	直接排入环境的	排入污水处理厂的	工业废水处理量/万吨	废水治理设施数/套	废水治理设施治理能力（万吨/日）	废水治理设施运行费用/万元
东南诸河	浙	江	126 924	53 979	72 945	175 684	6 389	991	435 147.5
	安	徽	1 042	834	208	856	87	5	1 274.1
	福	建	101 973	88 124	13 849	164 382	3 287	642	131 177.3
	合	计	229 939	142 937	87 002	340 922	9 763	1 638	567 598.9
西北诸河	河	北	1 349	1 088	262	265	8	5	1 131.7
	内蒙古		4 259	3 227	1 032	2 283	88	45	3 788.2
	西	藏	0	0	0	0	0	0	0
	甘	肃	8 861	8 763	98	9 239	197	53	14 009.9
	青	海	5 109	5 109	0	2 236	78	10	2 753.6
	新	疆	34 718	22 980	11 738	57 778	917	283	98 753.8
	合	计	54 297	41 167	13 130	71 802	1 288	397	120 437.2
西南诸河	云	南	28 629	28 369	260	44 671	875	205	35 337.8
	西	藏	400	400	0	722	33	7	712.1
	青	海	0	0	0	0	0	0	0
	合	计	29 029	28 769	260	45 393	908	212	36 049.9

十大水系接纳工业废水及处理情况（二）

（2013）

单位：吨

流域	地区名称	工业废水中污染物产生量				
		化学需氧量	氨氮	石油类	挥发酚	氰化物
总 计		22 148 280.8	1 391 192.7	261 789.7	69 996.5	5 772.0
辽河	内蒙古	90 205.2	4 622.3	342.8	227.2	4.8
	辽 宁	410 535.7	42 905.2	17 657.4	1 372.5	167.5
	吉 林	93 675.7	2 446.0	389.6	662.8	8.4
	合 计	594 416.6	49 973.5	18 389.8	2 262.6	180.7
海河	北 京	60 648.4	2 451.0	3 259.8	900.3	5.1
	天 津	153 419.7	8 553.4	969.0	2.8	1.5
	河 北	1 483 420.3	89 042.0	22 639.2	10 362.5	682.7
	山 西	294 775.1	16 650.0	2 844.1	3 828.0	110.1
	内蒙古	2 688.5	65.1	1.5	5.8	230.4
	山 东	363 799.3	11 315.5	879.3	305.7	17.1
	河 南	506 736.4	13 184.2	3 199.6	1 175.3	87.2
	合 计	2 865 487.7	141 261.1	33 792.5	16 580.4	1 134.2
淮河	江 苏	766 224.0	22 830.1	3 260.2	659.4	66.1
	安 徽	352 604.9	32 404.3	1 236.6	45.4	13.8
	山 东	1 780 962.8	84 117.5	23 017.0	4 143.9	200.3
	河 南	546 223.2	23 679.3	3 821.4	979.3	48.0
	合 计	3 446 014.8	163 031.2	31 335.2	5 827.9	328.2
松花江	内蒙古	191 716.6	20 118.5	200.3	5.2	0.4
	吉 林	295 349.9	7 349.0	2 211.4	220.0	10.6
	黑龙江	1 015 001.3	65 576.5	42 065.1	4 982.2	51.0
	合 计	1 502 067.8	93 044.0	44 476.8	5 207.5	62.0
珠江	福 建	11 231.0	900.1	120.5	5.4	3.4
	江 西	1 615.0	6 683.1	1.3	0	…
	湖 南	27 130.1	957.4	23.4	…	1.8
	广 东	1 343 356.8	50 014.9	4 711.9	185.8	666.4
	广 西	1 220 535.3	52 677.8	4 135.4	1 987.2	33.7
	海 南	94 608.4	2 352.5	108.3	…	…
	贵 州	39 901.2	4 000.9	972.3	405.6	14.7
	云 南	108 765.8	12 037.4	1 432.5	1 848.2	71.5
	合 计	2 847 143.6	129 624.2	11 505.6	4 432.2	791.5

十大水系接纳工业废水及处理情况（二）（续表）

（2013）

单位：吨

流域	地区名称	工业废水中污染物产生量				
		化学需氧量	氨氮	石油类	挥发酚	氰化物
长江	上 海	231 856.0	10 092.4	5 197.6	777.0	176.7
	江 苏	1 036 008.3	48 828.1	8 726.1	3 001.3	285.4
	浙 江	350 350.9	8 448.6	677.9	204.0	2.7
	安 徽	355 513.9	13 705.8	6 713.4	1 677.5	278.1
	福 建	0	0	0	0	0
	江 西	491 697.9	27 560.5	6 006.5	2 230.3	31.1
	河 南	278 476.1	3 523.4	650.0	0.7	1.0
	湖 北	527 492.4	31 890.4	3 799.9	1 059.8	51.2
	湖 南	584 207.2	57 892.3	5 652.0	989.7	211.6
	广 东	0	0	0	0	0
	广 西	2 363.1	41.8	…	…	…
	重 庆	327 201.4	24 216.3	2 922.9	962.0	91.8
	四 川	646 747.1	72 202.8	5 469.9	1 791.7	154.2
	贵 州	158 790.4	43 800.2	1 022.8	1 108.4	22.8
	云 南	75 357.2	21 054.9	614.4	659.1	24.4
	西 藏	0.4	…	…	…	…
	陕 西	51 775.8	5 533.2	142.8	1.7	0.7
	甘 肃	9 183.3	43.2	4.5	…	…
	青 海	0	0	0	0	0
	合 计	5 127 021.5	368 833.8	47 600.6	14 463.2	1 331.8
黄河	山 西	237 178.2	22 645.1	7 309.4	6 316.8	331.3
	内蒙古	236 368.7	45 747.8	3 384.5	2 103.2	53.9
	山 东	127 580.2	2 764.2	2 827.1	1 941.8	232.2
	河 南	264 262.8	17 501.3	7 503.0	605.3	23.5
	四 川	256.9	15.4	…	…	…
	陕 西	386 678.2	32 327.6	7 032.4	6 089.6	188.9
	甘 肃	109 674.6	16 665.1	2 232.8	179.7	2.7
	青 海	54 406.2	1 210.9	596.9	4.6	…
	宁 夏	401 115.2	62 372.5	2 873.1	615.0	26.8
	合 计	1 817 521.0	201 249.9	33 759.3	17 855.9	859.5

十大水系接纳工业废水及处理情况（二）（续表）

（2013）

单位：吨

流域	地区名称	工业废水中污染物产生量				
		化学需氧量	氨氮	石油类	挥发酚	氰化物
东南诸河	浙 江	1 791 647.9	69 786.8	23 497.5	194.8	751.2
	安 徽	5 216.1	215.7	16.6	…	…
	福 建	764 888.5	34 205.7	4 108.4	380.7	107.0
	合 计	2 561 752.5	104 208.2	27 622.5	575.4	858.2
西北诸河	河 北	8 534.3	159.4	5.5	…	…
	内蒙古	31 888.4	2 222.1	810.9	811.1	42.6
	西 藏	0	0	0	0	0
	甘 肃	79 511.0	7 568.3	206.6	83.2	1.1
	青 海	32 527.7	1 864.2	764.8	1.5	9.0
	新 疆	750 412.1	124 584.5	11 126.3	1 890.3	191.6
	合 计	902 873.6	136 398.6	13 749.3	2 786.1	244.2
西南诸河	云 南	479 488.2	3 423.7	389.3	5.3	10.1
	西 藏	4 493.5	144.7	4.0	…	…
	青 海	0	0	0	0	0
	合 计	483 981.7	3 568.3	393.3	5.3	10.1

十大水系接纳工业废水及处理情况（三）

（2013）

单位：吨

流域	地区名称	工业废水中污染物产生量					
		汞	镉	六价铬	总铬	铅	砷
总　计		20. 936	2 243. 096	3 250. 531	6 990. 343	2 962. 439	11 228. 723
辽河	内蒙古	0. 098	65. 055	0. 016	0. 018	65. 016	105. 840
	辽　宁	0. 184	40. 791	20. 523	30. 044	40. 812	38. 335
	吉　林	0. 203	0. 019	0. 014	0. 018	0. 120	0. 032
	合　计	0. 485	105. 866	20. 554	30. 081	105. 948	144. 207
海河	北　京	…	0. 001	11. 643	12. 328	0. 216	3. 291
	天　津	0. 077	0. 065	27. 576	37. 818	2. 483	0. 125
	河　北	0. 040	1. 289	309. 374	375. 532	9. 337	42. 062
	山　西	0. 494	0. 750	2. 372	2. 718	0. 031	…
	内蒙古	…	…	19. 362	19. 362	0. 010	0. 005
	山　东	0. 121	247. 212	1. 235	16. 322	242. 331	640. 121
	河　南	0. 368	89. 122	3. 064	15. 414	47. 568	22. 762
	合　计	1. 100	338. 439	374. 626	479. 495	301. 975	708. 367
淮河	江　苏	0. 002	0. 035	68. 172	111. 654	30. 431	0. 942
	安　徽	…	0. 004	2. 741	5. 797	2. 228	2. 101
	山　东	0. 924	39. 632	41. 663	88. 671	40. 992	278. 804
	河　南	0. 168	1. 210	5. 289	110. 930	2. 717	8. 273
	合　计	1. 094	40. 881	117. 866	317. 052	76. 368	290. 120
松花江	内蒙古	0. 079	0. 139	0. 005	0. 072	1. 384	2. 133
	吉　林	0. 004	0. 012	4. 184	4. 195	0. 041	2. 218
	黑龙江	0. 567	0. 130	7. 555	7. 602	1. 352	9. 433
	合　计	0. 650	0. 282	11. 744	11. 868	2. 777	13. 785
珠江	福　建	0. 007	3. 164	0. 706	0. 713	18. 687	0. 399
	江　西	0. 004	0. 057	0. 038	0. 113	2. 843	0. 326
	湖　南	0. 004	0. 551	…	4. 477	12. 067	7. 018
	广　东	0. 105	1. 645	146. 994	1 558. 396	15. 981	10. 053
	广　西	2. 370	73. 052	21. 050	52. 971	86. 853	129. 333
	海　南	…	0. 014	…	15. 119	0. 004	0. 005
	贵　州	0. 022	0. 032	0. 983	0. 983	0. 359	0. 615
	云　南	0. 060	23. 715	2. 949	2. 994	28. 544	103. 226
	合　计	2. 572	102. 230	172. 721	1 635. 766	165. 337	250. 974

十大水系接纳工业废水及处理情况（三）（续表）

（2013）

单位：吨

流域	地区名称	工业废水中污染物产生量					
		汞	镉	六价铬	总铬	铅	砷
长江	上　海	0.010	0.070	69.586	89.702	0.740	0.202
	江　苏	0.024	2.332	167.138	285.722	3.961	4.983
	浙　江	…	1.427	37.102	162.787	3.477	0.534
	安　徽	0.004	16.173	12.696	14.657	58.425	654.612
	福　建	0	0	0	0	0	0
	江　西	0.927	462.557	44.865	53.526	109.400	5 707.129
	河　南	0.001	0.005	0.679	9.828	0.121	0.127
	湖　北	0.259	7.563	40.241	51.941	18.627	115.430
	湖　南	2.787	162.690	33.633	46.239	378.059	333.133
	广　东	0	0	0	0	0	0
	广　西	…	…	…	…	0.004	0.013
	重　庆	…	0.028	107.302	146.155	7.410	1.646
	四　川	0.569	40.018	59.450	128.031	46.231	378.892
	贵　州	0.071	0.399	0.312	2.045	0.419	20.992
	云　南	0.535	157.416	0.091	0.707	219.691	730.036
	西　藏	0	0	0	0	0	0
	陕　西	0.313	66.402	3.684	3.839	85.355	33.842
	甘　肃	0.011	5.149	0.001	0.001	7.845	8.005
	青　海	0	0	0	0	0	0
	合　计	5.513	922.231	576.779	995.179	939.764	7 989.576
黄河	山　西	0.009	12.514	10.362	10.575	9.345	54.845
	内蒙古	1.914	15.842	0.865	0.964	37.654	119.370
	山　东	0.082	0.001	2.972	3.251	0.115	0.006
	河　南	0.999	179.650	5.634	48.414	315.686	88.550
	四　川	0	0	0	0	0	0
	陕　西	0.049	2.035	10.570	12.614	2.790	20.425
	甘　肃	0.059	23.125	0.446	28.795	26.820	119.285
	青　海	0.503	12.401	0.043	0.054	11.957	59.845
	宁　夏	0.711	0.086	0.799	0.958	1.296	73.280
	合　计	4.326	245.653	31.690	105.625	405.663	535.606

十大水系接纳工业废水及处理情况（三）（续表）

单位：吨

流域	地区名称	工业废水中污染物产生量					
		汞	镉	六价铬	总铬	铅	砷
东南诸河	浙江	0.006	2.583	1 543.020	2 628.876	4.852	31.998
	安徽	0.008	0.033	0.087	0.457	0.043	0.012
	福建	0.070	0.745	391.478	754.929	7.855	34.114
	合 计	0.084	3.361	1 934.585	3 384.263	12.750	66.124
西北诸河	河北	0	0	0	0	0	0
	内蒙古	0.324	12.239	0.013	0.013	11.086	15.210
	西藏	0	0	0	0	0	0
	甘肃	0.195	305.458	1.851	7.329	659.243	951.431
	青海	0.006	7.276	…	…	10.088	4.600
	新疆	4.214	2.001	7.538	17.232	5.248	20.449
	合 计	4.739	326.974	9.402	24.574	685.665	991.690
西南诸河	云南	0.314	157.150	0.064	0.078	265.447	222.512
	西藏	0.059	0.029	…	6.362	0.744	15.762
	青海	0	0	0	0	0	0
	合 计	0.372	157.179	0.064	6.440	266.191	238.274

十大水系接纳工业废水及处理情况（四）

（2013）

单位：吨

流域	地区名称	工业废水中污染物排放量				
		化学需氧量	氨氮	石油类	挥发酚	氰化物
总　计		**3 194 538.6**	**245 840.5**	**17 389.2**	**1 259.1**	**162.0**
辽河	内蒙古	29 296.2	2 935.9	137.3	…	…
	辽　宁	84 692.0	7 133.1	951.5	7.2	3.2
	吉　林	16 513.8	907.5	119.0	2.2	2.1
	合　计	**130 502.0**	**10 976.5**	**1 207.8**	**9.3**	**5.2**
海河	北　京	6 055.0	330.4	48.3	0.1	0.1
	天　津	26 214.7	3 338.7	118.2	1.8	1.0
	河　北	172 311.9	14 228.3	884.0	26.2	8.6
	山　西	30 495.2	3 510.2	496.3	317.6	8.4
	内蒙古	670.1	13.7	0.3	…	…
	山　东	20 283.5	1 501.5	31.4	4.6	0.5
	河　南	42 954.8	2 611.2	122.5	0.8	4.0
	合　计	**298 985.2**	**25 533.9**	**1 701.0**	**351.2**	**22.5**
淮河	江　苏	100 708.6	7 412.1	624.4	21.9	6.0
	安　徽	35 372.1	4 935.7	410.8	2.5	2.0
	山　东	105 153.1	8 509.9	432.1	32.8	3.1
	河　南	81 889.1	6 012.4	809.7	128.7	11.4
	合　计	**323 122.9**	**26 870.2**	**2 276.9**	**186.0**	**22.5**
松花江	内蒙古	20 406.2	1 085.6	49.7	5.2	…
	吉　林	48 643.1	3 310.2	210.6	2.1	0.3
	黑龙江	96 123.3	5 977.1	289.6	6.4	1.5
	合　计	**165 172.6**	**10 372.9**	**549.9**	**13.6**	**1.8**
珠江	福　建	2 189.5	523.0	21.7	…	0.1
	江　西	318.0	218.6	0.1	…	…
	湖　南	10 826.6	706.1	8.4	…	0.7
	广　东	234 281.2	14 599.3	643.2	12.5	9.0
	广　西	174 104.7	7 198.8	261.9	12.8	4.9
	海　南	12 526.0	913.2	8.5	…	…
	贵　州	17 621.9	549.6	138.8	0.1	0.2
	云　南	28 067.7	1 794.9	57.3	0.9	0.7
	合　计	**479 935.6**	**26 503.4**	**1 140.0**	**26.4**	**15.5**

十大水系接纳工业废水及处理情况（四）（续表）
（2013）

单位：吨

流域	地 区 名 称	工业废水中污染物排放量				
		化学需氧量	氨氮	石油类	挥发酚	氰化物
长 江	上 海	25 503.2	1 934.0	619.6	2.0	3.0
	江 苏	108 466.3	6 981.3	690.2	18.3	8.6
	浙 江	33 131.0	3 207.2	30.4	0.8	0.2
	安 徽	50 015.7	2 597.5	336.1	3.6	5.1
	福 建	0	0	0	0	0
	江 西	92 948.0	8 768.9	728.2	14.1	7.4
	河 南	11 659.1	1 226.2	33.3	0.1	…
	湖 北	128 087.6	13 608.9	1 002.6	17.8	6.8
	湖 南	129 862.1	22 340.5	566.9	17.9	9.5
	广 东	0	0	0	0	0
	广 西	1 312.4	22.1	…	…	…
	重 庆	51 533.7	3 265.7	335.8	9.5	1.5
	四 川	106 412.2	4 969.0	634.2	3.3	2.3
	贵 州	45 592.3	2 959.2	296.5	0.2	0.4
	云 南	14 627.4	644.4	117.5	0.3	0.4
	西 藏	0.4	…	…	…	…
	陕 西	14 150.6	2 417.8	17.7	…	0.1
	甘 肃	3 150.7	16.9	4.5	…	…
	青 海	0	0	0	0	0
	合 计	816 452.7	74 959.6	5 413.9	88.0	45.2
黄 河	山 西	48 311.7	4 108.3	484.5	303.2	20.9
	内蒙古	24 738.0	5 578.3	587.4	3.9	1.3
	山 东	7 290.3	212.5	61.8	1.2	0.1
	河 南	34 179.6	2 448.9	260.3	9.6	2.1
	四 川	80.3	6.0	…	…	…
	陕 西	81 249.3	5 928.7	640.0	2.5	3.9
	甘 肃	41 119.0	7 258.5	207.9	0.9	…
	青 海	25 407.9	757.2	64.9	0.5	…
	宁 夏	103 083.3	8 540.8	153.0	6.8	2.7
	合 计	365 459.5	34 839.3	2 459.9	328.6	31.1

十大水系接纳工业废水及处理情况（四）（续表）

（2013）

单位：吨

流域	地区名称	工业废水中污染物排放量				
		化学需氧量	氨氮	石油类	挥发酚	氰化物
东南诸河	浙 江	140 952.2	7 844.7	711.1	5.8	6.4
	安 徽	1 747.2	118.4	3.2	…	…
	福 建	78 979.5	5 627.0	478.9	1.8	4.3
	合 计	221 678.9	13 590.0	1 193.2	7.6	10.7
西北诸河	河 北	2 127.2	42.5	5.5	…	…
	内蒙古	16 866.7	1 769.6	164.9	231.3	0.3
	西 藏	0	0	0	0	0
	甘 肃	46 210.9	5 360.6	81.9	1.2	…
	青 海	16 663.1	1 318.3	276.0	0.7	…
	新 疆	185 914.7	11 788.5	727.5	15.1	7.1
	合 计	267 782.6	20 279.5	2 090.8	248.3	7.5
西南诸河	云 南	124 826.7	1 869.2	190.6	0.1	…
	西 藏	619.9	46.0	0.5	…	…
	青 海	0	0	0	0	0
	合 计	125 446.6	1 915.2	191.1	0.1	…

十大水系接纳工业废水及处理情况（五）

（2013）

单位：吨

流域	地区名称	工业废水中污染物排放量					
		汞	镉	六价铬	总铬	铅	砷
	总　计	0.783	17.875	58.139	161.863	74.084	111.607
辽河	内蒙古	0.034	0.064	…	…	1.009	1.203
	辽　宁	0.004	0.020	0.235	0.643	0.166	0.228
	吉　林	0.001	0.001	0.014	0.018	0.036	0.021
	合　计	0.039	0.085	0.250	0.661	1.210	1.452
海河	北　京	…	0.001	0.320	0.422	0.054	0.012
	天　津	0.006	0.003	0.133	0.356	0.086	0.012
	河　北	…	0.003	2.588	5.352	0.230	0.054
	山　西	0.002	0.750	0.006	0.086	0.004	…
	内蒙古	…	…	…	…	0.010	0.005
	山　东	0.007	0.978	0.030	6.215	0.445	1.835
	河　南	…	0.100	0.090	2.718	0.464	0.059
	合　计	0.015	1.835	3.168	15.149	1.292	1.977
淮河	江　苏	0.001	0.003	0.960	3.305	0.834	0.008
	安　徽	…	0.001	0.098	0.407	0.272	2.096
	山　东	0.008	0.057	0.479	1.562	0.395	0.548
	河　南	0.002	0.402	0.686	6.561	1.454	0.616
	合　计	0.012	0.463	2.223	11.836	2.956	3.268
松花江	内蒙古	0.037	0.020	0.005	0.072	0.680	1.035
	吉　林	0.001	0.003	0.043	0.065	0.010	0.546
	黑龙江	…	0.001	0.097	0.108	0.025	0.024
	合　计	0.039	0.025	0.145	0.245	0.714	1.605
珠江	福　建	0.002	0.053	0.008	0.012	0.620	0.137
	江　西	…	0.030	0.006	0.064	0.118	0.022
	湖　南	0.001	0.061	…	4.477	1.808	1.781
	广　东	0.012	0.352	5.170	22.098	2.507	1.022
	广　西	0.105	1.067	0.146	1.755	6.257	8.283
	海　南	…	0.001	…	0.122	0.003	0.001
	贵　州	0.016	0.017	0.005	0.005	0.222	0.440
	云　南	…	0.123	0.003	0.017	0.205	0.206
	合　计	0.137	1.704	5.338	28.551	11.741	11.893

十大水系接纳工业废水及处理情况（五）（续表）

（2013）

单位：吨

流域	地 区 名 称	工业废水中污染物排放量					
		汞	镉	六价铬	总铬	铅	砷
长 江	上 海	0.008	0.001	1.067	2.381	0.105	0.043
	江 苏	0.001	0.011	3.043	5.523	0.218	0.206
	浙 江	…	0.014	0.479	3.516	0.152	0.005
	安 徽	0.001	0.097	0.154	0.214	1.161	2.689
	福 建	0	0	0	0	0	0
	江 西	0.071	1.922	17.180	17.389	5.449	9.281
	河 南	…	0.001	0.031	0.351	0.030	0.028
	湖 北	0.021	0.508	9.043	9.845	2.525	6.063
	湖 南	0.224	6.640	1.005	6.786	22.395	40.738
	广 东	0	0	0	0	0	0
	广 西	0	0	0	0	0	0
	重 庆	…	0.003	0.218	0.423	0.081	0.033
	四 川	0.010	0.075	0.729	1.794	1.079	2.067
	贵 州	0.001	0.004	0.024	0.040	0.016	0.037
	云 南	0.001	0.774	…	0.039	2.894	3.108
	西 藏	0	0	0	0	0	0
	陕 西	0.024	0.523	0.015	0.020	1.257	0.489
	甘 肃	0.006	0.150	0.001	0.001	2.232	1.291
	青 海	0	0	0	0	0	0
	合 计	0.368	10.723	32.990	48.321	39.594	66.077
黄 河	山 西	0.009	0.030	0.135	0.147	0.169	0.192
	内蒙古	0.014	0.281	0.029	0.047	1.235	2.089
	山 东	…	…	…	0.039	…	…
	河 南	0.011	0.438	0.079	20.414	1.996	0.473
	四 川	0	0	0	0	0	0
	陕 西	0.001	0.033	0.207	1.616	0.155	0.210
	甘 肃	0.036	0.239	0.239	0.592	2.531	0.683
	青 海	0.007	0.269	0.007	0.011	0.271	0.758
	宁 夏	0.005	…	0.126	0.237	0.010	0.050
	合 计	0.083	1.290	0.821	23.103	6.366	4.455

十大水系接纳工业废水及处理情况（五）（续表）

（2013）

单位：吨

流域	地区名称	工业废水中污染物排放量					
		汞	镉	六价铬	总铬	铅	砷
东南诸河	浙 江	0.001	0.180	8.309	15.079	0.241	0.259
	安 徽	0.001	0.007	0.002	0.016	0.010	0.012
	福 建	0.010	0.180	3.081	11.383	2.449	1.281
	合 计	0.013	0.367	11.392	26.478	2.700	1.551
西北诸河	河 北	0	0	0	0	0	0
	内蒙古	0.004	0.152	0.003	0.004	0.742	1.713
	西 藏	0	0	0	0	0	0
	甘 肃	0.052	0.897	0.175	5.534	3.286	2.306
	青 海	0.001	0.029	…	…	0.355	0.721
	新 疆	0.016	0.011	1.631	1.974	0.072	0.825
	合 计	0.074	1.090	1.809	7.513	4.456	5.565
西南诸河	云 南	0.005	0.293	0.003	0.008	3.055	4.821
	西 藏	…	…	…	…	…	8.942
	青 海	0	0	0	0	0	0
	合 计	0.005	0.293	0.003	0.008	3.055	13.764

十大水系工业废水污染防治投资情况
（2013）

单位：个

流域	地 区 名 称	工业废水治理施工项目数/个	工业废水治理竣工项目数/个	工业废水治理项目完成投资/万元	废水治理竣工项目新增处理能力/（万吨/日）
	总　计	1 490	1 649	1 248 822.3	345.9
辽河	内蒙古	5	4	10 760.0	0.5
	辽　宁	3	13	25 596.3	7.0
	吉　林	8	7	6 457.0	0.2
	合　计	16	24	42 813.3	7.7
海河	北　京	7	9	8 427.9	3.2
	天　津	14	13	6 435.8	3.4
	河　北	41	50	59 636.9	7.3
	山　西	10	15	17 713.5	3.9
	内蒙古	0	0	0	0
	山　东	4	7	5 558.1	1.8
	河　南	10	27	26 649.0	4.8
	合　计	86	121	124 421.2	24.4
淮河	江　苏	34	40	47 036.0	13.2
	安　徽	13	25	14 126.0	2.5
	山　东	75	93	91 028.9	30.9
	河　南	13	20	8 751.0	1.0
	合　计	135	178	160 941.9	47.7
松花江	内蒙古	4	3	21 572.5	…
	吉　林	5	5	3 027.6	3.4
	黑龙江	8	11	17 508.7	2.4
	合　计	17	19	42 108.9	5.9
珠江	福　建	5	5	1 001.2	1.1
	江　西	0	0	0	0
	湖　南	2	0	613.5	0.6
	广　东	152	184	43 911.7	20.0
	广　西	52	52	66 085.2	25.2
	海　南	2	2	572.0	…
	贵　州	7	15	2 786.3	2.3
	云　南	29	24	11 300.0	9.2
	合　计	249	282	126 269.9	58.5

十大水系工业废水污染防治投资情况（续表）

（2013）

单位：个

流域	地名	区称	工业废水治理施工项目数/个	工业废水治理竣工项目数/个	工业废水治理项目完成投资/万元	废水治理竣工项目新增处理能力/（万吨/日）
	上	海	30	28	7 809.1	2.8
	江	苏	135	134	55 464.6	10.3
	浙	江	57	56	23 098.7	10.3
	安	徽	15	17	5 001.3	14.2
	福	建	0	0	0	0
	江	西	17	22	32 334.1	2.6
	河	南	5	9	10 416.9	4.4
	湖	北	28	42	15 872.6	5.6
	湖	南	33	37	52 408.8	21.5
长	广	东	0	0	0	0
江	广	西	0	1	150.0	0.2
	重	庆	19	27	6 399.1	0.5
	四	川	73	85	29 791.4	22.5
	贵	州	13	19	20 080.7	3.3
	云	南	15	15	3 961.4	1.1
	西	藏	0	0	0	0
	陕	西	2	3	2 629.7	0.3
	甘	肃	9	6	6 560.1	0.9
	青	海	0	0	0	0
	合	计	451	501	271 978.4	100.5
	山	西	43	52	25 301.0	14.8
	内蒙古		16	18	20 794.1	2.2
	山	东	7	6	4 693.7	3.8
	河	南	9	6	2 294.9	0.7
黄	四	川	0	0	0	0
河	陕	西	23	36	57 996.5	7.0
	甘	肃	6	6	8 663.2	0.9
	青	海	1	1	2 985.0	...
	宁	夏	15	8	18 947.0	5.1
	合	计	120	133	141 675.4	34.6

十大水系工业废水污染防治投资情况（续表）

（2013）

单位：个

流域	地区名称	工业废水治理施工项目数/个	工业废水治理竣工项目数/个	工业废水治理项目完成投资/万元	废水治理竣工项目新增处理能力/（万吨/日）
东南诸河	浙 江	275	230	127 535.5	26.4
	安 徽	0	0	0	0
	福 建	78	89	138 469.4	27.2
	合 计	353	319	266 004.9	53.6
西北诸河	河 北	0	0	0	0
	内蒙古	1	1	550.0	…
	西 藏	0	0	0	0
	甘 肃	10	11	6 559.4	0.9
	青 海	0	0	0	0
	新 疆	17	18	37 086.5	4.7
	合 计	28	30	44 195.9	5.6
西南诸河	云 南	28	26	19 962.7	6.8
	西 藏	7	16	8 450.0	0.7
	青 海	0	0	0	0
	合 计	35	42	28 412.7	7.6

十大水系农业污染排放情况

（2013）

单位：万吨

流域	地区名称	农业污染物排放（流失）总量			
		化学需氧量	氨氮	总氮	总磷
总　计		1 125.8	77.9	425.0	50.0
辽河	内蒙古	16.1	0.4	3.6	0.4
	辽　宁	85.0	3.4	18.6	2.6
	吉　林	11.6	0.5	2.6	0.3
	合　计	112.7	4.2	24.9	3.3
海河	北　京	7.5	0.5	2.6	0.3
	天　津	11.0	0.6	2.8	0.3
	河　北	88.6	4.3	36.6	4.4
	山　西	6.9	0.4	2.9	0.3
	内蒙古	2.7	…	0.6	…
	山　东	34.3	1.7	21.1	2.0
	河　南	15.5	1.1	7.9	1.2
	合　计	166.5	8.5	74.5	8.6
淮河	江　苏	27.7	2.6	12.4	1.4
	安　徽	25.2	2.5	12.0	1.2
	山　东	86.9	5.0	32.4	4.4
	河　南	46.3	3.8	22.6	2.7
	合　计	186.0	13.9	79.4	9.7
松花江	内蒙古	15.8	0.3	3.7	0.3
	吉　林	37.6	1.2	8.8	1.1
	黑龙江	104.0	3.3	24.4	2.4
	合　计	157.4	4.9	36.9	3.9
珠江	江　西	3.4	0.4	1.1	0.1
	湖　南	0.5	0.1	0.2	…
	广　东	0.4	0.1	0.2	…
	广　西	58.1	5.5	18.4	2.8
	贵　州	20.8	2.6	10.9	1.3
	云　南	10.1	0.9	2.9	0.4
	海　南	1.0	0.2	1.1	0.1
	福　建	2.4	0.4	1.9	0.2
	合　计	96.8	10.0	36.7	5.0

十大水系农业污染排放情况（续表）

（2013）

单位：万吨

流域	地区名称	农业污染物排放（流失）总量			
		化学需氧量	氨氮	总氮	总磷
长江	上　海	3.1	0.3	1.5	0.2
	江　苏	9.9	1.2	5.5	0.7
	浙　江	5.8	0.7	2.5	0.3
	安　徽	11.5	1.1	4.9	0.7
	福　建	0	0	0	0
	江　西	22.7	2.8	9.2	1.2
	河　南	6.9	0.5	4.5	0.5
	湖　北	46.1	4.5	19.2	2.5
	湖　南	55.3	6.1	19.0	2.2
	广　东	0	0	0	0
	广　西	0.3	0.1	0.2	...
	重　庆	12.1	1.3	5.2	0.7
	四　川	52.9	5.6	21.7	2.4
	贵　州	5.2	0.6	3.5	0.4
	云　南	2.1	0.4	2.6	0.2
	西　藏	0	0	0	0
	陕　西	4.1	0.6	2.3	0.3
	甘　肃	2.3	0.1	0.8	0.1
	青　海	0	0	0	0
	合　计	240.4	25.9	102.7	12.2
黄河	山　西	10.6	0.8	5.3	0.6
	内蒙古	15.0	0.3	4.2	0.3
	山　东	8.3	0.4	3.1	0.4
	河　南	9.5	0.7	4.3	0.5
	四　川	0	0	0	0
	陕　西	15.1	0.9	7.0	0.6
	甘　肃	6.2	0.3	2.3	0.2
	青　海	1.9	0.1	0.6	0.1
	宁　夏	10.1	0.2	3.0	0.3
	合　计	76.8	3.6	29.7	3.0

十大水系农业污染排放情况（续表）

（2013）

单位：万吨

流域	地 区名 称	农业污染物排放（流失）总量			
		化学需氧量	氨氮	总氮	总磷
东南诸河	浙 江	13.9	1.9	6.7	0.8
	安 徽	0.4	0.1	0.2	…
	福 建	17.4	2.8	8.5	1.3
	合 计	31.7	4.7	15.4	2.2
西北诸河	河 北	0.9	0	0.6	…
	内蒙古	11.2	0.1	2.4	0.2
	西 藏	0	0	0	0
	甘 肃	5.7	0.2	1.6	0.1
	青 海	0.3	…	0.1	…
	新 疆	36.4	1.3	16.5	1.3
	合 计	54.5	1.6	21.1	1.7
西南诸河	云 南	2.7	0.4	3.1	0.3
	西 藏	0.4	…	0.6	…
	青 海	0	0	0	0
	合 计	3.1	0.5	3.7	0.4

十大水系城镇生活污染排放及处理情况（一）

（2013）

流域	地 区 名 称	城镇人口/ 万人	生活用水总量/ 万吨	污水处理厂数/ 座	污水处理厂设计 处理能力/ （万吨/日）	生活污水实际 处理量/ 万吨
	总　计	73 726	5 725 663	5 364	16 574	3 879 495
辽 河	内蒙古	322	19 369	26	86	14 156
	辽　宁	2 917	187 027	158	771	133 544
	吉　林	314	18 467	12	48	12 117
	合　计	3 553	224 864	196	906	159 817
海 河	北　京	1 825	158 674	138	434	126 043
	天　津	1 214	73 554	60	272	59 890
	河　北	3 492	222 776	252	843	165 737
	山　西	654	41 622	67	118	28 049
	内蒙古	28	1 534	6	7	866
	山　东	668	41 898	54	187	38 020
	河　南	726	54 334	39	168	42 236
	合　计	8 607	594 392	616	2 028	460 841
淮 河	江　苏	2 225	165 400	218	427	94 095
	安　徽	1 478	98 821	91	229	61 443
	山　东	4 176	293 422	322	1 123	231 268
	河　南	2 381	189 737	114	474	138 596
	合　计	10 260	747 380	745	2 254	525 402
松 花 江	内蒙古	258	15 362	21	52	7 221
	吉　林	1 177	69 996	47	240	49 114
	黑龙江	2 188	129 816	107	348	71 506
	合　计	3 623	215 174	175	640	127 842
珠 江	福　建	109	10 333	6	24	7 306
	江　西	18	1 248	2	2	610
	湖　南	34	2 564	2	3	887
	广　东	7 277	833 353	471	2 075	590 852
	广　西	2 088	162 049	113	379	98 627
	海　南	472	38 275	56	108	21 087
	贵　州	188	11 239	28	22	5 271
	云　南	507	29 625	27	65	15 167
	合　计	10 693	1 088 687	705	2 678	739 807

十大水系城镇生活污染排放及处理情况（一）（续表）

（2013）

流域	地区名称	城镇人口/万人	生活用水总量/万吨	污水处理厂数/座	污水处理厂设计处理能力/（万吨/日）	生活污水实际处理量/万吨
	上 海	2 156	196 900	56	771	151 305
	江 苏	2 871	266 805	408	1 071	205 782
	浙 江	717	60 086	77	330	43 406
	安 徽	1 372	119 120	100	299	90 146
	福 建	0	0	0	0	0
	江 西	2 184	162 460	135	337	93 044
	河 南	397	28 018	14	50	14 777
	湖 北	3 161	244 543	172	666	173 258
	湖 南	3 177	247 646	159	540	152 094
长	广 东	0	0	0	0	0
江	广 西	27	2 139	4	4	733
	重 庆	1 754	127 455	149	286	92 771
	四 川	3 682	291 527	340	574	166 658
	贵 州	1 165	77 149	90	154	48 730
	云 南	743	72 488	42	164	51 354
	西 藏	1	57	0	0	0
	陕 西	333	18 340	28	48	8 687
	甘 肃	67	3 010	4	3	598
	青 海	9	530	0	0	0
	合 计	23 813	1 918 272	1 778	5 297	1 293 341
	山 西	1 260	74 625	118	213	45 827
	内蒙古	705	41 228	55	157	28 166
	山 东	389	27 133	30	67	19 161
	河 南	679	52 433	64	161	41 651
黄	四 川	3	176	0	0	0
河	陕 西	1 598	101 482	117	298	65 381
	甘 肃	767	40 680	39	114	17 666
	青 海	222	11 952	14	31	7 757
	宁 夏	340	26 601	20	75	17 653
	合 计	5 964	376 309	457	1 115	243 263

十大水系城镇生活污染排放及处理情况（一）（续表）

（2013）

流域	地区名称	城镇人口/万人	生活用水总量/万吨	污水处理厂数/座	污水处理厂设计处理能力/（万吨/日）	生活污水实际处理量/万吨
东南诸河	浙 江	2 803	241 629	284	768	149 681
	安 徽	63	4 923	13	15	4 428
	福 建	2 185	168 702	224	442	104 076
	合 计	5 051	415 254	521	1 224	258 186
西北诸河	河 北	25	1 319	3	6	1 123
	内蒙古	154	7 605	26	31	3 076
	西 藏	2	97	0	0	0
	甘 肃	213	11 755	23	66	7 796
	青 海	46	4 066	5	8	2 401
	新 疆	1 013	79 311	67	243	41 957
	合 计	1 453	104 153	124	355	56 353
西南诸河	云 南	609	35 842	45	62	12 605
	西 藏	97	5 271	2	14	2 038
	青 海	4	66	0	0	0
	合 计	710	41 178	47	76	14 643

十大水系城镇生活污染排放及处理情况（二）

（2013）

流域	地区名称	城镇生活污水排放量/万吨	城镇生活化学需氧量产生量/吨	城镇生活氨氮产生量/吨	城镇生活化学需氧量排放量/吨	城镇生活氨氮排放量/吨
	总　计	4 851 058	17 927 151	2 277 511	8 898 128	1 413 567
辽河	内蒙古	16 676	75 121	9 390	21 863	7 229
	辽　宁	156 106	676 907	91 595	312 225	61 720
	吉　林	15 786	71 398	9 411	44 089	7 567
	合　计	188 568	823 426	110 396	378 177	76 516
海河	北　京	134 991	527 173	64 729	89 868	14 189
	天　津	65 469	304 702	41 599	84 886	15 671
	河　北	199 834	754 669	101 498	230 879	48 759
	山　西	33 762	142 461	19 546	62 579	11 568
	内蒙古	1 330	6 307	817	2 576	484
	山　东	39 631	159 323	20 502	60 516	10 344
	河　南	49 183	161 355	20 480	60 011	11 575
	合　计	524 199	2 055 990	269 172	591 315	112 590
淮河	江　苏	145 526	558 649	72 911	366 933	52 781
	安　徽	89 788	346 160	41 104	240 056	30 478
	山　东	252 953	983 086	132 458	311 959	62 197
	河　南	161 882	525 289	66 234	246 979	41 885
	合　计	650 149	2 413 184	312 706	1 165 927	187 340
松花江	内蒙古	12 770	56 465	7 670	32 034	5 357
	吉　林	59 194	274 979	35 335	148 350	24 647
	黑龙江	105 244	481 048	64 340	309 467	48 229
	合　计	177 207	812 492	107 345	489 852	78 233
珠江	福　建	8 838	26 558	3 449	14 485	2 488
	江　西	1 060	4 396	517	3 700	432
	湖　南	1 859	8 179	991	7 169	907
	广　东	691 295	1 801 900	234 873	903 799	144 901
	广　西	133 823	506 271	60 448	364 345	46 776
	海　南	29 374	126 574	15 083	79 316	12 960
	贵　州	8 764	38 680	4 917	30 467	4 019
	云　南	24 171	130 289	14 999	94 648	12 143
	合　计	899 184	2 642 847	335 277	1 497 929	224 626

十大水系城镇生活污染排放及处理情况（二）（续表）

（2013）

流域	地名 区称 称	城镇生活污水排放量/万吨	城镇生活化学需氧量产生量/吨	城镇生活氨氮产生量/吨	城镇生活化学需氧量排放量/吨	城镇生活氨氮排放量/吨
	上 海	177 210	597 936	76 315	173 181	40 322
	江 苏	228 000	755 303	95 801	191 765	41 473
	浙 江	51 639	216 294	25 639	52 273	13 135
	安 徽	100 683	330 984	38 967	186 000	26 266
	福 建	0	0	0	0	0
	江 西	137 557	561 892	62 082	395 303	49 926
	河 南	24 075	78 131	9 817	44 014	6 666
	湖 北	208 836	782 731	93 134	455 112	64 560
长	湖 南	212 651	758 300	92 052	531 729	71 280
	广 东	0	0	0	0	0
江	广 西	1 818	5 859	792	4 954	612
	重 庆	108 937	435 221	55 683	218 601	36 211
	四 川	242 417	938 663	114 530	590 444	75 356
	贵 州	61 347	254 210	32 605	167 602	22 229
	云 南	61 585	190 559	21 883	60 832	11 509
	西 藏	49	306	36	269	34
	陕 西	16 401	72 377	9 435	50 769	7 215
	甘 肃	2 540	13 820	1 897	12 778	1 677
	青 海	358	2 050	262	1 990	270
	合 计	1 636 104	5 994 636	730 932	3 137 615	468 742
	山 西	56 441	295 010	39 711	142 778	23 753
	内蒙古	32 723	166 594	20 109	82 714	11 392
	山 东	20 540	92 387	12 157	40 811	7 216
	河 南	46 510	149 488	19 537	43 770	9 855
黄	四 川	158	762	93	762	93
	陕 西	80 765	376 740	48 256	175 397	28 352
河	甘 肃	32 898	158 609	20 991	113 914	15 852
	青 海	10 034	50 342	6 285	28 919	5 053
	宁 夏	22 810	76 982	9 685	16 783	6 373
	合 计	302 879	1 366 914	176 824	645 848	107 939

十大水系城镇生活污染排放及处理情况（二）（续表）

（2013）

流域	地区 名称		城镇生活污水 排放量/ 万吨	城镇生活化学需 氧量产生量/ 吨	城镇生活氨氮 产生量/ 吨	城镇生活化学需 氧量排放量/ 吨	城镇生活氨氮 排放量/ 吨
东 南 诸 河	浙	江	203 332	754 158	100 201	323 980	56 788
	安	徽	4 621	15 927	1 817	10 332	1 315
	福	建	145 420	534 364	69 463	331 162	49 857
	合	计	353 373	1 304 449	171 481	665 474	107 961
西 北 诸 河	河	北	1 119	5 370	734	2 397	417
	内蒙古		6 402	35 010	4 476	21 103	3 139
	西	藏	82	442	57	420	54
	甘	肃	9 331	42 546	5 920	18 686	3 396
	青	海	3 007	9 768	1 554	5 207	1 170
	新	疆	65 949	232 151	28 497	118 091	21 796
	合	计	85 889	325 287	41 238	165 903	29 973
西 南 诸 河	云	南	28 879	162 036	19 089	138 889	16 984
	西	藏	4 473	25 009	2 939	20 259	2 558
	青	海	152	880	113	941	104
	合	计	33 504	187 926	22 140	160 089	19 646

重点流域接纳工业废水及处理情况（一）

（2013）

流域	地区名称	工业废水排放量/万吨	直接排入环境的	排入污水处理厂的	工业废水处理量/万吨	废水治理设施数/套	废水治理设施治理能力/（万吨/日）	废水治理设施运行费用/万元
	总计	1 085 040	840 940	244 099	2 851 164	37 052	15 421	3 544 674.2
辽河	内蒙古	9 838	6 998	2 840	8 032	194	42	9 964.1
	辽 宁	40 793	29 352	11 442	194 242	1 334	1 063	182 320.5
	吉 林	2 814	2 578	236	2 844	112	17	5 873.9
	合 计	53 445	38 927	14 518	205 117	1 640	1 122	198 158.5
海河	北 京	7 643	1 959	5 684	8 820	432	52	25 375.6
	天 津	18 692	7 015	11 677	29 161	1 088	151	171 400.4
	河 北	99 461	67 862	31 599	680 307	4 310	3 688	432 883.3
	山 西	17 690	16 497	1 193	36 853	1 093	219	60 169.8
	内蒙古	463	463	0	588	34	5	1 941.0
	山 东	68 840	46 741	22 099	120 448	1 711	728	233 925.1
	河 南	28 285	26 961	1 324	54 151	788	302	58 223.1
	合 计	241 074	167 498	73 576	930 328	9 456	5 144	983 918.3
淮河	江 苏	61 927	51 382	10 545	108 809	1 994	593	154 107.1
	安 徽	30 628	26 248	4 380	35 799	865	284	64 687.3
	山 东	45 042	36 956	8 086	56 317	1 186	301	94 931.8
	河 南	63 761	52 774	10 988	60 541	1 211	406	76 631.9
	合 计	201 358	167 359	33 999	261 466	5 256	1 584	390 358.1
松花江	内蒙古	3 499	1 652	1 847	4 279	77	14	3 720.3
	吉 林	24 664	18 665	6 000	29 445	404	124	47 344.2
	黑龙江	39 796	33 647	6 149	92 522	1 032	671	285 126.6
	合 计	67 959	53 964	13 995	126 246	1 513	809	336 191.1

重点流域接纳工业废水及处理情况（一）（续表）

（2013）

流域	地名	区称	工业废水排放量/万吨	直接排入环境的	排入污水处理厂的	工业废水处理量/万吨	废水治理设施数/套	废水治理设施治理能力（万吨/日）	废水治理设施运行费用/万元
长江中下游	上	海	90 850	36 141	54 709	130 038	3 484	637	341 340.4
	江	苏	40 916	27 306	13 610	142 177	1 581	567	211 840.7
	安	徽	31 215	30 476	739	137 223	1 098	487	121 465.1
	江	西	67 857	65 850	2 007	167 116	2 521	1 069	186 689.6
	河	南	6 926	6 926	0	8 234	227	57	16 469.9
	湖	北	65 446	56 951	8 495	164 863	1 583	947	136 710.4
	湖	南	89 965	87 214	2 751	260 146	2 959	1 137	181 767.4
	广	西	1 050	1 005	45	1 248	82	6	603.1
	合	计	394 225	311 868	82 357	1 011 044	13 535	4 908	1 196 886.6
黄河中上游	山	西	30 166	26 625	3 540	88 436	1 899	579	146 178.2
	内蒙古		14 786	4 892	9 895	81 936	650	397	82 233.8
	河	南	30 719	28 091	2 628	41 252	1 055	276	67 696.1
	陕	西	24 663	19 150	5 512	42 722	1 265	293	82 388.0
	甘	肃	9 536	6 891	2 645	19 008	326	103	19 786.0
	青	海	3 438	3 438	0	23 138	127	68	11 438.9
	宁	夏	13 671	12 237	1 434	20 471	330	137	29 440.6
	合	计	126 978	101 324	25 655	316 963	5 652	1 853	439 161.6

重点流域接纳工业废水及处理情况（二）

（2013）

单位：吨

| 流域 | 地区名称 | 工业废水中污染物产生量 | | | | |
		化学需氧量	氨氮	石油类	挥发酚	氰化物
	总　计	11 711 682. 8	740 623. 5	182 243. 2	54 784. 5	3 320. 4
辽河	内蒙古	81 149. 0	4 158. 6	248. 6	227. 2	4. 8
	辽　宁	261 699. 1	14 950. 8	14 202. 5	1 150. 5	157. 5
	吉　林	46 925. 2	509. 6	148. 1	…	…
	合　计	389 773. 3	19 619. 0	14 599. 2	1 377. 7	162. 4
海河	北　京	52 196. 4	1 997. 1	3 254. 3	900. 3	5. 1
	天　津	153 419. 7	8 553. 4	969. 0	2. 8	1. 5
	河　北	1 376 620. 9	84 299. 5	20 659. 4	9 440. 2	633. 1
	山　西	284 480. 9	15 715. 4	2 853. 9	3 828. 0	110. 1
	内蒙古	2 735. 7	77. 0	1. 8	5. 8	230. 4
	山　东	1 007 617. 3	52 113. 9	14 732. 7	2 914. 8	150. 9
	河　南	491 228. 5	12 348. 0	3 103. 1	1 155. 6	87. 2
	合　计	3 368 299. 3	175 104. 4	45 574. 2	18 247. 6	1 218. 4
淮河	江　苏	669 273. 5	20 927. 2	2 691. 0	656. 5	66. 0
	安　徽	350 063. 8	32 268. 5	1 144. 3	45. 4	13. 3
	山　东	491 869. 3	20 401. 5	3 596. 4	1 259. 2	30. 3
	河　南	518 888. 3	23 274. 3	4 114. 3	979. 3	47. 9
	合　计	2 030 094. 9	96 871. 5	11 545. 9	2 940. 4	157. 6
松花江	内蒙古	178 321. 7	19 614. 6	95. 6	5. 2	…
	吉　林	279 308. 4	7 370. 4	2 151. 9	217. 0	10. 7
	黑龙江	926 316. 3	63 225. 2	42 000. 9	4 982. 1	47. 7
	合　计	1 383 946. 4	90 210. 2	44 248. 4	5 204. 3	58. 4

重点流域接纳工业废水及处理情况（二）（续表）

（2013）

单位：吨

流域	地 区 名 称	工业废水中污染物产生量				
		化学需氧量	氨氮	石油类	挥发酚	氰化物
长江中下游	上 海	463 711.1	20 184.8	10 395.2	1 554.1	353.4
	江 苏	383 800.7	19 635.7	4 532.3	2 819.0	168.0
	安 徽	257 930.4	9 248.6	5 494.3	1 677.4	275.1
	江 西	491 659.0	27 555.4	6 006.5	2 230.3	30.8
	河 南	281 243.3	3 490.6	529.4	0.1	⋯
	湖 北	422 789.4	26 086.6	2 740.1	1 043.1	47.5
	湖 南	562 028.9	58 427.5	5 660.5	982.0	213.4
	广 西	5 226.3	425.1	31.3	⋯	⋯
	合 计	2 868 389.0	165 054.2	35 389.5	10 306.0	1 088.2
黄河中上游	山 西	235 062.8	22 472.9	7 269.1	6 316.6	331.3
	内蒙古	233 439.1	45 975.2	3 513.4	2 903.4	61.7
	河 南	319 948.8	19 188.4	8 365.7	606.1	27.4
	陕 西	337 878.3	27 201.6	6 505.6	6 084.1	188.6
	甘 肃	109 674.6	16 665.1	2 232.8	179.7	2.7
	青 海	54 906.3	1 238.2	597.0	4.6	⋯
	宁 夏	380 269.8	61 022.8	2 402.5	614.0	23.9
	合 计	1 671 179.8	193 764.3	30 886.2	16 708.5	635.5

重点流域接纳工业废水及处理情况（三）

（2013）

单位：吨

流域	地区名称	工业废水中污染物产生量					
		汞	镉	六价铬	总铬	铅	砷
总 计		10.629	1 296.772	761.812	1 220.658	1 383.240	8 179.987
辽河	内蒙古	0.098	65.055	0.016	0.018	65.013	105.831
	辽 宁	0.101	0.024	14.720	24.125	0.192	0.188
	吉 林	0.203	0.017	…	…	0.114	…
	合 计	0.402	65.095	14.736	24.143	65.319	106.019
海河	北 京	…	0.001	11.488	12.171	0.216	1.802
	天 津	0.077	0.065	27.576	37.818	2.483	0.125
	河 北	0.040	1.289	260.484	305.617	8.988	42.062
	山 西	0.494	0.750	2.372	2.718	0.031	…
	内蒙古	…	…	19.362	19.362	0.010	0.005
	山 东	0.152	247.212	4.422	32.130	243.352	641.164
	河 南	0.368	89.122	3.064	15.414	47.568	22.762
	合 计	1.131	338.439	328.768	425.230	302.648	707.920
淮河	江 苏	0.001	0.016	66.973	108.409	30.166	0.922
	安 徽	…	0.004	2.734	5.789	2.228	2.101
	山 东	0.016	0.193	4.084	19.399	2.623	0.816
	河 南	0.168	1.210	5.289	110.930	2.717	8.273
	合 计	0.185	1.423	79.080	244.528	37.734	12.112
松花江	内蒙古	…	0.002	0.005	0.006	0.005	0.034
	吉 林	…	…	4.189	4.203	…	…
	黑龙江	0.567	0.129	7.545	7.592	1.226	9.395
	合 计	0.567	0.131	11.739	11.800	1.231	9.429

重点流域接纳工业废水及处理情况（三）（续表）

（2013）

单位：吨

流域	地 区 名 称	工业废水中污染物产生量					
		汞	镉	六价铬	总铬	铅	砷
长江中下游	上 海	0.020	0.140	139.171	179.405	1.481	0.405
	江 苏	0.025	0.070	50.123	73.659	1.699	1.285
	安 徽	0.004	16.115	12.639	14.600	57.188	654.344
	江 西	0.927	462.557	44.553	53.190	109.400	5 707.129
	河 南	0.001	0.006	0.646	9.780	0.121	5.240
	湖 北	0.017	7.463	25.950	29.479	15.389	113.445
	湖 南	2.790	163.228	33.581	50.645	390.123	338.125
	广 西	…	0.016	0.092	0.092	0.010	0.021
	合 计	3.784	649.595	306.755	410.851	575.411	6 819.993
黄河中上游	山 西	0.009	12.514	0.479	0.693	9.345	54.845
	内蒙古	2.231	11.836	0.860	0.958	31.748	106.995
	河 南	1.000	180.087	7.625	60.142	316.945	89.718
	陕 西	0.049	2.040	10.483	12.506	2.785	20.425
	甘 肃	0.059	23.125	0.446	28.795	26.820	119.285
	青 海	0.503	12.402	0.043	0.054	11.959	59.966
	宁 夏	0.707	0.086	0.799	0.958	1.296	73.280
	合 计	4.559	242.089	20.735	104.107	400.898	524.514

重点流域接纳工业废水及处理情况（四）

（2013）

单位：吨

流域	地区名称	工业废水中污染物排放量				
		化学需氧量	氨氮	石油类	挥发酚	氰化物
总 计		**1 587 453.9**	**148 127.6**	**11 031.6**	**1 153.2**	**116.8**
辽河	内蒙古	25 717.5	2 607.7	109.7	…	…
	辽 宁	46 215.5	2 624.9	383.8	3.6	2.8
	吉 林	4 879.5	207.8	8.2	…	…
	合 计	76 812.4	5 440.4	501.6	3.7	2.8
海河	北 京	5 219.6	290.6	43.2	0.1	0.1
	天 津	26 214.7	3 338.7	118.2	1.8	1.0
	河 北	161 489.0	13 319.5	822.0	24.6	6.6
	山 西	29 870.1	3 457.1	497.8	317.6	8.4
	内蒙古	680.7	14.5	0.3	…	…
	山 东	53 770.6	4 410.1	214.9	14.4	2.4
	河 南	41 373.5	2 516.0	104.5	0.8	4.0
	合 计	318 618.1	27 346.5	1 800.9	359.4	22.5
淮河	江 苏	88 846.4	6 735.1	518.2	20.0	6.0
	安 徽	34 496.4	4 879.4	371.6	2.5	2.0
	山 东	32 000.5	1 927.0	110.4	3.6	0.3
	河 南	81 078.6	5 929.5	808.3	128.7	11.4
	合 计	236 421.7	19 471.0	1 808.6	154.9	19.7
松花江	内蒙古	17 619.1	995.2	29.0	5.2	…
	吉 林	36 600.1	3 155.8	191.4	1.9	0.3
	黑龙江	77 723.5	5 427.8	271.2	6.3	1.4
	合 计	131 942.7	9 578.8	491.5	13.4	1.8

336

重点流域接纳工业废水及处理情况（四）（续表）

（2013）

单位：吨

流域	地 区名 称	工业废水中污染物排放量				
		化学需氧量	氨氮	石油类	挥发酚	氰化物
长江中下游	上 海	51 006.2	3 867.8	1 239.2	4.0	5.9
	江 苏	42 485.0	2 731.9	333.0	11.0	2.5
	安 徽	40 153.8	2 209.1	299.0	3.6	5.1
	江 西	92 915.0	8 766.1	728.2	14.1	7.4
	河 南	11 338.7	1 230.9	31.1	0.1	…
	湖 北	98 473.8	10 894.2	789.6	17.6	6.7
	湖 南	134 175.8	22 810.9	563.3	15.3	10.2
	广 西	2 197.0	251.7	16.3	…	…
	合 计	472 745.2	52 762.6	3 999.7	65.7	37.7
黄河中上游	山 西	48 459.1	4 043.5	483.5	303.2	20.9
	内蒙古	29 210.5	5 811.2	649.5	232.3	1.7
	河 南	38 429.5	2 801.6	319.8	10.2	4.0
	陕 西	67 533.6	4 575.9	562.4	2.5	3.9
	甘 肃	41 119.0	7 258.5	207.9	0.9	…
	青 海	25 496.3	772.1	65.0	0.5	…
	宁 夏	100 665.8	8 265.4	141.3	6.6	1.8
	合 计	350 913.8	33 528.3	2 429.4	556.2	32.3

重点流域接纳工业废水及处理情况（五）

（2013）

流域	地 区 名 称	工业废水中污染物排放量					
		汞	镉	六价铬	总铬	铅	砷
	总 计	0.457	12.918	29.766	89.232	45.049	71.754
辽 河	内蒙古	0.034	0.064	…	…	1.009	1.203
	辽 宁	0.003	0.004	0.133	0.521	0.025	0.023
	吉 林	0.001	0.001	…	…	0.034	…
	合 计	0.038	0.068	0.133	0.522	1.068	1.226
海 河	北 京	…	0.001	0.165	0.265	0.054	0.009
	天 津	0.006	0.003	0.133	0.356	0.086	0.012
	河 北	…	0.003	2.162	4.762	0.216	0.054
	山 西	0.002	0.750	0.006	0.086	0.004	…
	内蒙古	…	…	…	…	0.010	0.005
	山 东	0.007	0.978	0.190	6.660	0.469	1.857
	河 南	…	0.100	0.090	2.718	0.464	0.059
	合 计	0.015	1.835	2.747	14.847	1.301	1.997
淮 河	江 苏	…	0.001	0.931	2.640	0.820	0.006
	安 徽	…	0.001	0.098	0.407	0.272	2.096
	山 东	0.002	0.003	0.063	0.336	0.043	0.094
	河 南	0.002	0.402	0.686	6.561	1.454	0.616
	合 计	0.005	0.408	1.778	9.945	2.590	2.812
松 花 江	内蒙古	…	0.002	0.005	0.006	0.003	0.025
	吉 林	…	…	0.057	0.083	…	…
	黑龙江	…	0.001	0.097	0.108	0.014	0.013
	合 计	…	0.003	0.159	0.197	0.017	0.037

重点流域接纳工业废水及处理情况（五）（续表）

（2013）

单位：吨

流域	地区名称	工业废水中污染物排放量					
		汞	镉	六价铬	总铬	铅	砷
长江中下游	上　海	0.016	0.002	2.134	4.762	0.211	0.086
	江　苏	0.002	0.006	1.897	2.755	0.115	0.186
	安　徽	0.001	0.095	0.149	0.208	1.120	2.676
	江　西	0.071	1.922	17.172	17.369	5.449	9.281
	河　南	…	0.001	0.020	0.333	0.027	0.031
	湖　北	0.002	0.469	1.782	2.435	1.920	6.022
	湖　南	0.224	6.688	1.004	11.262	24.203	42.430
	广　西	…	0.003	0.002	0.002	0.002	0.002
	合　计	0.316	9.185	24.159	39.127	33.046	60.714
黄河中上游	山　西	0.009	0.030	0.133	0.144	0.169	0.192
	内蒙古	0.016	0.281	0.023	0.041	1.236	2.101
	河　南	0.011	0.566	0.099	22.019	2.681	0.944
	陕　西	0.001	0.034	0.162	1.550	0.129	0.210
	甘　肃	0.036	0.239	0.239	0.592	2.531	0.683
	青　海	0.007	0.269	0.007	0.011	0.272	0.788
	宁　夏	0.002	…	0.126	0.237	0.010	0.050
	合　计	0.082	1.419	0.789	24.595	7.028	4.969

重点流域工业废水污染防治投资情况

（2013）

流域	地 区名 称	工业废水治理施工项目数/个	工业废水治理竣工项目数/个	工业废水治理项目完成投资/万元	废水治理竣工项目新增处理能力/（万吨/日）
	总 计	534	655	580 456.3	159.6
辽河	内蒙古	5	4	10 760.0	0.5
	辽 宁	1	9	23 233.3	6.7
	吉 林	8	7	6 001.0	0.1
	合 计	14	20	39 994.3	7.3
海河	北 京	7	9	8 427.9	3.2
	天 津	14	13	6 435.8	3.4
	河 北	38	46	49 266.3	7.2
	山 西	10	15	17 713.5	3.9
	内蒙古	0	0	...	...
	山 东	28	37	42 218.6	10.7
	河 南	10	25	22 698.0	3.9
	合 计	107	145	146 760.2	32.2
淮河	江 苏	30	36	42 735.0	9.5
	安 徽	13	25	14 126.0	2.5
	山 东	17	32	25 653.0	14.9
	河 南	12	17	8 536.0	1.0
	合 计	72	110	91 050.0	27.9
松花江	内蒙古	4	3	21 572.5	...
	吉 林	5	5	3 027.6	3.4
	黑龙江	6	9	16 587.5	2.4
	合 计	15	17	41 187.6	5.9

重点流域工业废水污染防治投资情况（续表）

（2013）

流域	地区名称	工业废水治理施工项目数/个	工业废水治理竣工项目数/个	工业废水治理项目完成投资/万元	废水治理竣工项目新增处理能力/（万吨/日）
长江中下游	上　海	60	56	15 618.2	5.6
	江　苏	71	72	39 065.1	5.4
	安　徽	8	7	1 854.5	9.3
	江　西	17	22	32 334.1	2.6
	河　南	3	9	7 996.9	4.4
	湖　北	20	27	11 740.2	4.8
	湖　南	34	35	51 372.3	21.6
	广　西	0	1	150.0	0.2
	合　计	213	229	160 131.3	53.9
黄河中上游	山　西	42	52	25 693.0	14.8
	内蒙古	16	18	20 794.1	2.2
	河　南	11	17	5 887.9	2.6
	陕　西	22	32	18 362.8	6.8
	甘　肃	6	6	8 663.2	0.9
	青　海	1	1	2 985.0	…
	宁　夏	15	8	18 947.0	5.1
	合　计	113	134	101 333.0	32.4

重点流域农业污染排放情况

（2013）

单位：万吨

流域	地区名称	农业污染物排放（流失）总量			
		化学需氧量	氨氮	总氮	总磷
	总　计	752.5	46.5	278.6	32.0
辽河	内蒙古	14.9	0.4	3.4	0.3
	辽　宁	68.3	2.6	15.0	1.9
	吉　林	6.6	0.2	1.5	0.2
	合　计	89.8	3.2	19.9	2.5
海河	北　京	7.5	0.5	2.6	0.3
	天　津	11.0	0.6	2.8	0.3
	河　北	85.4	4.2	35.4	4.2
	山　西	7.0	0.4	3.0	0.3
	内蒙古	3.2	…	0.7	0.1
	山　东	49.3	2.4	26.7	2.7
	河　南	15.2	1.1	7.7	1.2
	合　计	178.5	9.1	78.8	9.1
淮河	江　苏	26.3	2.5	11.8	1.3
	安　徽	25.0	2.5	12.0	1.2
	山　东	33.2	2.2	13.1	1.6
	河　南	46.8	3.8	22.9	2.7
	合　计	131.4	11.0	59.8	6.8
松花江	内蒙古	11.4	0.3	2.7	0.2
	吉　林	38.9	1.4	9.2	1.2
	黑龙江	86.6	2.6	20.2	2.0
	合　计	136.9	4.2	32.1	3.4

重点流域农业污染排放情况（续表）

（2013）

单位：万吨

流域	地区名称	农业污染物排放（流失）总量			
		化学需氧量	氨氮	总氮	总磷
长江中下游	上　海	6.2	0.6	3.1	0.4
	江　苏	5.0	0.7	2.8	0.3
	安　徽	6.5	0.8	3.6	0.5
	江　西	22.7	2.8	9.2	1.2
	河　南	6.3	0.5	4.3	0.5
	湖　北	42.1	4.1	17.3	2.2
	湖　南	54.5	6.0	18.7	2.1
	广　西	0.5	0.1	0.3	…
	合　计	143.9	15.5	59.4	7.2
黄河中上游	山　西	10.6	0.8	5.3	0.6
	内蒙古	15.7	0.3	4.5	0.4
	河　南	13.1	0.9	6.2	0.8
	陕　西	14.6	0.8	6.8	0.6
	甘　肃	6.2	0.3	2.3	0.2
	青　海	2.0	0.1	0.6	0.1
	宁　夏	9.8	0.2	3.0	0.3
	合　计	72.0	3.4	28.6	2.9

重点流域城镇生活污染排放及处理情况（一）

（2013）

流域	地区名称	城镇人口/万人	生活用水总量/万吨	污水处理厂数/座	污水处理厂设计处理能力/（万吨/日）	生活污水实际处理量/万吨
	总　计	40 402	2 886 692	2 594	9 252	2 152 209
辽河	内蒙古	306	18 499	24	80	13 434
	辽　宁	2 047	131 712	110	569	109 174
	吉　林	141	8 613	5	25	6 087
	合　计	2 494	158 823	139	674	128 695
海河	北　京	1 825	158 674	135	422	123 906
	天　津	1 214	73 554	60	272	59 390
	河　北	3 384	215 573	235	801	157 468
	山　西	657	41 815	68	116	28 127
	内蒙古	33	1 799	7	7	897
	山　东	1 525	105 838	112	415	93 351
	河　南	697	52 262	34	153	40 978
	合　计	9 335	649 514	651	2 187	504 116
淮河	江　苏	2 067	153 605	192	364	84 399
	安　徽	1 460	97 613	88	216	54 349
	山　东	1 456	96 078	100	355	93 596
	河　南	2 249	175 334	105	367	102 389
	合　计	7 232	522 630	485	1 302	334 733
松花江	内蒙古	175	10 413	13	30	4 493
	吉　林	1 128	67 321	42	222	46 034
	黑龙江	1 954	115 746	73	329	69 015
	合　计	3 257	193 479	128	581	119 541

重点流域城镇生活污染排放及处理情况（一）（续表）

（2013）

流域	地　区 名　称	城镇人口/ 万人	生活用水总量/ 万吨	污水处理厂数/ 座	污水处理厂设计 处理能力/ （万吨/日）	生活污水实际 处理量/ 万吨
长 江 中 下 游	上　海	2 085	190 444	112	1 542	304 554
	江　苏	1 183	106 414	140	347	95 189
	安　徽	905	75 395	79	178	53 485
	江　西	2 178	162 019	134	336	92 126
	河　南	334	23 636	13	46	13 823
	湖　北	2 706	209 645	115	496	132 989
	湖　南	3 140	245 252	157	524	151 935
	广　西	38	3 029	5	7	1 241
	合　计	12 569	1 015 834	755	3 477	845 342
黄 河 中 上 游	山　西	1 240	73 460	119	212	45 776
	内蒙古	719	42 156	60	160	28 236
	河　南	745	57 339	67	162	40 252
	陕　西	1 479	94 176	114	276	62 167
	甘　肃	767	40 680	39	114	17 627
	青　海	234	12 502	17	33	8 038
	宁　夏	330	26 097	20	75	17 685
	合　计	5 514	346 410	436	1 031	219 781

重点流域城镇生活污染排放及处理情况（二）

（2013）

流域	地区名称	城镇生活污水排放量/万吨	城镇生活化学需氧量产生量/吨	城镇生活氨氮产生量/吨	城镇生活化学需氧量排放量/吨	城镇生活氨氮排放量/吨
	总　计	2 471 880	9 687 217	1 228 939	4 688 727	758 134
辽河	内蒙古	15 979	71 381	8 923	20 770	6 978
	辽　宁	108 170	473 748	63 913	203 414	45 302
	吉　林	7 414	32 051	4 202	16 510	3 415
	合　计	131 563	577 180	77 038	240 694	55 696
海河	北　京	134 991	527 173	64 729	89 868	14 189
	天　津	65 469	304 702	41 599	84 886	15 671
	河　北	193 482	731 469	98 425	224 247	47 416
	山　西	33 827	143 288	19 685	64 237	11 790
	内蒙古	1 555	7 480	971	3 711	616
	山　东	93 682	355 569	47 731	106 049	18 700
	河　南	47 438	155 088	19 693	56 962	11 030
	合　计	570 444	2 224 769	292 833	629 961	119 412
淮河	江　苏	135 243	522 049	67 758	348 159	49 243
	安　徽	88 686	341 854	40 594	236 832	30 018
	山　东	84 619	344 931	46 215	156 689	28 961
	河　南	150 836	496 886	62 529	243 257	40 534
	合　计	459 384	1 705 720	217 097	984 937	148 756
松花江	内蒙古	8 664	38 940	5 179	23 712	3 711
	吉　林	56 938	262 305	33 941	144 098	23 611
	黑龙江	93 903	433 884	58 105	268 992	42 459
	合　计	159 505	735 130	97 224	436 802	69 781

重点流域城镇生活污染排放及处理情况（二）（续表）

（2013）

流域	地 名	区 称	城镇生活污水排放量/万吨	城镇生活化学需氧量产生量/吨	城镇生活氨氮产生量/吨	城镇生活化学需氧量排放量/吨	城镇生活氨氮排放量/吨
长江中下游	上	海	171 400	578 332	73 813	167 503	39 000
	江	苏	87 491	329 611	40 490	119 929	23 336
	安	徽	61 973	216 643	25 513	137 861	19 306
	江	西	137 209	559 934	61 930	394 204	49 784
	河	南	20 333	65 733	8 155	38 551	5 710
	湖	北	178 795	675 482	79 980	403 921	55 467
	湖	南	210 394	750 316	91 182	526 817	70 669
	广	西	2 575	8 074	1 088	6 460	836
	合	计	870 169	3 184 125	382 152	1 795 247	264 107
黄河中上游	山	西	55 520	290 245	39 014	141 400	23 400
	内蒙古		33 464	169 975	20 535	85 109	11 732
	河	南	50 425	166 009	21 598	54 191	11 452
	陕	西	75 605	347 817	44 432	160 574	26 497
	甘	肃	32 898	158 609	20 991	113 914	15 852
	青	海	10 521	52 901	6 620	30 904	5 351
	宁	夏	22 381	74 738	9 403	14 995	6 100
	合	计	280 815	1 260 293	162 594	601 087	100 384

湖泊水库接纳工业废水及处理情况（一）

（2013）

流域	地区名称	工业废水排放量/万吨	直接排入环境的	排入污水处理厂的	工业废水处理量/万吨	废水治理设施数/套	废水治理设施治理能力/（万吨/日）	废水治理设施运行费用/万元
总　计		316 318	235 275	81 043	687 826	12 241	3 668	877 524.0
滇　池	云　南	806	653	153	2 019	163	17	9 267.7
巢　湖	安　徽	5 741	5 090	651	25 271	299	163	16 939.0
洞庭湖	江　西	2 069	2 069	0	26 022	129	341	15 531.6
	湖　北	608	608	0	353	13	5	781.8
	湖　南	89 965	87 214	2 751	260 146	2 959	1 137	181 767.4
	广　西	1 050	1 005	45	1 248	82	6	603.1
	合　计	93 692	90 897	2 795	287 770	3 183	1 490	198 683.9
鄱阳湖	安　徽	31	31	0	35	20	1	154.1
	江　西	64 912	62 905	2 007	140 383	2 382	726	169 679.1
	合　计	64 943	62 936	2 007	140 417	2 402	726	169 833.2
太　湖	上　海	1 746	111	1 635	1 212	173	6	6 943.2
	江　苏	101 725	56 186	45 540	125 046	3 299	661	304 086.5
	浙　江	36 901	9 037	27 864	79 146	1 876	420	138 918.8
	合　计	140 373	65 334	75 039	205 404	5 348	1 087	449 948.5
丹江口	河　南	3 912	3 912	0	11 752	273	63	13 136.2
	湖　北	1 977	1 717	261	2 083	147	16	5 072.5
	陕　西	4 874	4 737	137	13 110	426	105	14 643.0
	合　计	10 763	10 366	398	26 945	846	184	32 851.7

湖泊水库接纳工业废水及处理情况（二）

（2013）

单位：吨

流域	地区名称	工业废水中污染物产生量				
		化学需氧量	氨氮	石油类	挥发酚	氰化物
总　计		2 228 616.9	127 669.9	18 403.1	3 742.5	373.6
滇　池	云　南	9 152.3	592.2	77.6	189.7	1.7
巢　湖	安　徽	90 308.5	4 859.2	1 045.7	0.1	3.3
洞庭湖	江　西	22 605.7	274.3	1 194.0	297.3	5.8
	湖　北	16 146.4	69.3	3.6	…	…
	湖　南	562 028.9	58 427.5	5 660.5	982.0	213.4
	广　西	5 226.3	425.1	31.3	…	…
	合　计	606 007.3	59 196.2	6 889.4	1 279.3	219.2
鄱阳湖	安　徽	1 110.1	5.0	132.7	…	…
	江　西	458 012.8	26 849.2	4 805.7	1 932.9	25.0
	合　计	459 122.9	26 854.2	4 938.4	1 932.9	25.0
太　湖	上　海	6 951.9	370.4	12.2	…	12.4
	江　苏	615 411.1	19 485.7	4 117.9	134.2	107.6
	浙　江	349 138.5	8 393.0	677.7	204.0	2.7
	合　计	971 501.4	28 249.1	4 807.8	338.3	122.8
丹江口	河　南	25 057.5	810.3	70.5	0.4	1.0
	湖　北	16 171.3	1 889.2	432.6	0.1	…
	陕　西	51 295.8	5 219.6	141.2	1.6	0.7
	合　计	92 524.6	7 919.0	644.2	2.1	1.6

湖泊水库接纳工业废水及处理情况（三）

（2013）

单位：吨

流域	地区名称	工业废水中污染物产生量					
		汞	镉	六价铬	总铬	铅	砷
总　计		4.032	821.564	256.273	518.223	710.614	6 535.411
滇　池	云　南	…	125.622	0.006	0.006	118.986	451.443
巢　湖	安　徽	…	0.058	1.412	1.414	0.535	0.265
洞庭湖	江　西	…	0.003	…	…	0.035	0.036
	湖　北	…	…	…	…	…	0.098
	湖　南	2.790	163.228	33.581	50.645	390.123	338.125
	广　西	…	0.016	0.092	0.092	0.010	0.021
	合　计	2.790	163.246	33.673	50.737	390.168	338.280
鄱阳湖	安　徽	…	…	…	…	0.209	…
	江　西	0.927	462.555	44.553	53.190	109.365	5 707.060
	合　计	0.927	462.555	44.553	53.190	109.574	5 707.060
太　湖	上　海	…	…	16.940	16.941	0.104	0.001
	江　苏	…	2.263	110.773	220.679	3.403	3.875
	浙　江	…	1.427	36.982	162.638	3.477	0.534
	合　计		3.690	164.695	400.259	6.984	4.410
丹江口	河　南	…	…	0.232	0.375	0.007	0.005
	湖　北	0.001	0.010	8.067	8.450	0.030	0.630
	陕　西	0.313	66.383	3.634	3.791	84.330	33.318
	合　计	0.314	66.393	11.933	12.617	84.367	33.953

湖泊水库接纳工业废水及处理情况（四）

（2013）

单位：吨

流域	地区名称	工业废水中污染物排放量				
		化学需氧量	氨氮	石油类	挥发酚	氰化物
总　计		366 153.0	42 854.4	1 935.7	36.7	24.8
滇　池	云　南	1 290.8	67.7	46.4	0.1	0.1
巢　湖	安　徽	9 636.0	396.5	43.0	…	…
洞庭湖	江　西	4 294.3	121.0	24.2	0.3	0.3
	湖　北	1 685.6	50.2	2.6	…	…
	湖　南	134 175.8	22 810.9	563.3	15.3	10.2
	广　西	2 197.0	251.7	16.3	…	…
	合　计	142 352.7	23 233.8	606.5	15.6	10.5
鄱阳湖	安　徽	62.1	3.0	5.5	…	…
	江　西	87 856.2	8 549.6	702.8	13.8	7.1
	合　计	87 918.3	8 552.6	708.3	13.8	7.1
太　湖	上　海	736.4	107.0	9.1	…	0.2
	江　苏	62 994.1	3 895.2	340.8	6.3	6.5
	浙　江	33 121.6	3 208.2	30.3	0.8	0.2
	合　计	96 852.1	7 210.4	380.2	7.1	6.9
丹江口	河　南	4 260.1	330.1	2.4	0.1	…
	湖　北	9 730.9	718.5	131.2	…	…
	陕　西	14 112.1	2 344.8	17.7	…	0.1
	合　计	28 103.1	3 393.4	151.3	0.1	0.1

湖泊水库接纳工业废水及处理情况（五）

（2013）

单位：吨

流域	地 区 名 称	工业废水中污染物排放量					
		汞	镉	六价铬	总铬	铅	砷
总 计		0.321	9.488	27.329	43.021	31.938	53.557
滇 池	云 南	0	0.327	…	…	0.877	1.412
巢 湖	安 徽	…	0.002	0.022	0.024	0.029	0.012
洞庭湖	江 西	0	0	0	0	0.013	0
	湖 北	0	0	0	0	0	0.098
	湖 南	0.224	6.688	1.004	11.262	24.203	42.430
	广 西	…	0.003	0.002	0.002	0.002	0.002
	合 计	0.224	6.691	1.006	11.264	24.218	42.529
鄱阳湖	安 徽	…	…	0	0	0.023	…
	江 西	0.071	1.922	17.172	17.369	5.436	9.273
	合 计	0.071	1.922	17.172	17.369	5.459	9.273
太 湖	上 海	0	…	0.149	0.151	0.020	…
	江 苏	…	0.005	1.169	3.309	0.120	0.007
	浙 江	0	0.014	0.479	3.515	0.152	0.005
	合 计	…	0.020	1.797	6.975	0.291	0.013
丹江口	河 南	…	…	0.036	0.054	0.005	0.004
	湖 北	0.001	0.010	7.237	7.253	0.030	0.014
	陕 西	0.024	0.517	0.059	0.082	1.028	0.301
	合 计	0.026	0.527	7.331	7.388	1.063	0.318

湖泊水库工业污染防治投资情况
（2013）

流域	地名	区称	工业废水治理施工项目数/个	工业废水治理竣工项目数/个	工业废水治理项目完成投资/万元	废水治理竣工项目新增处理能力/（万吨/日）
总 计			187	199	133 782.7	45.4
滇 池	云 南		2	1	150.0	···
巢 湖	安 徽		6	12	5 847.8	5.0
	江 西		0	0	···	···
	湖 北		0	0	···	···
洞庭湖	湖 南		34	35	51 372.3	21.6
	广 西		0	1	150.0	0.2
	合 计		34	36	51 522.3	21.8
	安 徽					
鄱阳湖	江 西		17	22	32 334.1	2.6
	合 计		17	22	32 334.1	2.6
	上 海		2	2	102.0	···
太 湖	江 苏		64	62	16 399.4	4.9
	浙 江		57	58	23 487.4	10.3
	合 计		123	122	39 988.9	15.2
	河 南		1	0	600.0	0.1
丹江口	湖 北		2	3	710.0	0.4
	陕 西		2	3	2 629.7	0.3
	合 计		5	6	3 939.7	0.8

湖泊水库农业污染排放情况

（2013）

单位：万吨

流域	地区名称	农业污染物排放/流失总量			
		化学需氧量	氨氮	总氮	总磷
总　计		103.4	11.6	40.0	4.8
滇　池	云　南	0.4	0.1	0.2	…
巢　湖	安　徽	5.8	0.4	1.8	0.2
洞庭湖	江　西	0.6	0.1	0.2	…
	湖　北	0.4	0.1	0.3	…
	湖　南	54.5	6.0	18.7	2.1
	广　西	0.5	0.1	0.3	…
	合　计	56.0	6.2	19.5	2.2
鄱阳湖	安　徽	0	0	0	0
	江　西	22.0	2.7	8.9	1.2
	合　计	22.0	2.7	8.9	1.2
太　湖	上　海	0.2	…	0.1	…
	江　苏	4.8	0.5	2.7	0.4
	浙　江	5.9	0.7	2.5	0.3
	合　计	10.9	1.2	5.3	0.7
丹江口	河　南	2.7	0.2	1.2	0.1
	湖　北	1.1	0.1	0.6	0.1
	陕　西	4.5	0.7	2.4	0.3
	合　计	8.3	1.0	4.2	0.5

湖泊水库城镇生活污染排放及处理情况（一）

（2013）

流域	地 区名 称	城镇人口/万人	生活用水总量/万吨	污水处理厂数/座	污水处理厂设计处理能力/（万吨/日）	生活污水实际处理量/万吨
总　计		9 435	781 614	717	2 209	524 113
滇　池	云　南	340	47 169	10	84	28 542
巢　湖	安　徽	546	48 513	21	111	39 405
洞庭湖	江　西	96	6 186	3	11	2 781
	湖　北	33	849	1	2	526
	湖　南	3 140	245 252	157	524	151 935
	广　西	38	3 029	5	7	1 241
	合　计	3 307	255 317	166	543	156 482
鄱阳湖	安　徽	7	338	1	1	256
	江　西	2 066	154 610	130	325	89 040
	合　计	2 073	154 948	131	326	89 296
太　湖	上　海	79	7 229	12	26	6 876
	江　苏	1 570	152 569	254	684	123 866
	浙　江	872	74 080	75	322	52 902
	合　计	2 521	233 878	341	1 032	183 644
丹江口	河　南	142	10 796	6	15	4 554
	湖　北	164	12 585	15	49	11 122
	陕　西	342	18 408	27	49	8 580
	合　计	648	41 788	48	112	24 255

湖泊水库城镇生活污染排放及处理情况（二）

（2013）

流域	地 区 名 称	城镇生活污水 排放量/ 万吨	城镇生活化学需氧 量产生量/ 吨	城镇生活氨氮 产生量/ 吨	城镇生活化学需 氧量排放量/ 吨	城镇生活氨氮 排放量/ 吨
总 计		674 596	2 371 351	284 639	1 207 023	179 360
滇 池	云 南	40 094	92 575	10 366	3 541	3 324
巢 湖	安 徽	42 806	133 093	15 699	58 580	8 264
洞庭湖	江 西	5 945	23 430	2 730	19 440	2 407
	湖 北	722	7 677	982	6 434	888
	湖 南	210 394	750 316	91 182	526 817	70 669
	广 西	2 575	8 074	1 088	6 460	836
	合 计	219 636	789 496	95 982	559 150	74 800
鄱阳湖	安 徽	320	1 775	271	1 198	153
	江 西	130 300	531 081	58 778	371 166	46 969
	合 计	130 620	532 855	59 049	372 364	47 122
太 湖	上 海	6 506	21 952	2 802	6 358	1 480
	江 苏	134 576	396 167	51 461	61 574	16 548
	浙 江	63 246	263 002	31 050	54 053	13 888
	合 计	204 328	681 121	85 312	121 985	31 916
丹江口	河 南	9 329	30 215	3 723	17 449	2 681
	湖 北	11 337	37 742	4 729	20 448	3 649
	陕 西	16 447	74 253	9 777	53 506	7 603
	合 计	37 112	142 211	18 229	91 403	13 934

三峡地区接纳工业废水及处理情况（一）

（2013）

流域	地区名称		工业废水排放量/万吨	直接排入环境的	排入污水处理厂的	工业废水处理量/万吨	废水治理设施数/套	废水治理设施治理能力/（万吨/日）	废水治理设施运行费用/万元
	总　计		131 070	119 546	11 524	283 731	7 371	1 664	339 938.1
库　区		湖　北	3 435	2 942	494	2 668	55	15	2 118.8
		重　庆	15 579	14 419	1 160	16 555	942	146	37 374.5
	合　计		19 014	17 361	1 653	19 223	997	161	39 493.3
影响区		湖　北	7 835	6 721	1 114	6 468	88	23	7 900.1
		重　庆	17 093	17 008	85	16 968	594	90	19 912.2
		四　川	11 038	10 760	278	10 251	663	77	27 976.9
		贵　州	1 332	1 331	1	774	68	7	4 176.6
	合　计		37 298	35 820	1 478	34 462	1 413	198	59 965.8
上游区		重　庆	384	384	0	511	84	4	1 585.3
		四　川	53 826	45 806	8 019	161 651	3 203	976	171 703.6
		贵　州	13 135	13 032	103	30 097	840	145	29 074.5
		云　南	7 413	7 143	270	37 786	834	181	38 115.6
	合　计		74 758	66 366	8 392	230 046	4 961	1 305	240 479.0

三峡地区接纳工业废水及处理情况（二）

（2013）

单位：吨

流域	地区名称		工业废水中污染物产生量				
			化学需氧量	氨氮	石油类	挥发酚	氰化物
	总　计		1 187 163.2	144 578.6	10 706.2	3 479.2	264.8
库　区		湖　北	15 135.7	330.1	10.3	…	…
		重　庆	170 468.4	20 231.3	2 015.5	788.6	71.0
	合　计		185 604.1	20 561.4	2 025.8	788.6	71.0
影响区		湖　北	19 436.7	2 245.0	450.6	16.4	1.0
		重　庆	150 983.3	3 037.3	884.7	173.4	20.0
		四　川	191 027.7	40 874.4	569.3	61.3	3.9
		贵　州	32 133.1	1 144.7	…	…	…
	合　计		393 580.8	47 301.4	1 904.6	251.0	24.9
上游区		重　庆	1 870.1	849.1	3.5	…	0.8
		四　川	455 976.3	31 343.8	4 900.6	1 730.4	122.0
		贵　州	73 004.6	23 701.1	924.7	55.1	23.4
		云　南	77 127.4	20 821.9	946.9	654.0	22.7
	合　计		607 978.3	76 715.9	6 775.8	2 439.6	168.9

三峡地区接纳工业废水及处理情况（三）

（2013）

单位：吨

流域	地 区名 称	工业废水中污染物产生量					
		汞	镉	六价铬	总铬	铅	砷
总 计		1.287	233.391	168.066	277.882	339.221	1 171.787
库 区	湖 北	…	0.006	0.001	0.051	0.009	0.005
	重 庆	…	0.011	62.444	100.877	7.013	0.108
	合 计	…	0.016	62.446	100.928	7.022	0.113
影响区	湖 北	…	…	…	0	0.001	…
	重 庆	0	0.018	44.412	44.824	0.396	1.538
	四 川	0.004	0.037	12.760	13.650	0.099	6.356
	贵 州	0	0	0	0	0	0
	合 计	0.004	0.055	57.172	58.474	0.496	7.895
上游区	重 庆	…	0	0.445	0.445	0	0
	四 川	0.565	39.981	46.690	114.381	46.131	372.536
	贵 州	0.042	0.013	1.176	2.905	0.024	20.680
	云 南	0.676	193.326	0.136	0.749	285.547	770.563
	合 计	1.283	233.320	48.448	118.480	331.702	1 163.780

三峡地区接纳工业废水及处理情况（四）

（2013）

单位：吨

流域	地 区名 称	工业废水中污染物排放量				
		化学需氧量	氨氮	石油类	挥发酚	氰化物
总 计		217 354.3	12 019.1	1 525.5	13.2	4.2
库 区	湖 北	5 940.4	271.3	6.6	…	…
	重 庆	27 441.6	1 808.7	131.5	0.6	0.1
	合 计	33 382.0	2 079.9	138.2	0.6	0.1
影响区	湖 北	6 360.0	1 173.7	28.7	…	…
	重 庆	22 968.4	872.7	200.9	8.9	0.9
	四 川	31 305.2	1 572.5	175.5	2.4	1.4
	贵 州	5 861.2	263.4	…	…	…
	合 计	66 494.8	3 882.4	405.1	11.3	2.2
上游区	重 庆	847.8	556.6	2.4	…	0.6
	四 川	75 187.2	3 402.5	458.8	1.0	0.9
	贵 州	27 004.3	1 517.9	265.7	…	0.1
	云 南	14 438.1	579.8	255.3	0.2	0.3
	合 计	117 477.4	6 056.8	982.3	1.3	1.9

三峡地区接纳工业废水及处理情况（五）

（2013）

单位：吨

流域	地区名称	工业废水中污染物排放量					
		汞	镉	六价铬	总铬	铅	砷
总　计		0.025	0.925	0.955	2.326	4.794	5.713
库　区	湖　北	…	0.004	0.001	0.051	0.007	0.004
	重　庆	0	0.002	0.088	0.270	0.072	…
	合　计	…	0.006	0.090	0.322	0.079	0.004
影响区	湖　北	…	…	0	0	…	0
	重　庆	0	0.001	0.045	0.051	0.009	0.033
	四　川	0.003	…	0.234	0.270	0.016	0.021
	贵　州	0	0	0	0	0	0
	合　计	0.003	0.001	0.279	0.321	0.025	0.054
上游区	重　庆	…	0	0.084	0.100	0	0
	四　川	0.007	0.075	0.496	1.524	1.063	2.045
	贵　州	0.014	0.009	0.006	0.021	0.017	0.037
	云　南	0.001	0.834	…	0.039	3.611	3.573
	合　计	0.022	0.918	0.586	1.683	4.691	5.655

三峡地区工业污染防治投资情况

（2013）

流域	地区名称	工业废水治理施工项目数/个	工业废水治理竣工项目数/个	工业废水治理项目完成投资/万元	废水治理竣工项目新增处理能力/（万吨/日）
总　计		112	144	56 569.3	32.7
库　区	湖　北	0	2	437.0	0.1
	重　庆	13	19	4 906.4	0.2
	合　计	13	21	5 343.4	0.3
影响区	湖　北	1	3	665.0	0.1
	重　庆	2	3	653.9	0.2
	四　川	28	29	11 082.9	5.9
	贵　州	0	3	12 738.0	0.3
	合　计	31	38	25 139.8	6.5
上游区	重　庆	0	1	230.8	0.1
	四　川	45	56	18 708.5	16.7
	贵　州	6	10	2 475.5	2.9
	云　南	17	18	4 671.4	6.3
	合　计	68	85	26 086.1	26.0

三峡地区农业污染排放情况

（2013）

单位：万吨

流域	地 区 名 称	农业污染物排放/流失总量			
		化学需氧量	氨氮	总氮	总磷
总 计		72.1	7.9	33.1	3.7
库区	湖 北	0.9	0.1	0.5	0.1
	重 庆	5.9	0.6	2.6	0.3
	合 计	6.8	0.7	3.1	0.4
影响区	湖 北	0.6	0.1	0.3	…
	重 庆	5.5	0.6	2.3	0.3
	四 川	9.8	1.1	3.7	0.4
	贵 州	0.2	…	0.1	…
	合 计	16.1	1.8	6.5	0.8
上游区	重 庆	0.7	0.1	0.3	…
	四 川	43.1	4.5	18.0	2.0
	贵 州	3.1	0.4	2.5	0.3
	云 南	2.4	0.4	2.8	0.2
	合 计	49.3	5.4	23.5	2.6

三峡地区城镇生活污染排放及处理情况（一）

（2013）

流域	地区名称		城镇人口/万人	生活用水总量/万吨	污水处理厂数/座	污水处理厂设计处理能力/（万吨/日）	生活污水实际处理量/万吨
	总　计		7 145	554 783	625	1 156	349 984
库区	湖　北		63	4 421	23	15	3 321
	重　庆		1 194	87 901	96	232	71 818
	合　计		1 257	92 322	119	247	75 139
影响区	湖　北		105	7 458	11	34	8 349
	重　庆		481	33 934	50	50	16 548
	四　川		673	40 415	30	60	18 768
	贵　州		45	2 341	4	8	2 331
	合　计		1 303	84 149	95	152	45 995
上游区	重　庆		40	2 633	3	4	1 059
	四　川		3 012	251 287	310	514	147 890
	贵　州		808	56 646	55	104	32 733
	云　南		725	67 747	43	136	47 167
	合　计		4 585	378 313	411	758	228 849

三峡地区城镇生活污染排放及处理情况（二）

（2013）

流域	地区名称		城镇生活污水排放量/万吨	城镇生活化学需氧量产生量/吨	城镇生活氨氮产生量/吨	城镇生活化学需氧量排放量/吨	城镇生活氨氮排放量/吨
	总　计		463 172	1 771 402	218 468	1 023 660	143 824
库区	湖　北		3 759	14 153	1 769	6 948	1 231
	重　庆		74 879	296 435	37 926	124 742	22 564
	合　计		78 638	310 588	39 695	131 691	23 796
影响区	湖　北		6 364	23 671	2 970	11 925	2 298
	重　庆		29 118	119 277	15 260	84 369	12 131
	四　川		34 020	168 516	20 588	124 814	15 884
	贵　州		2 034	10 415	1 302	8 138	734
	合　计		71 537	321 880	40 121	229 247	31 048
上游区	重　庆		2 401	9 836	1 258	8 370	1 073
	四　川		208 554	770 909	94 035	466 392	59 565
	贵　州		44 756	173 562	22 041	117 540	16 176
	云　南		57 286	184 628	21 318	70 420	12 166
	合　计		312 997	1 138 934	138 652	662 722	88 980

入海陆源接纳工业废水及处理情况（一）

（2013）

流域	地名	区称	工业废水排放量/万吨	直接排入环境的	排入污水处理厂的	工业废水处理量/万吨	废水治理设施数/套	废水治理设施治理能力/（万吨/日）	废水治理设施运行费用/万元
	总	计	360 741	234 240	126 500	573 289	14 196	3 053	1 126 864.7
渤	天	津	9 029	1 890	7 139	9 569	317	55	50 048.7
	河	北	6 402	3 203	3 199	89 868	248	485	49 820.6
	辽	宁	24 488	23 044	1 444	12 109	415	194	33 694.3
海	山	东	28 676	17 228	11 448	51 070	577	284	103 486.2
	合	计	68 596	45 365	23 230	162 617	1 557	1 018	237 049.8
黄	辽	宁	3 759	3 302	457	2 804	225	23	4 292.4
	江	苏	20 674	14 183	6 491	29 035	1 033	160	68 514.9
	山	东	15 306	5 396	9 909	56 614	741	235	47 756.8
海	合	计	39 739	22 881	16 858	88 453	1 999	418	120 564.1
东	上	海	30 125	17 111	13 015	54 371	888	262	114 556.9
	浙	江	67 279	22 842	44 436	65 573	3 488	431	296 176.9
	福	建	73 266	60 290	12 976	74 289	1 655	289	80 642.8
海	广	东	2 074	1 945	130	1 444	114	18	4 387.7
	合	计	172 745	102 188	70 556	195 676	6 145	1 000	495 764.3
南	广	东	71 107	56 453	14 654	84 126	4 101	507	246 279.3
	广	西	5 134	4 825	309	39 385	151	85	21 527.5
	海	南	3 420	2 527	893	3 032	243	24	5 679.8
海	合	计	79 661	63 805	15 856	126 543	4 495	617	273 486.6

入海陆源接纳工业废水及处理情况（二）

（2013）

单位：吨

| 流域 | 地 区
名 称 | 工业废水中污染物产生量 | | | | |
		化学需氧量	氨氮	石油类	挥发酚	氰化物
	总 计	3 381 310.1	153 792.7	51 816.8	3 269.6	1 099.7
渤 海	天 津	73 758.1	3 666.9	468.8	0.7	1.0
	河 北	135 319.2	2 808.2	3 672.0	1 249.9	218.2
	辽 宁	76 988.7	22 125.4	3 331.8	188.1	5.0
	山 东	388 392.1	9 292.8	8 718.7	115.8	1.2
	合 计	674 458.1	37 893.3	16 191.3	1 554.5	225.3
黄 海	辽 宁	34 957.2	1 639.8	92.0	…	0.3
	江 苏	365 528.3	7 967.6	1 709.1	141.0	33.0
	山 东	126 213.0	5 972.1	1 436.9	10.9	25.7
	合 计	526 698.5	15 579.4	3 237.9	151.9	59.0
东 海	上 海	175 000.4	6 899.2	4 852.4	776.8	158.7
	浙 江	710 531.1	54 405.5	22 049.5	182.7	326.0
	福 建	466 560.5	17 608.1	3 300.4	353.8	78.7
	广 东	9 643.2	285.2	0.3	…	…
	合 计	1 361 735.2	79 198.0	30 202.6	1 313.3	563.4
南 海	广 东	654 981.6	17 325.5	1 019.0	8.2	242.7
	广 西	137 931.6	2 384.0	1 153.9	241.7	9.2
	海 南	25 505.2	1 412.4	12.0	…	…
	合 计	818 418.4	21 121.9	2 184.9	249.9	252.0

入海陆源接纳工业废水及处理情况（三）

（2013）

单位：吨

流域	地区名称	工业废水中污染物产生量					
		汞	镉	六价铬	总铬	铅	砷
总　计		1.2	59.0	1 359.6	2 907.9	81.8	249.7
渤海	天　津	0.1	0.1	24.2	33.8	2.3	0.1
	河　北	0	0	1.6	8.7	0	0
	辽　宁	0.1	34.3	1.9	2.3	33.3	23.0
	山　东	…	…	6.2	16.6	1.3	104.7
	合　计	0.2	34.4	34.0	61.4	36.9	127.8
黄海	辽　宁	0	…	1.8	1.9	…	…
	江　苏	…	…	59.5	62.9	21.2	…
	山　东	0.9	23.7	19.7	30.8	17.8	104.4
	合　计	0.9	23.8	81.0	95.6	39.0	104.4
东海	上　海	…	0.1	27.7	39.9	0.4	0.1
	浙　江	…	0.2	780.4	1 627.7	0.5	15.2
	福　建	…	0.2	378.8	741.5	2.8	0.6
	广　东	…	…	…	…	0.6	…
	合　计	…	0.4	1 186.9	2 409.1	4.3	15.9
南海	广　东	0.1	0.4	57.3	318.6	1.6	1.5
	广　西	0	0	0.3	8.2	0	…
	海　南	…	…	…	15.1	…	…
	合　计	0.1	0.4	57.7	341.9	1.6	1.5

入海陆源接纳工业废水及处理情况（四）

（2013）

单位：吨

流域	地区名称	工业废水中污染物排放量				
		化学需氧量	氨氮	石油类	挥发酚	氰化物
	总　计	402 073. 6	27 123. 0	2 266. 9	35. 4	12. 6
渤海	天　津	15 361. 2	1 972. 4	25. 7	0. 7	1. 0
	河　北	13 057. 1	864. 6	80. 8	0. 3	…
	辽　宁	21 039. 4	2 518. 3	483. 4	2. 7	0. 1
	山　东	22 507. 8	1 938. 1	68. 8	8. 0	0. 1
	合　计	71 965. 4	7 293. 4	658. 7	11. 7	1. 2
黄海	辽　宁	10 935. 5	846. 0	34. 8	…	…
	江　苏	33 341. 2	2 584. 4	260. 6	16. 0	1. 2
	山　东	11 800. 0	929. 4	68. 9	0. 5	0. 7
	合　计	56 076. 7	4 359. 7	364. 4	16. 5	1. 8
东海	上　海	20 193. 3	1 133. 4	589. 9	1. 8	2. 6
	浙　江	83 100. 2	5 309. 2	197. 6	4. 7	2. 8
	福　建	42 465. 0	2 602. 3	242. 3	0. 1	1. 6
	广　东	2 701. 2	198. 7	0. 3	…	…
	合　计	148 459. 7	9 243. 6	1 030. 2	6. 6	6. 9
南海	广　东	104 371. 2	5 599. 4	160. 2	0. 6	2. 5
	广　西	15 219. 6	278. 3	45. 0	0. 1	0. 1
	海　南	5 981. 0	348. 6	8. 4	…	…
	合　计	125 571. 9	6 226. 4	213. 7	0. 6	2. 6

入海陆源接纳工业废水及处理情况（五）

（2013）

单位：吨

流域	地区名称	工业废水中污染物排放量					
		汞	镉	六价铬	总铬	铅	砷
	总　计	**0.032**	**0.231**	**10.946**	**30.490**	**1.665**	**0.644**
渤 海	天　津	0.006	0.002	0.092	0.295	0.081	0.012
	河　北	0	0	0.004	0.071	0	0
	辽　宁	…	…	0.060	0.078	0.009	…
	山　东	…	0.001	0.038	3.035	0.004	0.001
	合　计	**0.006**	**0.002**	**0.194**	**3.480**	**0.095**	**0.013**
黄 海	辽　宁	0	…	0.061	0.066	…	…
	江　苏	0.001	0.006	1.790	2.066	0.379	0.009
	山　东	0.006	0	0.185	0.464	0.003	0.063
	合　计	**0.007**	**0.006**	**2.036**	**2.595**	**0.382**	**0.072**
东 海	上　海	0.008	0.001	0.474	1.212	0.075	0.031
	浙　江	…	0.149	4.276	8.090	0.051	0.011
	福　建	0.001	0.019	2.830	11.070	0.592	0.179
	广　东	…	0.001	…	…	0.028	0.001
	合　计	**0.009**	**0.170**	**7.579**	**20.372**	**0.747**	**0.222**
南 海	广　东	0.010	0.052	1.131	2.667	0.438	0.335
	广　西	0	0	0.006	1.254	0	…
	海　南	…	0.001	…	0.122	0.003	0.001
	合　计	**0.010**	**0.053**	**1.137**	**4.043**	**0.442**	**0.337**

入海陆源工业废水污染防治投资情况

（2013）

单位：个

流域	地名	区称	工业废水治理施工项目数/个	工业废水治理竣工项目数/个	工业废水治理项目完成投资/万元	废水治理竣工项目新增处理能力/（万吨/日）
	总　计		368	356	316 307.3	49.6
渤	天　津		1	3	4 490.0	0.2
	河　北		2	2	7 029.8	1.2
	辽　宁		1	3	5 710.0	3.0
海	山　东		20	27	35 315.0	7.9
	合　计		24	35	52 544.8	12.2
黄	辽　宁		0	0	…	…
	江　苏		59	54	40 216.3	3.9
	山　东		14	13	11 364.9	2.1
海	合　计		73	67	51 581.2	6.0
	上　海		18	17	5 759.1	2.3
东	浙　江		125	98	79 056.2	9.5
	福　建		39	38	93 807.2	10.6
海	广　东		0	0	…	…
	合　计		182	153	178 622.5	22.4
南	广　东		83	92	19 964.3	8.7
	广　西		5	7	13 082.6	0.3
	海　南		1	2	512.0	…
海	合　计		89	101	33 558.9	9.0

入海陆源农业污染排放情况

（2013）

单位：万吨

流域	地区名称	农业污染物排放/流失总量			
		化学需氧量	氨氮	总氮	总磷
	总　计	108.2	9.5	39.3	5.5
渤海	天　津	0.6	…	0.1	…
	河　北	7.5	0.4	3.4	0.4
	辽　宁	11.6	0.6	2.6	0.4
	山　东	13.2	0.7	4.8	0.7
	合　计	32.8	1.8	11.0	1.6
黄海	辽　宁	6.8	0.3	1.5	0.3
	江　苏	11.3	1.2	5.0	0.5
	山　东	12.9	0.7	4.2	0.7
	合　计	31.1	2.1	10.7	1.5
东海	上　海	2.6	0.3	1.2	0.2
	浙　江	6.2	0.9	3.3	0.4
	福　建	9.2	1.8	3.9	0.5
	广　东	1.9	0.2	0.6	0.1
	合　计	20.0	3.2	9.0	1.2
南海	广　东	14.2	1.5	5.4	0.8
	广　西	1.7	0.1	0.7	0.1
	海　南	8.3	0.7	2.4	0.3
	合　计	24.3	2.4	8.6	1.2

入海陆源城镇生活污染排放及处理情况（一）

（2013）

流域	地区名称	城镇人口/万人	生活用水总量/万吨	污水处理厂数/座	污水处理厂设计处理能力/（万吨/日）	生活污水实际处理量/万吨
总　计		9 085	753 621	769	3 072	536 495
渤海	天　津	276	13 494	22	59	7 516
	河　北	230	17 188	16	67	13 375
	辽　宁	423	29 572	25	78	21 346
	山　东	370	26 042	43	118	20 686
	合　计	1 297	86 295	106	322	62 923
黄海	辽　宁	296	19 150	14	64	18 220
	江　苏	582	42 938	99	102	19 096
	山　东	935	68 555	73	220	54 452
	合　计	1 813	130 643	186	386	91 767
东海	上　海	858	78 358	25	629	60 213
	浙　江	1 372	121 218	97	419	81 817
	福　建	1 249	97 247	101	284	62 886
	广　东	107	6 567	24	12	2 931
	合　计	3 586	303 390	247	1 344	207 847
南海	广　东	2 017	206 385	174	878	159 461
	广　西	164	12 814	7	40	8 761
	海　南	207	14 094	49	102	5 737
	合　计	2 388	233 293	230	1 020	173 958

入海陆源城镇生活污染排放及处理情况（二）

（2013）

流域	地区名称	城镇生活污水排放量/万吨	城镇生活化学需氧量产生量/吨	城镇生活氨氮产生量/吨	城镇生活化学需氧量排放量/吨	城镇生活氨氮排放量/吨
	总　计	652 164	2 265 904	296 505	1 020 831	174 010
渤	天　津	10 121	67 852	9 259	35 293	6 768
	河　北	15 476	45 555	6 562	9 287	3 172
	辽　宁	25 327	102 740	13 784	50 950	7 775
海	山　东	22 782	87 259	11 680	26 781	5 113
	合　计	73 706	303 405	41 285	122 310	22 829
黄	辽　宁	16 048	65 743	9 009	29 047	4 385
	江　苏	37 900	144 360	18 805	86 734	14 154
海	山　东	58 162	222 370	29 758	46 998	11 629
	合　计	112 110	432 473	57 572	162 779	30 169
东	上　海	70 522	237 953	30 370	68 919	16 046
	浙　江	100 420	369 715	47 376	159 824	29 298
	福　建	83 864	305 721	39 698	172 234	27 912
海	广　东	5 064	24 087	3 122	25 037	3 317
	合　计	259 870	937 477	120 566	426 014	76 573
南	广　东	185 001	494 430	65 520	230 856	35 443
	广　西	10 550	37 646	4 620	26 281	2 692
海	海　南	10 926	60 474	6 943	52 590	6 304
	合　计	206 478	592 549	77 082	309 728	44 439

12

大气污染防治重点区域废气排放统计

ANNUAL STATISTIC REPORT ON ENVIRONMENT IN CHINA
2013

大气防治重点区域工业废气排放及处理情况（一）

（2013）

区域	地区名称	工业废气排放量/亿米³	废气污染物产生量/吨			废气污染物排放量/吨		
			二氧化硫	氮氧化物	烟（粉）尘	二氧化硫	氮氧化物	烟（粉）尘
总计		361 126	28 909 321	9 672 664	360 244 359	8 516 657	7 585 831	4 559 470
京津冀	北　京	3 692	156 984	95 394	4 229 328	52 041	75 927	27 182
	天　津	8 080	1 425 856	349 931	6 703 176	207 793	250 646	62 766
	河　北	79 121	3 368 736	1 305 414	52 508 510	1 173 147	1 105 634	1 187 198
	合　计	90 894	4 951 576	1 750 739	63 441 014	1 432 981	1 432 207	1 277 146
长三角	上　海	13 344	538 840	328 829	6 608 157	172 867	262 346	67 174
	江　苏	49 797	3 198 660	1 362 931	40 764 970	909 478	985 313	455 568
	浙　江	24 565	1 701 631	755 404	25 826 955	579 104	573 498	296 586
	合　计	87 706	5 439 132	2 447 164	73 200 083	1 661 449	1 821 157	819 328
珠三角	广　东	17 310	1 377 949	593 533	11 801 244	418 419	432 826	169 361
	合　计	17 310	1 377 949	593 533	11 801 244	418 419	432 826	169 361
辽宁中部城市群	辽　宁	20 049	911 026	416 966	18 297 111	497 828	393 964	372 532
	合　计	20 049	911 026	416 966	18 297 111	497 828	393 964	372 532
山东城市群	山　东	47 160	5 714 807	1 372 935	63 837 220	1 445 348	1 171 600	542 371
	合　计	47 160	5 714 807	1 372 935	63 837 220	1 445 348	1 171 600	542 371
武汉及其周边城市群	湖　北	13 148	1 514 942	316 838	12 985 107	317 651	254 659	146 886
	合　计	13 148	1 514 942	316 838	12 985 107	317 651	254 659	146 886
长株潭城市群	湖　南	4 565	535 604	163 451	9 411 332	109 087	79 074	51 376
	合　计	4 565	535 604	163 451	9 411 332	109 087	79 074	51 376
成渝城市群	重　庆	9 532	1 404 981	302 028	20 346 826	494 415	247 905	179 842
	四　川	14 516	1 601 973	399 138	21 491 981	583 791	344 636	165 117
	合　计	24 048	3 006 955	701 166	41 838 806	1 078 206	592 540	344 959
海峡西岸城市群	福　建	16 183	961 931	491 625	19 458 436	342 040	330 628	240 583
	合　计	16 183	961 931	491 625	19 458 436	342 040	330 628	240 583
山西中北部城市群	山　西	16 607	1 423 896	402 597	12 552 782	361 390	313 415	220 279
	合　计	16 607	1 423 896	402 597	12 552 782	361 390	313 415	220 279
陕西关中城市群	陕　西	9 670	1 610 245	564 571	16 776 789	430 822	360 876	225 380
	合　计	9 670	1 610 245	564 571	16 776 789	430 822	360 876	225 380
甘宁城市群	甘　肃	5 159	720 156	175 948	6 268 434	174 374	134 537	56 293
	宁　夏	2 168	350 693	91 103	4 567 803	92 369	84 321	27 170
	合　计	7 327	1 070 849	267 051	10 836 237	266 743	218 858	83 463
新疆乌鲁木齐城市群	新　疆	6 459	390 409	184 028	5 808 198	154 695	184 028	65 806
	合　计	6 459	390 409	184 028	5 808 198	154 695	184 028	65 806

大气防治重点区域工业废气排放及处理情况（二）

（2013）

区域	地区名称	废气治理设施数/套	废气治理设施处理能力/（万米³/小时）	废气治理设施运行费用/万元	脱硫治理设施数/套	脱硫治理设施处理能力/（千克/小时）	脱硫治理设施运行费用/万元
总　计		138 287	841 679	8 925 373	14 185	18 889 598	3 568 423
京津冀	北　京	3 316	10 729	90 066	436	27 059	27 587
	天　津	4 262	21 118	256 506	1 027	159 835	124 620
	河　北	18 172	179 690	1 331 603	1 327	721 945	388 178
	合　计	25 750	211 537	1 678 174	2 790	908 838	540 385
长三角	上　海	4 896	29 883	471 487	238	215 238	136 694
	江　苏	17 964	104 344	1 269 572	1 676	8 470 467	511 141
	浙　江	17 512	57 354	945 787	1 440	998 342	446 203
	合　计	40 372	191 581	2 686 846	3 354	9 684 048	1 094 039
珠三角	广　东	14 048	49 288	644 370	1 427	252 130	299 974
	合　计	14 048	49 288	644 370	1 427	252 130	299 974
辽宁中部城市群	辽　宁	7 231	52 866	345 370	522	823 875	99 956
	合　计	7 231	52 866	345 370	522	823 875	99 956
山东城市群	山　东	16 436	120 289	1 308 652	2 752	3 166 035	587 993
	合　计	16 436	120 289	1 308 652	2 752	3 166 035	587 993
武汉及其周边城市群	湖　北	3 079	22 250	337 761	230	470 240	132 997
	合　计	3 079	22 250	337 761	230	470 240	132 997
长株潭城市群	湖　南	1 374	4 275	73 407	81	171 817	30 899
	合　计	1 374	4 275	73 407	81	171 817	30 899
成渝城市群	重　庆	4 439	23 084	231 696	227	509 633	95 065
	四　川	7 287	36 383	334 087	528	1 556 916	130 845
	合　计	11 726	59 467	565 783	755	2 066 549	225 910
海峡西岸城市群	福　建	8 112	35 192	410 418	174	644 723	141 282
	合　计	8 112	35 192	410 418	174	644 723	141 282
山西中北部城市群	山　西	5 364	42 978	360 683	1 794	166 455	162 281
	合　计	5 364	42 978	360 683	1 794	166 455	162 281
陕西关中城市群	陕　西	2 451	21 289	233 276	192	410 334	120 060
	合　计	2 451	21 289	233 276	192	410 334	120 060
甘宁城市群	甘　肃	1 256	9 708	86 517	29	26 697	26 754
	宁　夏	436	4 111	88 174	32	52 961	52 975
	合　计	1 692	13 819	174 691	61	79 657	79 729
新疆乌鲁木齐城市群	新　疆	652	16 849	105 943	53	44 898	52 919
	合　计	652	16 849	105 943	53	44 898	52 919

大气防治重点区域工业废气排放及处理情况（三）

（2013）

区域	地区名称	脱硝治理设施数/套	脱硝治理设施处理能力/（千克/小时）	脱硝治理设施运行费用/万元	除尘治理设施数/套	除尘治理设施处理能力/（千克/小时）	除尘治理设施运行费用/万元
总 计		1 367	3 450 746	1 011 036	98 326	280 314 675	3 466 611
京津冀	北 京	58	6 896	22 488	2 309	1 603 692	26 160
	天 津	24	15 889	23 760	2 452	1 507 805	79 565
	河 北	120	98 984	102 087	15 879	20 908 270	758 041
	合 计	202	121 769	148 335	20 640	24 019 767	863 767
长三角	上 海	48	798 664	47 862	2 651	20 387 490	144 173
	江 苏	179	172 236	182 649	11 297	66 761 056	413 039
	浙 江	191	63 196	132 125	11 154	48 368 891	269 418
	合 计	418	1 034 096	362 636	25 102	135 517 437	826 630
珠三角	广 东	141	31 452	120 995	6 278	3 047 703	131 354
	合 计	141	31 452	120 995	6 278	3 047 703	131 354
辽宁中部城市群	辽 宁	22	10 901	8 388	6 480	29 815 597	215 107
	合 计	22	10 901	8 388	6 480	29 815 597	215 107
山东城市群	山 东	113	50 272	82 428	12 588	27 898 870	543 949
	合 计	113	50 272	82 428	12 588	27 898 870	543 949
武汉及其周边城市群	湖 北	40	34 796	40 699	2 458	9 825 759	149 128
	合 计	40	34 796	40 699	2 458	9 825 759	149 128
长株潭城市群	湖 南	18	4 465	7 218	1 141	5 128 778	27 259
	合 计	18	4 465	7 218	1 141	5 128 778	27 259
成渝城市群	重 庆	38	19 495	27 363	3 275	8 491 715	93 947
	四 川	195	2 040 526	22 721	5 959	6 904 134	137 600
	合 计	233	2 060 021	50 084	9 234	15 395 849	231 546
海峡西岸城市群	福 建	66	33 599	56 621	6 922	4 412 883	171 688
	合 计	66	33 599	56 621	6 922	4 412 883	171 688
山西中北部城市群	山 西	41	20 098	66 425	3 198	2 917 205	125 128
	合 计	41	20 098	66 425	3 198	2 917 205	125 128
陕西关中城市群	陕 西	55	40 543	37 138	2 103	19 406 256	69 290
	合 计	55	40 543	37 138	2 103	19 406 256	69 290
甘宁城市群	甘 肃	14	6 061	15 353	1 189	1 102 921	41 911
	宁 夏	4	2 673	14 716	400	760 144	20 483
	合 计	18	8 734	30 069	1 589	1 863 065	62 394
新疆乌鲁木齐城市群	新 疆	0	0	0	593	1 065 506	49 371
	合 计	0	0	0	593	1 065 506	49 371

大气防治重点区域工业废气排放及处理情况（四）

（2013）

区域	地区名称	工业锅炉数/台				工业锅炉蒸吨数（蒸吨）				工业窑炉数/座
		35蒸吨及以上	20（含）～35蒸吨	10（含）～20蒸吨	10蒸吨以下	35蒸吨及以上	20（含）～35蒸吨	10（含）～20蒸吨	10蒸吨以下	
总　计		7 676	3 464	7 731	38 946	2 164 298	75 748	84 804	126 257	45 537
京津冀	北　京	329	239	402	992	32 759	4 958	4 580	3 635	242
	天　津	424	302	369	1 389	52 181	6 400	3 991	4 128	866
	河　北	774	367	863	4 477	758 151	8 025	9 412	15 062	6 696
	合　计	1 527	908	1 634	6 858	843 091	19 383	17 983	22 825	7 804
长三角	上　海	153	65	153	1 648	60 497	1 432	1 632	5 258	1 179
	江　苏	979	257	688	4 850	209 780	4 812	6 843	14 681	4 416
	浙　江	660	268	918	5 864	148 610	8 202	9 806	18 897	4 316
	合　计	1 792	590	1 759	12 362	418 887	14 446	18 281	38 835	9 911
珠三角	广　东	374	199	787	3 784	105 605	4 407	8 846	12 973	2 469
	合　计	374	199	787	3 784	105 605	4 407	8 846	12 973	2 469
辽宁中部城市群	辽　宁	700	393	749	2 032	91 042	8 463	8 337	7 023	4 752
	合　计	700	393	749	2 032	91 042	8 463	8 337	7 023	4 752
山东城市群	山　东	1 829	484	851	3 348	299 582	10 148	9 419	11 561	4 965
	合　计	1 829	484	851	3 348	299 582	10 148	9 419	11 561	4 965
武汉及其周边城市群	湖　北	133	44	156	784	43 903	744	1 585	2 518	1 000
	合　计	133	44	156	784	43 903	744	1 585	2 518	1 000
长株潭城市群	湖　南	41	20	45	513	14 462	384	485	1 498	844
	合　计	41	20	45	513	14 462	384	485	1 498	844
成渝城市群	重　庆	116	62	120	1 157	30 781	1 324	1 323	2 944	1 768
	四　川	223	163	314	2 412	43 904	3 726	3 594	6 507	5 524
	合　计	339	225	434	3 569	74 684	5 050	4 917	9 451	7 292
海峡西岸城市群	福　建	170	118	411	2 660	70 531	2 510	4 850	8 671	2 597
	合　计	170	118	411	2 660	70 531	2 510	4 850	8 671	2 597
山西中北部城市群	山　西	264	158	260	1 580	77 099	3 450	2 818	5 380	1 520
	合　计	264	158	260	1 580	77 099	3 450	2 818	5 380	1 520
陕西关中城市群	陕　西	221	149	294	819	52 901	3 087	3 255	2 925	1 070
	合　计	221	149	294	819	52 901	3 087	3 255	2 925	1 070
甘宁城市群	甘　肃	86	84	233	368	20 713	1 750	2 700	1 665	975
	宁　夏	93	71	79	120	20 265	1 501	889	528	82
	合　计	179	155	312	488	40 978	3 251	3 589	2 193	1 057
新疆乌鲁木齐城市群	新　疆	107	21	39	149	31 534	425	440	404	256
	合　计	107	21	39	149	31 534	425	440	404	256

大气防治重点区域工业废气污染防治投资情况（一）

（2013）

区域	地区名称	工业废气治理施工项目数/个			工业废气治理竣工项目数/个		
		脱硫治理项目	脱硝治理项目	其他废气治理项目	脱硫治理项目	脱硝治理项目	其他废气治理项目
总　计		376	390	922	410	408	977
京津冀	北　京	1	8	15	1	7	17
	天　津	16	3	15	18	5	16
	河　北	64	36	133	68	39	155
	合　计	81	47	163	87	51	188
长三角	上　海	2	0	37	5	1	34
	江　苏	29	46	93	32	51	121
	浙　江	34	42	221	38	44	187
	合　计	65	88	351	75	96	342
珠三角	广　东	33	11	92	39	17	119
	合　计	33	11	92	39	17	119
辽宁中部城市群	辽　宁	7	15	16	6	15	25
	合　计	7	15	16	6	15	25
山东城市群	山　东	114	97	163	108	68	162
	合　计	114	97	163	108	68	162
武汉及其周边城市群	湖　北	9	13	8	20	13	11
	合　计	9	13	8	20	13	11
长株潭城市群	湖　南	1	3	17	1	7	20
	合　计	1	3	17	1	7	20
成渝城市群	重　庆	3	5	17	2	9	24
	四　川	12	36	16	15	36	20
	合　计	15	41	33	17	45	44
海峡西岸城市群	福　建	14	17	43	18	30	31
	合　计	14	17	43	18	30	31
山西中北部城市群	山　西	14	19	15	19	25	18
	合　计	14	19	15	19	25	18
陕西关中城市群	陕　西	8	21	13	5	27	11
	合　计	8	21	13	5	27	11
甘宁城市群	甘　肃	2	3	3	2	7	3
	宁　夏	8	9	5	7	6	3
	合　计	10	12	8	9	13	6
新疆乌鲁木齐城市群	新　疆	5	6	0	6	1	0
	合　计	5	6	0	6	1	0

大气防治重点区域工业废气污染防治投资情况（二）

（2013）

区域	地区名称	工业废气治理项目完成投资/万元			废气治理竣工项目新增处理能力/（万米³/小时）
		脱硫治理项目	脱硝治理项目	其他废气治理项目	
总　计		884 198	1 987 845	687 159	57 471
京津冀	北　京	50	14 173	17 509	220
	天　津	38 022	28 974	13 296	1 221
	河　北	140 131	215 765	93 188	9 944
	合　计	178 203	258 912	123 992	11 385
长三角	上　海	1 300	15 000	2 756	460
	江　苏	109 494	284 994	107 181	9 025
	浙　江	41 810	204 269	89 387	4 416
	合　计	152 604	504 262	199 325	13 900
珠三角	广　东	29 468	70 092	53 983	2 289
	合　计	29 468	70 092	53 983	2 289
辽宁中部城市群	辽　宁	10 007	60 906	29 314	2 345
	合　计	10 007	60 906	29 314	2 345
山东城市群	山　东	211 302	348 602	151 321	12 469
	合　计	211 302	348 602	151 321	12 469
武汉及其周边城市群	湖　北	25 880	162 394	5 094	1 385
	合　计	25 880	162 394	5 094	1 385
长株潭城市群	湖　南	5 250	15 105	14 055	514
	合　计	5 250	15 105	14 055	514
成渝城市群	重　庆	8 667	56 632	6 609	2 334
	四　川	34 959	75 096	9 441	1 777
	合　计	43 626	131 728	16 050	4 110
海峡西岸城市群	福　建	48 386	106 726	78 867	3 031
	合　计	48 386	106 726	78 867	3 031
山西中北部城市群	山　西	22 481	137 122	8 433	2 951
	合　计	22 481	137 122	8 433	2 951
陕西关中城市群	陕　西	31 208	102 289	4 152	2 288
	合　计	31 208	102 289	4 152	2 288
甘宁城市群	甘　肃	29 515	31 802	967	298
	宁　夏	12 185	25 666	1 607	494
	合　计	41 700	57 468	2 574	792
新疆乌鲁木齐城市群	新　疆	84 083	32 240	0	9
	合　计	84 083	32 240	0	9

大气防治重点区域城镇生活废气排放情况

（2013）

区域	地区名称	生活煤炭消费量/万吨	生活天然气消费量/万米^3	二氧化硫排放量/吨	氮氧化物排放量/吨	烟尘排放量/吨
总　计		6 778	2 531 137	809 636	173 465	458 017
京津冀	北　京	412	625 243	34 967	13 638	28 258
	天　津	204	124 829	8 959	5 221	18 400
	河　北	922	83 387	111 524	23 319	77 015
	合　计	1 537	833 458	155 449	42 177	123 673
长三角	上　海	230	155 000	42 947	23 474	6 451
	江　苏	201	178 689	31 950	6 112	17 091
	浙　江	119	50 167	14 032	2 982	6 902
	合　计	550	383 856	88 929	32 567	30 444
珠三角	广　东	21	89 224	4 239	2 152	1 361
	合　计	21	89 224	4 239	2 152	1 361
辽宁中部城市群	辽　宁	443	18 143	41 991	11 048	40 445
	合　计	443	18 143	41 991	11 048	40 445
山东城市群	山　东	1 859	123 431	199 411	26 412	108 087
	合　计	1 859	123 431	199 411	26 412	108 087
武汉及其周边城市群	湖　北	452	15 195	33 719	7 214	22 142
	合　计	452	15 195	33 719	7 214	22 142
长株潭城市群	湖　南	57	51 335	5 669	1 006	4 945
	合　计	57	51 335	5 669	1 006	4 945
成渝城市群	重　庆	146	194 906	53 261	4 487	4 401
	四　川	388	397 690	55 566	8 316	7 667
	合　计	534	592 596	108 828	12 804	12 068
海峡西岸城市群	福　建	116	9 982	18 950	2 356	9 767
	合　计	116	9 982	18 950	2 356	9 767
山西中北部城市群	山　西	573	67 974	60 921	11 919	51 566
	合　计	573	67 974	60 921	11 919	51 566
陕西关中城市群	陕　西	433	172 052	62 856	17 271	37 306
	合　计	433	172 052	62 856	17 271	37 306
甘宁城市群	甘　肃	71	32 991	9 644	2 648	3 375
	宁　夏	34	70 796	5 697	1 237	3 016
	合　计	105	103 787	15 341	3 885	6 391
新疆乌鲁木齐城市群	新　疆	98	70 104	13 333	2 654	9 823
	合　计	98	70 104	13 333	2 654	9 823

大气防治重点区域机动车废气排放情况

（2013）

区域	地区名称	机动车污染物排放总量/吨			
		总颗粒物	氮氧化物	一氧化碳	碳氢化合物
总　计		262 090	3 003 542	16 418 744	1 981 147
京津冀	北　京	3 806	76 472	690 086	77 094
	天　津	6 267	55 669	452 373	51 310
	河　北	49 057	523 476	2 511 253	318 511
	合　计	59 130	655 617	3 653 712	446 915
长三角	上　海	7 218	94 119	424 910	61 876
	江　苏	27 040	346 175	1 814 921	215 756
	浙　江	16 059	176 297	1 235 693	142 315
	合　计	50 318	616 591	3 475 525	419 947
珠三角	广　东	31 070	318 806	1 700 118	193 672
	合　计	31 070	318 806	1 700 118	193 672
辽宁中部城市群	辽　宁	12 553	131 258	659 987	82 243
	合　计	12 553	131 258	659 987	82 243
山东城市群	山　东	46 214	453 122	2 269 575	279 449
	合　计	46 214	453 122	2 269 575	279 449
武汉及其周边城市群	湖　北	8 931	99 392	553 990	64 393
	合　计	8 931	99 392	553 990	64 393
长株潭城市群	湖　南	3 391	47 085	264 706	29 386
	合　计	3 391	47 085	264 706	29 386
成渝城市群	重　庆	6 952	109 631	620 122	79 760
	四　川	11 772	159 854	944 494	111 343
	合　计	18 724	269 485	1 564 615	191 102
海峡西岸城市群	福　建	9 005	105 316	582 952	69 056
	合　计	9 005	105 316	582 952	69 056
山西中北部城市群	山　西	7 430	80 522	426 661	51 026
	合　计	7 430	80 522	426 661	51 026
陕西关中城市群	陕　西	5 480	102 195	595 872	68 256
	合　计	5 480	102 195	595 872	68 256
甘宁城市群	甘　肃	2 043	36 847	318 682	38 963
	宁　夏	3 962	35 474	146 198	19 177
	合　计	6 005	72 320	464 880	58 140
新疆乌鲁木齐城市群	新　疆	3 839	51 830	206 151	27 561
	合　计	3 839	51 830	206 151	27 561

大气防治重点区域生活垃圾处理厂废气排放情况

（2013）

区域	地区名称	焚烧废气中污染物排放量		
		二氧化硫/吨	氮氧化物/吨	烟尘/吨
总　计		809.3	2 205.6	540.5
京津冀	北　京	33.9	284.0	39.3
	天　津	18.5	148.0	10.0
	河　北	2.7	…	1.1
	合　计	55.0	432.0	50.4
长三角	上　海	…	…	…
	江　苏	163.1	193.4	42.0
	浙　江	108.7	96.4	137.0
	合　计	271.8	289.8	178.9
珠三角	广　东	212.0	1 147.7	224.2
	合　计	212.0	1 147.7	224.2
辽宁中部城市群	辽　宁	…	…	…
	合　计	…	…	…
山东城市群	山　东	42.7	99.9	28.4
	合　计	42.7	99.9	28.4
武汉及其周边城市群	湖　北	…	…	…
	合　计	…	…	…
长株潭城市群	湖　南	…	…	…
	合　计	…	…	…
成渝城市群	重　庆			
	四　川	223.9	232.5	57.0
	合　计	223.9	232.5	57.0
海峡西岸城市群	福　建	3.9	3.7	1.6
	合　计	3.9	3.7	1.6
山西中北部城市群	山　西	…	…	…
	合　计	…	…	…
陕西关中城市群	陕　西	…	…	…
	合　计	…	…	…
甘宁城市群	甘　肃	…	…	…
	宁　夏	…	…	…
	合　计	…	…	…
新疆乌鲁木齐城市群	新　疆	…	…	…
	合　计	…	…	…

大气防治重点区域危险（医疗）废物处置厂废气排放情况
（2013）

区域	地区 名称	焚烧废气中污染物排放量		
		二氧化硫/吨	氮氧化物/吨	烟尘/吨
总　计		624.9	1 256.2	623.9
京津冀	北　京	0.2	8.4	0.8
	天　津	61.8	35.8	13.4
	河　北	23.5	38.9	41.5
	合　计	85.5	83.0	55.7
长三角	上　海	33.7	416.1	81.7
	江　苏	88.2	247.3	219.1
	浙　江	119.6	103.0	61.8
	合　计	241.5	766.5	362.6
珠三角	广　东	44.2	130.8	29.5
	合　计	44.2	130.8	29.5
辽宁中部城市群	辽　宁	3.0	9.8	3.6
	合　计	3.0	9.8	3.6
山东城市群	山　东	166.2	93.0	24.5
	合　计	166.2	93.0	24.5
武汉及其周边城市群	湖　北	21.0	18.6	13.4
	合　计	21.0	18.6	13.4
长株潭城市群	湖　南	…	1.2	0.1
	合　计	…	1.2	0.1
成渝城市群	重　庆	9.5	19.4	9.1
	四　川	34.8	55.2	38.4
	合　计	44.3	74.6	47.5
海峡西岸城市群	福　建	9.5	39.9	7.2
	合　计	9.5	39.9	7.2
山西中北部城市群	山　西	0.4	11.6	1.0
	合　计	0.4	11.6	1.0
陕西关中城市群	陕　西	7.1	19.4	41.5
	合　计	7.1	19.4	41.5
甘宁城市群	甘　肃	0.3	2.0	1.8
	宁　夏	0.6	2.4	1.9
	合　计	0.9	4.4	3.8
新疆乌鲁木齐城市群	新　疆	1.3	3.4	33.6
	合　计	1.3	3.4	33.6

13

环境管理统计

ANNUAL STATISTIC REPORT ON ENVIRONMENT IN CHINA
2013

各地区环保机构情况（一）

（2013）

地 区 名 称	环保系统机构总数	国家级、省级机构数	环保行政主管部门机构数	环境监察机构数	环境监测站机构数	辐射监测机构数	科研机构数	宣教机构数	信息机构数	应急机构数	其他机构数
总 计	14 257	445	32	51	47	34	35	34	24	10	178
国家级	45	45	1	12	1	1	3	1	1	1	24
北 京	105	15	1	1	1	2	1	1	1	1	6
天 津	83	14	1	1	1	1	1	1	1	1	6
河 北	821	13	1	2	1	1	1	1	1	0	5
山 西	616	19	1	1	2	1	1	1	1	0	11
内蒙古	416	12	1	3	1	0	1	1	0	0	5
辽 宁	509	20	1	1	1	2	1	1	0	0	13
吉 林	252	11	1	1	1	1	1	1	1	1	3
黑龙江	446	16	1	1	7	1	1	1	1	0	3
上 海	68	8	1	1	1	1	1	1	1	0	1
江 苏	589	20	1	4	2	1	1	1	0	1	9
浙 江	611	12	1	1	2	1	1	1	1	0	4
安 徽	409	11	1	1	1	1	1	1	1	0	4
福 建	870	13	1	1	3	1	1	1	1	1	3
江 西	431	11	1	1	1	1	1	2	1	0	3
山 东	752	12	1	1	2	1	2	2	0	0	3
河 南	788	9	1	1	1	1	1	1	0	0	3
湖 北	505	10	1	1	1	1	1	1	1	0	3
湖 南	501	15	1	1	3	1	1	1	1	1	5
广 东	747	11	1	0	1	1	1	1	1	0	5
广 西	383	13	1	1	2	1	1	1	0	1	5
海 南	181	8	1	1	0	1	1	2	1	0	1
重 庆	937	11	1	1	0	1	1	1	1	0	5
四 川	726	13	1	1	1	1	1	1	1	0	6
贵 州	425	14	1	1	1	1	1	1	0	1	7
云 南	494	11	1	1	1	1	1	1	1	0	4
西 藏	121	8	1	1	1	1	0	0	1	0	3
陕 西	409	15	1	2	1	2	1	1	1	0	6
甘 肃	422	15	1	1	1	1	1	1	1	1	7
青 海	126	10	1	1	2	1	1	1	1	0	2
宁 夏	69	13	1	2	1	1	1	1	0	0	6
新 疆	400	17	1	2	2	1	2	1	1	0	7

各地区环保机构情况（二）

（2013）

单位：个

地 区 名 称	地市级环保机构数	环保行政主管部门机构数	环境监察机构数	环境监测站机构数	辐射监测机构数	科研机构数	宣教机构数	信息机构数	应急机构数	其他机构数
总 计	2 252	333	357	361	108	193	123	128	61	588
北 京	0	0	0	0	0	0	0	0	0	0
天 津	0	0	0	0	0	0	0	0	0	0
河 北	89	11	18	12	3	12	7	7	2	17
山 西	116	11	13	16	7	11	8	10	4	36
内蒙古	85	12	10	13	2	7	4	2	1	34
辽 宁	147	14	17	16	0	13	13	8	4	62
吉 林	67	9	10	11	7	3	7	3	3	14
黑龙江	92	13	15	16	8	12	6	4	0	18
上 海	0	0	0	0	0	0	0	0	0	0
江 苏	102	13	16	14	10	8	9	5	5	22
浙 江	86	11	12	11	0	6	5	2	2	37
安 徽	96	16	16	16	2	9	3	13	1	20
福 建	70	9	11	13	4	9	5	9	2	8
江 西	76	11	12	11	4	10	6	3	0	19
山 东	137	17	21	19	4	10	8	4	5	49
河 南	116	17	23	18	14	11	6	2	5	20
湖 北	81	13	14	13	7	9	3	4	1	17
湖 南	94	14	14	15	5	12	2	5	0	27
广 东	155	21	11	23	2	16	10	12	2	58
广 西	67	14	14	14	3	9	1	3	6	3
海 南	10	3	2	2	0	1	0	2	0	0
重 庆	0	0	0	0	0	0	0	0	0	0
四 川	95	21	21	24	4	2	2	9	1	11
贵 州	68	9	11	9	4	2	4	5	7	17
云 南	87	16	16	18	2	6	4	1	1	23
西 藏	19	7	4	5	0	0	0	0	0	3
陕 西	72	10	13	10	3	3	6	1	0	26
甘 肃	106	14	15	16	10	7	1	8	8	27
青 海	29	8	8	6	0	1	0	1	1	4
宁 夏	20	5	5	6	0	1	0	1	0	2
新 疆	70	14	15	14	3	3	3	4	0	14

各地区环保机构情况（三）

（2013）

单位：个

地 区 名 称	县级环保 机构数	环保行政主管 部门机构数	环境监察 机构数	环境监测 站机构数	辐射监测 机构数	科研机 构数	宣教机 构数	信息机 构数	应急机 构数	其他机 构数	乡镇环保 机构数
总　计	8 866	2 811	2 515	2 346	75	96	98	98	61	766	2 694
北　京	90	16	25	17	4	3	1	0	0	24	0
天　津	67	16	16	16	0	0	0	0	0	19	2
河　北	557	172	166	151	1	8	8	8	1	42	162
山　西	417	119	128	112	8	3	5	9	0	33	64
内蒙古	318	102	94	92	1	0	1	0	1	27	1
辽　宁	334	100	89	77	0	2	3	0	0	63	8
吉　林	167	57	50	49	0	0	5	1	0	5	7
黑龙江	335	114	101	92	0	0	0	0	0	28	3
上　海	60	17	16	17	5	2	2	0	0	1	0
江　苏	359	99	101	91	8	8	3	2	11	36	108
浙　江	327	90	78	74	1	4	3	3	8	66	186
安　徽	288	104	89	68	0	1	0	7	0	19	14
福　建	276	83	81	77	2	2	1	5	7	18	511
江　西	336	100	99	100	1	2	0	1	0	33	8
山　东	501	137	125	120	4	5	7	6	7	90	102
河　南	491	159	132	123	15	14	3	3	7	35	172
湖　北	334	99	90	83	6	12	6	2	1	35	80
湖　南	368	121	111	104	3	4	0	0	1	24	24
广　东	396	117	47	106	5	22	20	21	2	56	185
广　西	293	105	92	79	0	0	0	1	4	12	10
海　南	54	20	16	13	0	0	0	0	0	5	109
重　庆	150	38	40	41	3	1	12	3	4	8	776
四　川	567	183	180	172	0	1	3	9	0	19	51
贵　州	335	88	87	87	2	1	11	15	6	38	8
云　南	376	129	112	120	0	1	3	1	0	10	20
西　藏	94	74	9	11	0	0	0	0	0	0	0
陕　西	310	107	107	95	0	0	0	0	0	1	12
甘　肃	247	86	87	63	6	0	0	0	1	4	54
青　海	87	43	32	9	0	0	0	0	0	3	0
宁　夏	35	17	9	8	0	0	0	0	0	1	1
新　疆	297	99	106	79	0	0	1	1	0	11	16

各地区环保机构情况（四）
（2013）

单位：人

地 区 名 称	环保系统 人员总数	国家级、 省级人 员数	环保行政 主管部门 人员数	环境监 察机构 人员数	环境监 测站人 员数	辐射监 测机构 人员数	科研机 构人员 数	宣教机 构人员 数	信息机 构人员 数	应急机 构人员 数	其他机 构人员 数
总　计	212 048	17 681	3 158	1 824	3 283	1 511	3 240	532	310	144	3 670
国家级	2 951	2 951	357	519	185	481	575	38	35	39	713
北　京	2 764	928	123	73	163	58	261	47	15	9	179
天　津	1 835	621	108	51	162	28	143	25	22	6	76
河　北	18 639	436	109	62	81	30	46	16	12	0	80
山　西	11 617	488	98	50	104	23	47	9	12	0	145
内蒙古	5 998	317	69	45	59	0	22	13	0	0	109
辽　宁	9 285	522	70	27	105	42	95	10	0	0	173
吉　林	5 355	365	84	16	94	24	69	18	13	9	38
黑龙江	5 104	429	85	33	171	19	74	11	8	0	28
上　海	2 251	658	104	63	181	37	194	29	15	0	35
江　苏	10 597	721	132	51	167	39	191	14	0	24	103
浙　江	7 000	445	76	38	130	84	74	7	6	0	30
安　徽	5 711	368	80	40	88	20	75	12	11	0	42
福　建	4 614	408	88	38	89	54	42	10	6	18	63
江　西	5 371	386	70	45	76	40	96	11	15	0	33
山　东	14 431	488	95	34	131	25	65	38	0	0	100
河　南	21 815	373	108	50	66	30	40	23	0	0	56
湖　北	7 511	333	85	37	63	15	75	19	8	0	31
湖　南	10 000	1 003	69	24	176	10	129	18	4	10	563
广　东	11 657	815	146	0	167	50	39	17	9	0	387
广　西	4 013	310	54	23	96	38	37	16	0	8	38
海　南	1 543	232	91	26	0	13	84	2	12	0	4
重　庆	5 662	566	143	68	0	29	189	23	28	0	86
四　川	9 545	537	106	32	110	70	149	13	14	0	43
贵　州	4 034	386	74	53	68	24	77	9	0	9	72
云　南	5 557	510	99	55	95	30	150	15	11	0	55
西　藏	890	172	56	17	40	13	0	0	8	0	38
陕　西	5 931	457	101	84	82	61	43	15	16	0	55
甘　肃	4 635	442	99	29	84	25	77	13	10	12	93
青　海	941	223	52	26	84	29	0	6	8	0	18
宁　夏	937	305	58	56	62	32	0	21	0	0	76
新　疆	3 854	486	69	59	104	38	82	14	12	0	108

各地区环保机构情况（五）

（2013）

单位：人

地区名称	地市级环保机构人员数	环保行政主管部门人员数	环境监察机构人员数	环境监测站人员数	辐射监测机构人员数	科研机构人员数	宣教机构人员数	信息机构人员数	应急机构人员数	其他机构人员数
总　计	47 016	9 993	9 861	16 433	694	3 310	873	813	402	4 637
北　京	0	0	0	0	0	0	0	0	0	0
天　津	0	0	0	0	0	0	0	0	0	0
河　北	2 722	545	651	821	43	285	80	62	12	223
山　西	2 051	377	434	569	23	175	43	82	59	289
内蒙古	1 636	302	242	552	66	128	34	8	14	290
辽　宁	3 428	499	626	1 009	0	507	99	69	32	587
吉　林	1 120	287	283	294	43	45	42	11	17	98
黑龙江	1 488	347	305	496	36	129	43	15	0	117
上　海	0	0	0	0	0	0	0	0	0	0
江　苏	2 836	528	740	1 065	67	189	56	25	45	121
浙　江	1 721	298	391	629	0	129	30	15	8	221
安　徽	1 649	339	411	623	9	68	14	65	5	115
福　建	1 311	200	331	500	20	136	31	45	10	38
江　西	1 259	254	305	393	28	110	39	17	0	113
山　东	3 190	725	583	1 046	25	193	81	31	35	471
河　南	3 469	681	957	1 183	126	200	53	19	47	203
湖　北	1 711	330	362	639	41	153	23	26	5	132
湖　南	2 172	452	423	846	12	221	18	30	0	170
广　东	3 884	1 147	417	1 146	14	299	61	56	8	736
广　西	1 487	314	328	693	17	74	10	24	14	13
海　南	248	57	50	92	0	6	0	43	0	0
重　庆	0	0	0	0	0	0	0	0	0	0
四　川	2 352	556	509	1 030	28	85	29	39	29	47
贵　州	1 037	231	199	394	21	13	30	43	37	69
云　南	1 384	375	253	518	13	75	15	4	3	128
西　藏	285	197	26	37	0	0	0	0	0	25
陕　西	1 515	279	325	615	23	27	31	12	0	203
甘　肃	1 261	284	233	503	32	28	3	41	19	118
青　海	223	68	60	77	0	4	0	8	3	3
宁　夏	349	83	100	152	0	0	0	8	0	6
新　疆	1 228	238	317	511	7	31	8	15	0	101

各地区环保机构情况（六）

（2013）

单位：人

地 区名 称	县级环保机构人员数	环保行政主管部门人员数	环境监察机构人员数	环境监测站人员数	辐射监测机构人员数	科研机构人员数	宣教机构人员数	信息机构人员数	应急机构人员数	其他机构人员数	乡镇环保机构人员数
总　计	137 099	39 694	51 011	38 168	313	895	501	412	206	5 899	10 252
北　京	1 836	411	426	333	20	18	16	0	0	612	0
天　津	1 196	309	249	447	0	0	0	0	0	191	18
河　北	14 207	4 667	5 247	3 315	7	143	122	84	7	615	1 274
山　西	8 394	1 437	3 578	2 886	46	45	24	58	0	320	684
内蒙古	4 026	1 296	1 493	1 046	3	0	2	0	4	182	19
辽　宁	5 278	1 192	2 107	1 258	0	17	5	0	0	699	57
吉　林	3 845	865	1 928	943	0	0	24	4	0	81	25
黑龙江	3 160	795	1 399	830	0	0	0	0	0	136	27
上　海	1 593	272	446	787	45	7	12	0	0	24	0
江　苏	6 854	1 941	2 332	2 317	32	57	7	1	24	143	186
浙　江	4 286	1 034	1 521	1 465	2	22	10	6	35	191	548
安　徽	3 653	987	1 647	887	0	0	0	18	0	114	41
福　建	2 774	638	1 116	935	3	0	2	35	7	38	121
江　西	3 688	1 294	1 261	983	0	20	0	3	0	127	38
山　东	9 781	3 173	2 963	2 794	19	34	86	31	32	649	972
河　南	16 738	3 859	7 916	4 255	67	151	25	9	45	411	1 235
湖　北	5 079	1 621	1 858	1 306	15	79	27	1	2	170	388
湖　南	6 660	2 022	2 572	1 892	8	25	0	0	1	140	165
广　东	5 867	2 640	606	1 801	19	181	56	94	10	460	1 091
广　西	2 197	805	730	617	0	0	0	2	20	23	19
海　南	779	325	233	129	0	0	0	0	0	92	284
重　庆	2 415	603	819	874	4	0	38	15	12	50	2 681
四　川	6 558	1 939	2 320	2 001	0	91	5	24	0	178	98
贵　州	2 575	842	921	640	6	2	23	24	7	110	36
云　南	3 598	1 573	869	1 089	0	3	14	0	0	50	65
西　藏	433	410	10	13	0	0	0	0	0	0	0
陕　西	3 958	801	2 102	1 051	0	0	0	0	0	4	1
甘　肃	2 793	995	1 241	518	17	0	0	0	0	22	139
青　海	495	193	182	114	0	0	0	0	0	6	0
宁　夏	276	74	85	109	0	0	0	0	0	8	7
新　疆	2 107	681	834	533	0	0	3	3	0	53	33

各地区环境信访与环境法制情况（一）

（2013）

单位：件

地　区　名　称	当年颁布地方性法规	当年颁布地方政府规章	现行有效的地方性法规总数	现行有效的地方政府规章总数	当年受理行政复议案件数	本级行政处罚案件数	电话/网络投诉数	电话/网络投诉办结数
总　　计	29	41	379	314	550	139 059	1 112 172	1 098 555
国家级	…	…	…	…	185	7	1 960	1 960
北　京	0	0	3	3	8	23 583	48 138	48 138
天　津	0	1	4	11	3	369	22 914	22 914
河　北	1	1	11	12	7	5 809	24 263	24 545
山　西	1	0	19	10	2	1 569	14 077	13 997
内蒙古	0	0	17	10	0	1 135	9 952	9 210
辽　宁	3	2	18	16	7	9 345	34 128	33 891
吉　林	0	0	10	9	0	1 652	13 966	13 921
黑龙江	0	0	11	18	4	38 434	18 489	18 407
上　海	0	0	3	10	8	2 145	54 195	53 938
江　苏	1	2	22	21	23	5 813	114 630	114 092
浙　江	1	0	20	18	22	9 979	55 026	54 250
安　徽	0	2	13	8	9	763	31 805	31 641
福　建	0	0	22	7	15	3 770	63 580	61 518
江　西	1	0	4	3	8	1 553	9 232	9 121
山　东	0	1	19	10	8	5 943	98 803	98 668
河　南	1	1	8	10	29	1 949	48 606	47 694
湖　北	0	1	12	12	7	2 232	34 262	33 881
湖　南	0	0	7	4	7	1 718	27 141	26 366
广　东	7	3	36	13	94	11 144	134 435	131 585
广　西	0	0	3	6	11	646	35 819	35 422
海　南	3	0	25	5	17	228	3 006	2 966
重　庆	0	0	3	6	42	1 453	86 819	86 819
四　川	1	0	15	9	13	1 629	34 644	34 631
贵　州	0	1	7	5	0	855	8 830	8 783
云　南	5	8	24	32	7	1 189	21 403	18 190
西　藏	0	2	1	5	0	15	160	160
陕　西	1	0	12	7	9	1 749	37 945	37 945
甘　肃	2	16	8	16	1	488	6 065	6 060
青　海	1	0	10	5	2	211	3 738	3 729
宁　夏	0	0	6	9	0	559	5 175	5 167
新　疆	0	0	6	4	2	1 125	8 966	8 946

各地区环境信访与环境法制情况（二）

（2013）

地 区 名 称	当年备案的 地方环境标 准数/件	累积备案的 地方环境标 准总数/件	来信总数/ 件	来访总数/ 批次	来访总数 （人数）/ 人	来信、来访 已办结数量/ 件	承办的人大 建议数/件	承办的政协 提案数/件
总 计	31	126	103 776	46 162	107 165	151 635	7 222	10 990
国家级	…	…	3 155	846	2 455	3 820	434	335
北 京	6	34	1 698	334	551	2 172	101	86
天 津	0	3	510	307	554	817	44	70
河 北	4	8	2 202	1 690	3 301	4 082	332	792
山 西	0	0	1 532	520	1 180	2 035	169	201
内蒙古	0	0	1 015	889	3 149	1 712	61	130
辽 宁	0	1	4 109	2 594	6 173	6 611	244	332
吉 林	0	0	804	1 205	3 246	2 414	86	124
黑龙江	0	2	2 169	2 250	5 926	5 001	134	161
上 海	0	9	1 742	575	7 050	2 267	98	151
江 苏	0	0	3 890	1 820	4 280	6 010	403	602
浙 江	1	5	5 888	2 203	4 409	8 801	727	912
安 徽	0	0	6 729	1 409	3 457	8 899	222	384
福 建	4	8	4 628	1 305	2 761	6 184	433	480
江 西	0	0	5 352	3 161	6 499	8 928	198	257
山 东	6	20	3 768	1 293	3 396	5 420	245	560
河 南	5	9	4 344	2 218	4 475	6 534	315	574
湖 北	0	0	5 672	2 027	4 581	7 016	369	543
湖 南	0	0	5 168	4 033	9 683	9 465	446	443
广 东	0	13	14 894	2 795	6 003	16 500	471	553
广 西	1	1	2 184	1 859	3 517	4 000	129	195
海 南	2	2	168	82	358	241	27	42
重 庆	2	8	3 931	1 425	2 582	5 417	220	360
四 川	0	0	7 445	3 051	6 281	11 261	297	486
贵 州	0	0	2 250	911	1 657	3 260	131	213
云 南	0	0	2 216	1 618	3 286	4 257	382	448
西 藏	0	0	208	53	84	287	13	560
陕 西	0	2	1 476	1 284	2 160	2 483	169	268
甘 肃	0	1	1 808	542	1 205	1 593	107	118
青 海	0	0	566	85	186	628	28	36
宁 夏	0	0	297	656	893	673	14	411
新 疆	0	0	1 958	1 122	1 827	2 847	173	163

各地区环境保护能力建设投资情况（一）

（2013）

单位：万元

地 区名 称	本级环保能力建设资金使用总额	监测能力建设	监察能力建设	核与辐射安全监管能力建设	固体废物管理能力建设	环境应急能力建设	环境信息能力建设	环境宣教能力建设	环境监管运行保障
总 计	1 319 412.3	571 722.2	164 614.6	29 677.9	21 757.9	44 490.3	62 919.5	43 848.7	353 878.6
国家级	23 788.0	5 319.0	730.0	3 063.0	…	1 462.0	…	988.0	11 226.0
北 京	30 298.2	9 282.4	1 100.9	1 139.5	349.4	362.8	586.1	2 096.2	15 353.6
天 津	11 027.2	4 779.3	1 667.0	100.0	243.0	541.5	240.0	242.3	3 213.9
河 北	66 285.2	35 743.3	10 169.9	628.5	257.4	2 274.9	6 249.7	1 101.1	9 761.0
山 西	44 639.0	19 968.6	6 670.7	717.6	478.6	7 026.2	3 696.9	281.6	5 757.2
内蒙古	49 228.9	21 687.8	6 748.1	230.0	1 717.8	2 724.7	1 154.9	712.1	3 978.5
辽 宁	65 749.3	26 939.7	7 097.4	2 303.0	65.0	2 682.7	4 477.3	1 647.5	20 536.7
吉 林	17 516.7	12 916.4	1 842.6	786.6	8.0	245.7	744.8	115.1	857.4
黑龙江	17 568.2	9 090.7	4 912.9	760.4	114.5	562.0	240.2	420.2	1 444.6
上 海	24 265.5	13 041.4	1 624.0	676.5	36.0	435.2	1 863.8	512.2	6 075.6
江 苏	100 227.3	41 208.0	11 913.5	2 245.4	5 207.9	3 376.0	6 125.9	3 162.4	26 971.2
浙 江	99 713.9	54 504.0	6 397.7	1 400.4	1 120.3	2 264.3	5 283.2	2 434.6	26 139.7
安 徽	31 754.1	13 254.2	4 262.1	348.9	1 347.1	1 287.4	935.3	584.0	9 662.2
福 建	58 290.7	23 765.2	8 451.1	477.0	662.2	677.0	965.0	1 686.0	19 458.7
江 西	25 218.9	12 680.6	3 014.4	174.1	129.6	410.0	707.1	774.3	7 315.7
山 东	39 538.6	16 401.9	5 826.2	1 264.7	109.4	3 143.1	1 625.8	708.7	10 418.1
河 南	56 234.8	37 612.7	8 567.4	960.5	170.5	699.3	727.7	822.4	6 674.3
湖 北	123 293.4	12 534.9	4 149.6	283.6	45.0	1 303.0	1 338.3	10 394.5	93 244.5
湖 南	35 160.6	15 214.7	7 414.0	736.8	270.3	868.8	2 332.2	1 076.4	6 351.4
广 东	95 956.3	38 950.3	15 985.8	679.4	3 470.1	1 703.9	5 324.9	7 899.5	16 445.1
广 西	17 849.1	11 225.3	1 359.3	296.5	80.7	600.8	305.6	229.9	3 533.8
海 南	5 916.8	3 726.6	204.2	837.0	4.0	53.9	89.3	77.0	908.3
重 庆	36 302.9	11 543.0	6 645.8	847.0	336.0	4 768.8	1 781.7	1 413.3	8 956.0
四 川	72 642.0	33 574.4	8 803.1	3 565.6	295.9	841.8	5 509.6	782.6	13 712.6
贵 州	23 468.5	13 819.0	4 004.6	373.5	1 654.0	370.0	1 049.9	502.6	1 573.7
云 南	32 466.8	11 430.2	6 197.2	880.3	2 196.7	506.8	2 613.5	902.9	7 639.6
西 藏	2 740.0	129.1	143.0	515.1	0.1	42.0	757.0	54.6	992.0
陕 西	31 479.2	17 360.4	3 848.5	1 535.5	53.9	630.2	1 253.7	753.6	6 043.3
甘 肃	17 691.6	9 698.6	3 674.3	1 059.8	290.5	229.0	278.1	358.9	2 100.7
青 海	9 994.2	5 480.0	1 025.9	614.0	140.0	23.0	361.0	181.0	2 169.3
宁 夏	11 746.1	7 819.2	425.6	45.5	792.2	1 323.0	469.6	538.0	333.0
新 疆	41 360.3	21 021.3	9 737.8	132.2	111.0	1 050.5	3 831.4	395.2	5 030.9

各地区环境保护能力建设投资情况（二）

（2013）

单位：万元

地 区 名 称	资金来源			
	国家拨付	省级拨付	地市拨付	县（区）拨付
总　计	159 839.0	341 428.6	350 555.4	467 589.3
国家级	23 788.0	…	…	…
北　京	0	14 759.6	2 497.4	13 041.2
天　津	1 750.4	4 696.8	801.4	3 778.6
河　北	17 577.1	8 458.4	23 684.9	16 564.8
山　西	6 048.1	17 188.5	8 094.8	13 307.6
内蒙古	4 167.4	31 439.1	8 574.8	5 047.6
辽　宁	12 080.6	23 511.0	25 464.4	4 693.3
吉　林	9 847.3	1 536.3	2 563.6	3 569.5
黑龙江	3 793.6	6 370.9	4 975.9	2 427.8
上　海	557.0	8 400.0	2 333.8	12 974.7
江　苏	1 704.0	14 755.3	35 043.8	48 724.2
浙　江	2 127.0	27 581.8	27 193.5	42 811.6
安　徽	4 603.2	8 323.4	11 741.9	7 085.6
福　建	4 234.3	2 085.9	30 074.6	21 895.9
江　西	2 525.6	7 899.0	6 871.8	7 922.5
山　东	2 917.8	10 130.8	13 662.6	12 827.4
河　南	6 492.8	18 456.0	16 362.9	14 923.1
湖　北	4 472.4	8 608.9	7 196.4	103 015.7
湖　南	3 835.0	14 420.5	8 697.4	8 207.7
广　东	2 228.5	14 069.2	25 413.3	54 245.3
广　西	3 195.8	4 020.8	7 052.2	3 580.3
海　南	96.0	1 700.8	2 666.1	1 453.9
重　庆	3 627.0	16 245.6	4 540.4	11 889.9
四　川	5 318.0	22 575.0	21 825.5	22 923.5
贵　州	4 281.6	8 518.8	2 980.5	7 687.6
云　南	3 779.0	9 084.9	7 059.9	12 543.0
西　藏	1 929.2	33.0	600.2	177.6
陕　西	4 146.0	15 057.3	10 540.5	1 735.4
甘　肃	6 829.8	5 565.0	3 551.9	1 744.9
青　海	2 269.0	6 464.5	453.0	807.7
宁　夏	5 661.0	5 571.5	81.6	432.0
新　疆	3 956.5	3 900.0	27 954.4	5 549.4

各地区环境污染源控制与管理情况

（2013）

地 区 名 称	清洁生产审核当年完成企业数/家	强制性审核当年完成数	应开展监测的重金属污染防控重点企业数/家	重金属排放达标的重点企业数	已发放危险废物经营许可证数/个	具有医疗废物经营范围的许可证数	受省级环保部门委托的机动车环保检验机构数/个	受省级环保部门委托的机动车环保检验机构人数/人
总　计	9 763	6 385	3 805	3 107	1 763	258	2 423	31 624
国家级	…	…	…	…	20	4	…	…
北　京	61	7	4	4	11	3	41	0
天　津	89	46	60	56	15	1	33	193
河　北	517	420	161	145	69	14	111	1 168
山　西	123	106	66	35	19	6	102	968
内蒙古	91	70	40	39	51	11	70	850
辽　宁	335	302	109	84	50	7	47	858
吉　林	112	93	28	27	64	4	95	591
黑龙江	50	42	29	22	35	7	108	709
上　海	541	214	196	182	33	2	94	858
江　苏	1 803	1 126	157	157	317	19	140	1 828
浙　江	1 516	460	116	101	100	11	90	1 840
安　徽	260	230	142	131	51	14	74	1 023
福　建	412	411	194	193	29	5	464	4 544
江　西	120	102	143	94	72	10	93	773
山　东	448	373	128	128	105	17	169	4 169
河　南	306	251	302	302	84	16	65	1 059
湖　北	74	45	74	57	36	11	11	161
湖　南	204	147	576	235	141	11	29	297
广　东	1 595	1 115	417	361	147	17	220	3 578
广　西	130	113	77	60	23	4	12	179
海　南	11	6	7	6	8	2	25	308
重　庆	198	194	89	89	16	2	93	1 394
四　川	267	121	107	103	59	18	8	203
贵　州	82	81	32	32	28	3	63	801
云　南	176	118	277	246	62	6	22	537
西　藏	4	0	33	9	1	1	2	30
陕　西	58	51	64	60	42	5	55	756
甘　肃	108	108	131	109	23	13	43	742
青　海	13	10	17	15	7	2	13	434
宁　夏	19	14	17	17	10	3	24	366
新　疆	40	9	12	8	35	9	7	407

各地区环境监测情况（一）

（2013）

地 区 名 称	监测用房 面积/米²	监测业务 经费/万元	环境监测— 监测仪器设 备台（套） 数/台（套）	环境监测—监 测仪器设备原 值总值/万元	环境空气 监测点位 数/个	国控监测 点位数	酸雨监测 点位/个	沙尘天气影 响环境质量 监测点位数/ 个
总　计	2 867 723	369 878	268 306	1 462 525	3 001	1 436	1 176	82
国家级	14 998	17 825	339	2 831	…	…	…	…
北　京	24 227	5 308	6 290	45 801	35	12	2	1
天　津	23 653	4 020	5 474	24 877	69	15	17	1
河　北	155 918	16 387	13 452	57 759	88	53	22	3
山　西	89 239	11 827	10 702	48 732	81	58	23	2
内蒙古	125 577	7 280	9 655	54 866	53	44	28	20
辽　宁	117 665	11 075	11 229	66 812	96	77	54	9
吉　林	73 457	9 824	5 252	25 781	174	33	20	5
黑龙江	70 616	6 120	5 487	35 181	77	57	30	0
上　海	56 867	6 886	5 692	63 205	68	10	22	0
江　苏	182 138	41 604	17 515	115 170	145	72	100	0
浙　江	128 857	25 983	14 884	98 567	159	47	90	0
安　徽	82 959	11 813	8 456	35 939	74	68	40	0
福　建	114 766	7 919	12 524	61 148	104	37	34	0
江　西	119 376	7 201	6 003	21 039	60	60	18	0
山　东	123 568	12 591	14 036	71 830	236	74	59	2
河　南	120 554	10 882	9 517	58 252	118	75	50	1
湖　北	88 222	12 435	7 804	31 677	87	51	51	0
湖　南	124 280	15 104	10 629	49 190	187	78	68	0
广　东	178 255	32 444	19 974	137 403	211	102	48	0
广　西	62 194	5 832	5 217	25 638	55	50	38	0
海　南	14 947	1 503	2 101	15 128	45	7	25	0
重　庆	60 345	7 166	10 114	40 223	106	17	47	0
四　川	197 705	20 206	19 622	93 713	174	94	123	0
贵　州	46 882	5 796	4 564	21 414	94	33	21	0
云　南	145 542	6 314	12 130	38 980	89	40	46	0
西　藏	18 788	1 097	627	4 447	15	18	2	0
陕　西	86 772	5 776	6 265	32 460	75	50	24	4
甘　肃	59 654	32 646	4 365	25 565	71	33	29	9
青　海	18 434	2 385	2 015	14 371	35	11	8	3
宁　夏	23 404	1 640	1 871	12 672	22	19	8	5
新　疆	117 864	4 989	4 501	31 854	98	41	29	17

各地区环境监测情况（二）

（2013）

单位：个

地区名称	地表水水质监测断面（点位）数	国控断面（点位）数	饮用水水源地监测点位数	地表水监测点位数	地下水监测点位数	近岸海域监测点位数	近岸海域环境功能区点位数	近岸海域环境质量点位数
总　计	9 414	972	912	565	347	882	581	301
北　京	199	8	5	1	4	0	0	0
天　津	582	11	1	1	0	31	21	10
河　北	273	32	38	8	30	28	20	8
山　西	107	15	25	2	23	0	0	0
内蒙古	123	34	60	8	52	0	0	0
辽　宁	274	37	45	15	30	135	107	28
吉　林	220	41	18	16	2	0	0	0
黑龙江	145	68	43	18	25	0	0	0
上　海	711	3	4	4	0	30	20	10
江　苏	842	83	49	45	4	35	19	16
浙　江	703	40	27	27	0	100	50	50
安　徽	274	73	36	22	14	0	0	0
福　建	365	17	25	22	3	101	66	35
江　西	222	38	28	28	0	0	0	0
山　东	577	50	50	25	25	136	95	41
河　南	280	47	46	15	31	0	0	0
湖　北	498	51	36	36	0	0	0	0
湖　南	244	39	29	29	0	0	0	0
广　东	493	20	81	75	6	119	67	52
广　西	149	17	40	33	7	70	48	22
海　南	119	12	6	5	1	97	68	29
重　庆	276	11	5	5	0	0	0	0
四　川	323	21	35	30	5	0	0	0
贵　州	134	8	16	15	1	0	0	0
云　南	641	69	53	53	0	0	0	0
西　藏	48	12	4	0	4	0	0	0
陕　西	138	23	25	9	16	0	0	0
甘　肃	61	12	21	7	14	0	0	0
青　海	65	8	13	3	10	0	0	0
宁　夏	52	5	15	1	14	0	0	0
新　疆	276	67	33	7	26	0	0	0

各地区环境监测情况（三）

（2013）

单位：个

地 区 名 称	开展环境噪声监测的监测点位数	区域环境噪声监测点位数	道路交通噪声监测点位数	功能区环境噪声监测点位数	开展生态监测的监测点位数	开展污染源监督性监测的重点企业数
总　计	245 011	183 901	50 526	10 584	163	61 454
北　京	707	185	518	4		384
天　津	1 350	771	507	72	1	571
河　北	9 035	5 973	2 511	551	1	3 421
山　西	9 625	7 554	1 851	220	12	979
内蒙古	9 457	6 105	3 021	331	17	1 192
辽　宁	10 751	8 105	2 128	518	17	2 404
吉　林	9 263	6 705	2 034	524	2	847
黑龙江	10 308	7 594	2 286	428	14	1 467
上　海	538	283	199	56	3	1 560
江　苏	15 649	11 610	3 413	626	14	4 298
浙　江	13 676	10 427	2 675	574	15	4 738
安　徽	5 461	3 930	1 213	318	2	1 522
福　建	13 604	10 708	2 426	470	4	3 489
江　西	5 161	3 635	1 076	450	1	1 566
山　东	20 544	16 333	3 501	710	11	6 120
河　南	11 263	8 518	2 017	728	15	1 905
湖　北	11 086	8 268	2 331	487	6	2 228
湖　南	11 481	8 814	2 281	386	3	2 556
广　东	15 332	11 193	3 510	629	2	8 295
广　西	6 306	5 236	852	218	4	1 235
海　南	2 082	1 711	358	13	1	305
重　庆	5 122	4 002	875	245	1	1 085
四　川	16 182	13 274	2 429	479	3	3 071
贵　州	5 977	4 977	919	81	2	874
云　南	6 875	5 230	1 475	170	1	1 684
西　藏	231	195	32	4	1	9
陕　西	6 559	4 330	1 479	750	2	1 102
甘　肃	4 517	3 093	1 223	201	1	944
青　海	556	439	103	14	3	391
宁　夏	1 469	1 009	344	116	1	639
新　疆	4 844	3 694	939	211	2	573

各地区污染源自动监控情况

（2013）

单位：个

地 区 名 称	已实施自动监控国家重点监控企业数	已实施自动监控国家重点监控企业中排放口数		已实施自动监控国家重点监控企业中监控设备与环保部门稳定联网数			
		水排放口数	气排放口数	COD 监控设备与环保部门稳定联网数	NH₃-N 监控设备与环保部门稳定联网数	SO₂ 监控设备与环保部门稳定联网数	NOₓ 监控设备与环保部门稳定联网数
总　计	9 779	7 812	7 145	5 382	3 822	5 489	5 445
北　京	33	24	11	22	22	11	11
天　津	65	45	98	33	33	81	80
河　北	552	508	676	487	424	676	676
山　西	793	586	851	153	151	294	293
内蒙古	253	150	348	142	118	384	384
辽　宁	301	238	247	192	179	232	213
吉　林	172	116	168	90	68	108	103
黑龙江	193	133	165	128	45	165	163
上　海	76	63	43	56	44	8	8
江　苏	776	635	356	635	429	356	356
浙　江	532	434	196	410	178	193	194
安　徽	247	207	211	164	114	130	133
福　建	281	246	130	141	100	130	130
江　西	198	152	142	100	75	72	59
山　东	848	569	559	99	64	469	468
河　南	597	610	459	323	242	459	459
湖　北	482	409	226	318	182	140	131
湖　南	432	373	214	272	208	180	181
广　东	512	448	237	226		204	205
广　西	526	434	245	344	294	161	167
海　南	47	35	28	29	11	24	24
重　庆	132	98	77	86	68	47	43
四　川	451	348	207	339	251	180	167
贵　州	145	157	158	123	128	158	158
云　南	196	179	164	110	94	91	84
陕　西	421	258	301	131	102	121	126
甘　肃	138	137	195	75	57	151	157
青　海	56	33	69	33	33	50	38
宁　夏	98	54	90	54	54	90	89
新　疆	226	133	274	67	54	124	145

各地区排污费征收情况

（2013）

地 区 名 称	排污费解缴入库户数/户	排污费解缴入库户金额/万元
总　计	352 316	2 048 126
北　京	1 890	3 146
天　津	3 285	18 697
河　北	19 515	166 499
山　西	9 975	151 279
内蒙古	4 223	100 501
辽　宁	22 370	141 254
吉　林	16 373	34 413
黑龙江	8 999	46 884
上　海	4 535	22 421
江　苏	31 153	199 967
浙　江	20 557	74 605
安　徽	9 804	54 315
福　建	13 340	33 691
江　西	7 285	84 080
山　东	13 710	158 321
河　南	18 056	95 414
湖　北	6 679	40 554
湖　南	15 786	57 008
广　东	56 083	92 938
广　西	6 753	27 562
海　南	544	4 516
重　庆	6 595	38 602
四　川	10 629	73 309
贵　州	7 977	54 499
云　南	7 409	35 773
西　藏	3 012	1 645
陕　西	4 814	60 185
甘　肃	5 131	23 283
青　海	808	6 947
宁　夏	1 804	16 933
新　疆	6 611	64 443

各地区自然生态保护与建设情况（一）

（2013）

地 区 名 称	自然保护区 个数/个	国家级	省级	自然保护区面积/ 公顷	国家级	省级
总　计	2 697	407	855	146 309 848	94 039 262	39 194 628
北　京	20	2	12	133 966	26 403	71 413
天　津	8	3	5	90 380	37 862	52 518
河　北	44	13	25	707 165	255 062	418 821
山　西	46	7	39	1 105 370	117 468	987 902
内蒙古	184	27	60	13 688 964	4 167 260	6 856 302
辽　宁	105	15	30	2 805 108	1 000 796	866 151
吉　林	44	19	17	2 430 085	1 093 215	1 298 846
黑龙江	226	33	83	6 806 078	2 912 310	2 560 839
上　海	4	2	2	93 821	66 175	27 646
江　苏	31	3	10	530 400	299 292	85 448
浙　江	33	10	9	199 035	146 542	14 885
安　徽	104	7	29	524 263	139 221	287 729
福　建	90	15	22	428 128	229 576	88 875
江　西	199	13	33	1 245 654	220 299	340 514
山　东	86	7	37	1 099 845	219 828	541 862
河　南	34	11	21	738 858	426 316	311 142
湖　北	70	15	25	1 019 256	355 761	406 582
湖　南	128	23	28	1 283 962	609 901	377 313
广　东	392	14	64	1 849 663	313 582	526 265
广　西	78	21	46	1 455 932	374 122	856 018
海　南	50	9	24	2 735 080	106 526	2 614 313
重　庆	57	6	18	844 567	276 664	231 325
四　川	167	29	63	8 977 554	2 923 820	3 028 471
贵　州	123	8	5	880 781	243 539	75 073
云　南	154	20	38	2 857 020	1 503 374	681 122
西　藏	47	9	14	41 368 882	37 153 065	4 209 486
陕　西	57	22	28	1 166 414	599 698	470 635
甘　肃	60	19	37	7 462 545	5 117 234	2 230 411
青　海	11	7	4	21 765 351	20 733 751	1 031 600
宁　夏	14	8	6	533 049	439 449	93 600
新　疆	31	10	21	19 482 672	11 931 151	7 551 521

各地区自然生态保护与建设情况（二）

（2013）

单位：个

地　区 名　称	生态市、县建设 个数	国家级生态市个数	国家级生态县（区） 个数	省级生态市个数	省级生态县（区） 个数
总　计	399	6	86	35	272
北　京	11	0	2	9	0
天　津	5	0	1	0	4
河　北	0	0	0	0	0
山　西	1	0	0	0	1
内蒙古	7	0	0	0	7
辽　宁	19	0	8	3	8
吉　林	1	0	0	0	1
黑龙江	22	0	0	1	21
上　海	1	0	1	0	0
江　苏	55	5	30	1	19
浙　江	73	0	16	7	50
安　徽	19	0	3	0	16
福　建	51	0	5	6	40
江　西	10	0	0	1	9
山　东	10	0	4	1	5
河　南	6	0	1	0	5
湖　北	1	0	0	0	1
湖　南	1	0	0	0	1
广　东	10	1	5	1	3
广　西	3	0	0	0	3
海　南	7	0	0	2	5
重　庆	6	0	0	0	6
四　川	42	0	7	0	35
贵　州	4	0	0	0	4
云　南	3	0	0	0	3
西　藏	0	0	0	0	0
陕　西	12	0	2	1	9
甘　肃	0	0	0	0	0
青　海	0	0	0	0	0
宁　夏	2	0	0	0	2
新　疆	17	0	1	2	14

各地区自然生态保护与建设情况（三）
（2013）

单位：个

地 区 名 称	国家级生态村镇 个数	国家级生态村 个数	国家级生态乡镇 个数	国家有机食品生产基地 数量
总 计	3 219	238	2 981	138
北 京	93	2	91	1
天 津	21	0	21	0
河 北	51	11	40	0
山 西	11	3	8	0
内蒙古	83	4	79	4
辽 宁	149	8	141	15
吉 林	41	3	38	6
黑龙江	68	3	65	1
上 海	53	2	51	2
江 苏	488	44	444	11
浙 江	459	9	450	30
安 徽	125	21	104	11
福 建	204	3	201	0
江 西	162	9	153	3
山 东	414	6	408	16
河 南	90	7	83	1
湖 北	59	22	37	0
湖 南	123	27	96	0
广 东	74	6	68	0
广 西	22	1	21	3
海 南	4	1	3	0
重 庆	6	0	6	3
四 川	144	7	137	3
贵 州	48	8	40	1
云 南	58	3	55	5
西 藏	0	0	0	0
陕 西	60	11	49	0
甘 肃	36	6	30	2
青 海	1	0	1	2
宁 夏	24	4	20	0
新 疆	48	7	41	18

各地区环境影响评价

（2013）

地 区 名 称	当年审批的建设 项目投资总额/ 万元	当年审批的建设 项目环保投资总 额/万元	当年审批的建设 项目环境影响评 价文件数量/个	编制报告书的 项目数量	填报报告表的 项目数量	编制登记表的 项目数量
总　计	3 354 806 656	99 793 840	476 241	34 113	190 363	251 765
国家级	191 629 800	8 838 592	243	223	14	6
北　京	60 445 762	1 686 455	8 769	427	3 104	5 238
天　津	71 060 709	1 935 356	3 330	519	2 414	397
河　北	201 978 936	4 508 139	18 155	1 247	8 379	8 529
山　西	98 128 957	1 742 657	9 120	1 037	4 361	3 722
内蒙古	93 181 522	7 215 377	11 733	817	3 520	7 396
辽　宁	89 100 892	1 602 175	19 024	1 182	7 178	10 664
吉　林	74 807 478	3 685 406	15 322	910	5 842	8 570
黑龙江	45 073 278	1 469 456	9 752	754	4 765	4 233
上　海	64 959 659	1 262 591	11 706	518	5 783	5 405
江　苏	368 025 955	9 034 834	40 131	3 365	18 188	18 578
浙　江	181 710 763	2 840 578	32 562	2 652	16 216	13 694
安　徽	163 096 465	7 637 359	20 078	1 651	9 040	9 387
福　建	117 714 361	3 173 808	16 682	1 682	8 338	6 662
江　西	60 745 062	1 887 038	7 205	915	3 620	2 670
山　东	231 039 960	7 457 341	34 475	2 113	16 399	15 963
河　南	129 246 787	2 154 633	14 537	1 322	6 721	6 494
湖　北	131 464 035	2 836 177	10 476	1 300	4 678	4 498
湖　南	83 720 520	3 252 643	10 871	1 276	3 324	6 271
广　东	269 544 030	6 003 359	50 139	1 988	23 439	24 712
广　西	50 983 273	2 015 846	15 170	671	4 419	10 080
海　南	35 926 880	540 949	2 928	225	1 067	1 636
重　庆	73 933 507	1 531 120	5 057	775	2 390	1 892
四　川	65 293 480	1 656 865	17 049	1 991	5 612	9 446
贵　州	67 625 817	2 843 455	15 398	716	4 509	10 173
云　南	80 410 167	2 347 755	31 110	1 195	5 325	24 590
西　藏	11 553 377	96 943	2 951	134	618	2 199
陕　西	69 893 962	1 463 432	9 712	724	2 028	6 960
甘　肃	25 164 376	606 032	5 801	449	1 879	3 473
青　海	24 515 940	964 486	2 227	132	574	1 521
宁　夏	44 458 636	1 218 448	3 037	298	1 109	1 630
新　疆	78 372 310	4 284 535	21 491	905	5 510	15 076

各地区建设项目竣工环境保护验收情况（一）

（2013）

地 区 名 称	当年完成环保验收项目数/个	环保验收一次合格项目数/个	经限期改正验收合格项目数/个	当年完成环保验收项目总投资/万元
总 计	150 392	145 363	5 029	818 100 429
国家级	284	279	5	160 203 089
北 京	3 557	3 430	127	10 977 989
天 津	1 283	1 282	1	17 131 546
河 北	5 448	5 289	159	21 001 199
山 西	1 789	1 757	32	11 915 293
内蒙古	2 411	2 385	26	17 203 508
辽 宁	7 691	7 539	152	40 078 897
吉 林	4 608	4 562	46	20 860 026
黑龙江	4 037	3 948	89	14 692 862
上 海	4 506	4 487	19	47 980 399
江 苏	9 728	9 344	384	102 562 974
浙 江	10 680	10 348	332	43 027 581
安 徽	5 280	5 111	169	25 653 694
福 建	6 407	5 746	661	31 518 806
江 西	1 714	1 527	187	8 576 296
山 东	11 980	11 307	673	55 145 297
河 南	4 680	4 636	44	14 443 580
湖 北	3 267	3 138	129	18 801 219
湖 南	3 144	3 032	112	10 466 650
广 东	17 923	17 080	843	43 344 245
广 西	5 149	4 978	171	5 997 696
海 南	706	703	3	3 012 035
重 庆	2 117	2 092	25	19 775 479
四 川	4 724	4 598	126	19 214 754
贵 州	5 315	5 265	50	2 600 543
云 南	8 788	8 756	32	12 889 866
西 藏	883	860	23	1 169 881
陕 西	2 635	2 542	93	10 989 199
甘 肃	1 691	1 677	14	4 919 947
青 海	421	421	0	1 905 593
宁 夏	408	406	2	6 795 751
新 疆	7 138	6 838	300	13 244 535

各地区建设项目竣工环境保护验收情况（二）

（2013）

单位：万元

地 区 名 称	当年完成工业企业验 收项目环保投资	废水治理环保 投资	废气治理环保 投资	噪声治理环保 投资	固体废物治理 环保投资	绿化及生态 环保投资
总　计	19 428 391	5 427 312	7 657 238	996 606	1 346 066	1 641 531
国家级	3 841 640	366 634	1 726 731	65 546	338 557	47 549
北　京	130 013	25 917	36 960	50 498	6 151	6 733
天　津	285 037	54 764	118 280	14 926	8 956	73 162
河　北	505 801	115 720	241 349	47 523	42 441	32 962
山　西	601 395	167 673	284 057	28 155	36 385	43 352
内蒙古	601 751	125 371	291 005	10 942	18 257	24 314
辽　宁	737 471	247 754	277 454	70 170	62 892	48 554
吉　林	267 207	100 120	106 210	7 976	2 818	47 850
黑龙江	256 106	124 415	74 180	14 292	13 130	15 757
上　海	590 840	55 425	149 878	11 470	16 324	178 761
江　苏	1 830 439	825 345	518 364	138 621	108 886	161 418
浙　江	1 012 387	359 832	406 854	64 095	59 187	84 526
安　徽	930 134	268 943	454 646	48 526	42 221	78 271
福　建	615 523	176 204	311 232	29 974	27 954	39 219
江　西	275 154	84 980	98 010	21 245	20 157	29 295
山　东	1 486 716	534 442	475 969	88 848	161 593	132 805
河　南	471 784	152 489	148 365	37 886	51 757	44 752
湖　北	342 654	114 272	125 908	17 821	15 337	24 481
湖　南	223 857	76 401	69 083	21 978	24 621	20 744
广　东	1 122 600	346 992	455 391	62 202	45 988	176 420
广　西	312 174	133 695	99 024	13 486	15 835	23 923
海　南	19 880	4 587	2 156	1 451	967	9 517
重　庆	180 850	66 989	55 039	17 576	12 818	23 880
四　川	567 897	243 481	148 232	24 826	82 217	57 538
贵　州	106 809	42 498	41 176	4 355	4 010	7 303
云　南	433 240	123 257	150 839	19 046	32 620	86 934
西　藏	10 450	1 433	4 207	84	550	2 530
陕　西	235 436	80 807	58 146	23 470	17 965	33 997
甘　肃	495 671	128 726	300 610	14 016	20 495	12 968
青　海	67 056	10 737	44 844	1 326	3 241	2 543
宁　夏	241 970	72 872	109 350	7 163	21 545	20 307
新　疆	628 449	194 537	273 689	17 113	30 191	49 166

各地区突发环境事件

（2013）

单位：次

地 区 名 称	突发环境事件次数	特别重大环境事件次数	重大环境事件次数	较大环境事件次数	一般环境事件次数
总 计	712	0	3	12	697
北 京	16	0	0	0	16
天 津	0	0	0	0	0
河 北	3	0	0	0	3
山 西	13	0	1	0	12
内蒙古	4	0	0	0	4
辽 宁	12	0	0	1	11
吉 林	1	0	0	0	1
黑龙江	0	0	0	0	0
上 海	251	0	0	1	250
江 苏	125	0	0	1	124
浙 江	26	0	0	1	25
安 徽	6	0	0	1	5
福 建	13	0	0	0	13
江 西	5	0	1	0	4
山 东	5	0	0	1	4
河 南	17	0	0	0	17
湖 北	7	0	0	2	5
湖 南	3	0	0	1	2
广 东	5	0	0	1	4
广 西	16	0	1	1	14
海 南	4	0	0	0	4
重 庆	11	0	0	0	11
四 川	14	0	0		14
贵 州	9	0	0	0	9
云 南	2	0	0	0	2
西 藏	0	0	0	0	0
陕 西	118	0	0	0	118
甘 肃	11	0	0	0	11
青 海	2	0	0	0	2
宁 夏	3	0	0	1	2
新 疆	10	0	0	0	10

各地区环境宣教情况

（2013）

地　区 名　称	当年开展的社会环境宣传教 育活动数/次	当年开展的社会环境宣传教育活 动人数/人	环境教育基地数/个
总　计	17 087	32 941 736	1 902
国家级	17	12 370	80
北　京	709	1 635 858	21
天　津	551	375 950	10
河　北	713	9 126 426	32
山　西	98	808 273	12
内蒙古	126	86 759	129
辽　宁	455	2 411 400	81
吉　林	6 195	391 896	45
黑龙江	135	245 486	121
上　海	518	544 743	73
江　苏	358	2 568 410	321
浙　江	674	893 610	120
安　徽	167	645 787	20
福　建	383	398 422	59
江　西	85	123 585	11
山　东	748	2 340 393	169
河　南	214	104 689	74
湖　北	249	296 706	31
湖　南	383	315 990	20
广　东	385	3 790 166	103
广　西	96	56 537	11
海　南	132	40 703	7
重　庆	638	2 275 311	143
四　川	730	436 190	30
贵　州	52	36 640	3
云　南	683	1 197 296	86
西　藏	59	6 945	6
陕　西	822	482 158	17
甘　肃	137	622 672	8
青　海	289	216 024	34
宁　夏	155	115 305	11
新　疆	131	339 036	14

14

附 表

ANNUAL STATISTIC REPORT ON ENVIRONMENT IN CHINA
2013

全国行政区划

（2013年底）

单位：个

省级区划名称	地级区划数	地级市	县级区划数	县级市	市辖区	县	自治县
全 国	333	286	2 853	368	872	1 442	117
北京市			16		14	2	
天津市			16		13	3	
河北省	11	11	172	22	37	107	6
山西省	11	11	119	11	23	85	
内蒙古自治区	12	9	102	11	22	17	
辽宁省	14	14	100	17	56	19	8
吉林省	9	8	60	21	20	16	3
黑龙江省	13	12	128	18	64	45	1
上海市			17		16	1	
江苏省	13	13	100	23	55	22	
浙江省	11	11	90	21	34	34	1
安徽省	16	16	105	6	43	56	
福建省	9	9	85	14	26	45	
江西省	11	11	100	10	20	70	
山东省	17	17	137	29	48	60	
河南省	17	17	159	21	50	88	
湖北省	13	12	103	24	38	38	2
湖南省	14	13	122	16	35	64	7
广东省	21	21	121	23	58	37	3
广西壮族自治区	14	14	110	7	36	55	12
海南省	3	3	20	6	4	4	6
重庆市			38		19	15	4
四川省	21	18	183	14	48	117	4
贵州省	9	6	88	7	13	56	11
云南省	16	8	129	12	13	75	29
西藏自治区	7	1	74	1	1	72	
陕西省	10	10	107	3	24	80	
甘肃省	14	12	86	4	17	58	7
青海省	8	2	43	3	5	28	7
宁夏回族自治区	5	5	22	2	9	11	
新疆维吾尔自治区	14	2	101	22	11	62	6
香港特别行政区							
澳门特别行政区							
台湾省							

注：数据摘自《中国统计摘要》（中国统计出版社），下同。

国民经济与社会发展总量指标摘要

指标	单位	2006 年	2007 年	2008 年	2009 年	2010 年	2011 年	2012 年	2013 年
人口									
年底总人口	万人	131 448	132 129	132 802	133 450	134 091	134 735	135 404	136 072
其中：城镇人口	万人	58 288	60 633	62 403	64 512	66 978	69 079	71 182	73 111
乡村人口	万人	73 160	71 496	70 399	68 938	67 113	65 656	64 222	62 961
国民经济核算									
国内生产总值	亿元	216 314.4	265 810.3	314 045.4	340 902.8	401 512.8	471 563.7	518 942.1	568 845.2
其中：第一产业	亿元	24 040.0	28 627.0	33 702.0	35 226.0	40 533.6	47 712.0	52 373.6	56 957.0
第二产业	亿元	103 719.5	125 831.4	149 003.4	157 638.8	187 383.2	220 591.6	235 162.0	249 684.4
第三产业	亿元	88 554.9	111 351.9	131 340.0	148 038.0	173 596.0	203 260.1	231 406.5	262 203.8
人均国内生产总值	元/人	16 500	20 169	23 708	25 608	30 015	35 083	38 459	41 908
支出法国内生产总值	亿元	222 711.9	266 556.7	315 977.2	348 771.3	402 818.7	465 998.7	529 238.4	586 673.0
其中：最终消费	亿元	113 103.8	132 232.9	153 422.5	169 274.8	194 115.0	224 740.8	261 832.8	292 165.6
资本形成总额	亿元	92 954.1	110 943.2	138 325.3	164 463.2	193 603.9	229 102.4	252 773.2	280 356.1
固定资产投资									
全社会固定资产投资总额	亿元	109 998.2	137 323.9	172 828.4	224 599	278 121.9	311 021.9	374 694.7	446 294.1
城镇	亿元	93 368.7	117 464.5	148 738.3	193 920	241 430.9	301 932.8	364 854.1	435 747.4
其中：房地产开发	亿元	19 422.9	25 288.8	31 203.2	36 242	48 259.4	61 739.8	71 803.8	86 013.4
农村	亿元	16 629.5	19 859.5	24 123.0	30 678	36 724.9			
主要农业、工业产品产量									
粮食	万吨	49 804	50 160.3	52 870.9	53 082	54 647.7	57 121	58 958.0	60 193.8
棉花	万吨	753.3	762.4	749.2	638	596.1	658.9	683.6	629.9
油料	万吨	2 640.3	2 568.7	2 952.8	3 154	3 230.1	3 306.8	3 436.8	3 517.0
肉类	万吨	7 089.0	6 865.7	7 278.3	7 650	7 925.8	7 957.8	8 387.2	8 535.0
原煤	亿吨	25.29	26.92	28.02	29.73	32.35	35.20	36.50	36.80
原油	万吨	18 476.57	18 632	19 043.1	18 949	20 241.4	20 287.6	20 748	20 947
发电量	亿千瓦·时	28 657.26	32 816	34 957.6	37 146.5	42 071.6	47 000.7	49 876	53 976
粗钢	万吨	41 914.9	48 929	50 305.8	57 218	63 723.0	68 388.3	72 388	77 904
水泥	万吨	123 676.5	136 117	142 355.7	164 398	188 191.2	208 500.0	220 984	241 613

国民经济与社会发展总量指标摘要（续表）

指标	单位	2006 年	2007 年	2008 年	2009 年	2010 年	2011 年	2012 年	2013 年
国内商业和对外贸易									
社会消费品零售总额	亿元	79 145.2	93 571.6	114 830.1	132 678.4	156 998	183 918.6	210 307.0	237 810
进出口总额	亿美元	17 604	21 765.7	25 632.6	22 075.4	29 740.0	36 420.6	38 671.2	41 589.9
其中：出口额	亿美元	9 689	12 204.6	14 306.9	12 016.1	15 777.5	18 986.0	20 487.1	22 090.0
进口额	亿美元	7 915	9 561.1	11 325.6	10 059.2	13 962.4	17 434.6	18 184.1	19 499.9
利用外资									
实际利用（使用）外资额	亿美元	735.2	783.4	952.6	918.0	1 088.2	1 177.0	1 132.9	1 187.2
其中：外商直接投资	亿美元	694.7	747.7	924.0	900.3	1 057.3	1 160.1	1 117.2	1 175.9
外商其他投资	亿美元	40.6	35.7	28.6	17.7	30.9	16.9	15.8	11.3
旅游									
入境过夜旅游者人数	万人次	4 991	5 472.0	5 304.9	5 087.5	5 566.5	5 758.1	5 772.5	5 568.6
国内旅游人数	亿人次	13.9	16.1	17.1	19.0	21.0	26.41	29.57	32.62
国际旅游外汇收入	亿美元	339.5	419.2	408.4	396.8	458.1	484.6	500.3	516.6
国内旅游总收入 （国内旅游总花费）	亿元	6 229.7	7 770.6	8 749.3	10 183.7	12 579.8	19 305.4	22 706.2	26 276.1
教育、科技、文化、卫生									
普通高等学校学生数	万人	1 738.8	1 884.9	2 021.0	2 144.7	2 231.8	2 308.5	2 391.3	2 468.1
普通中等学校学生数	万人	8 451.9	8 243.3	8 050.4	7 876.9	7 703.2	7 519.0	7 228.4	6 875.0
小学学生数	万人	10 711.5	10 564.0	10 331.5	10 071.5	9 940.7	9 926.4	9 695.9	9 360.5
研究与试验发展经费支出	亿元	3 003.1	3 710.2	4 570.0	5 791.9	7 063	8 610	10 298.4	11 846.6
技术市场成交额	亿元	1 818.2	2 227	2 665	3 039	3 907	4 764	6 437.1	7 469.0
图书总印数	亿册（张）	64.0	62.9	70.6	70.4	71.7	76.8	79.2	83.1
杂志总印数（期刊）	亿册	28.5	30.4	31.0	31.5	32.2	32.7		
报纸总印数	亿份	424.5	438.0	442.9	439.1	452.1	466.8		
医院、卫生院数 （医疗卫生机构数）	万个	91.8	91.2	89.1	91.7	93.7	95.4	95.0	97.4
医生数	万人	209.9	212.3	220.2	232.9	241.3	246.6	261.6	279.5
医院、卫生院床位数 （医疗卫生机构床位数）	万张	351.2	370.1	403.9	441.7	478.7	516.0	416.1	457.9
城市公用事业									
城市房屋集中供热面积	亿米^2	26.6				43.6	47.4	51.8	
年末供水综合生产能力	万米^3/日	26 962							
年末污水处理能力	万米^3/日	9 734							
年末城市道路长度	千米	241 351							

注：① 由于计算误差的影响，支出法国内生产总值不等于按生产法计算的国内生产总值；

② 本表价值量指标均按当年价格计算。

413

国民经济与社会发展结构指标摘要

单位：%

指标	2006 年	2007 年	2008 年	2009 年	2010 年	2011 年	2012 年	2013 年
总人口	100	100	100	100	100	100	100	100
其中：城镇	44.3	45.9	47.0	48.3	50.0	51.3	51.3	53.7
乡村	5 537	54.1	53.0	51.7	50.0	48.7	48.7	46.3
国内生产总值	100	100	100	100	100	100	100	100
其中：第一产业	11.1	10.8	10.7	10.3	10.1	10.1	33.6	10.0
第二产业	47.9	47.3	47.4	46.2	46.7	46.8	30.3	43.9
第三产业	40.9	41.9	41.8	43.4	43.2	43.1	36.1	46.1
固定资产投资总额	100	100	100	100	100	100	100	100
城镇	84.9	85.6	86.0	86.3	86.8	97.1	97.4	97.6
农村	15.1	14.4	14.0	13.7	13.2	2.9	2.6	2.4
财政收入	100	100	100	100	100	100	100	100
其中：中央	52.8	54.1	53.3	52.4	51.1	49.5	47.9	46.6
地方	47.2	45.9	46.7	47.6	48.9	50.5	52.1	53.4
财政支出	100	100	100	100	100	100	100	100
其中：中央	24.7	23.1	21.4	20.0	17.8	15.2	14.9	14.6
地方	75.3	76.9	78.6	80.0	82.2	84.8	85.1	85.4
农、林、牧、渔业产值	100	100	100	100	100	100	100	100
其中：农业	51.1	50.4	55.7	50.7	53.3	51.6	52.5	53.1
林业	3.9	3.8	3.8	3.9	3.7	3.8	3.9	4.0
牧业	29.4	33.0	29.7	32.3	30.0	31.7	30.4	29.3
渔业	9.6	9.1	10.9	9.3	9.3	9.3	9.7	9.9
农、林、牧、渔、服务业	6.0	3.7		3.8	3.7			
工业总产值	100	100	100	100	100	100	100	100
其中：轻工业	30.5							
重工业	69.5							

注：2002 年起轻重工业结构为国有及规模以上非国有工业企业口径。

自然资源状况

项目	单位	2006 年	2007 年	2008 年	2009 年	2010 年	2011 年	2012 年	2013 年
国土面积	万千米^2	960	960	960	960	960	960	960	960
海域面积	万千米^2	473	473	473	473	473	473	473	
大陆岸线长度	万千米	1.80	1.80	1.80	1.80	1.80	1.8	1.8	
岛屿面积	万千米^2	3.87	3.87	3.87	3.87	3.87	3.87	3.87	
降水量									
台湾中部山区	毫米	≥4 000	≥4 000	≥4 000	≥4 000	≥4 000	≥4 000	≥4 000	
华南沿海	毫米	1 600~2 000	1 600~2 000	1 600~2 000	1 600~2 000	1 600~2 000	1 600~2 000	1 600~2 000	
长江流域	毫米	1 000~1 500	1 000~1 500	1 000~1 500	1 000~1 500	1 000~1 500	1 000~1 500	1 000~1 500	
华北、东北	毫米	400~800	400~800	400~800	400~800	400~800	400~800	400~800	
西北内陆	毫米	100~200	100~200	100~200	100~200	100~200	100~200	100~200	
塔里木盆地、吐鲁番盆地和柴达木盆地	毫米	≤25	≤25	≤25	≤25	≤25	≤25	≤25	
耕地面积	万公顷	13 004	13 004	12 178	12 178	12 172	12 172	12 172	12 172
荒地面积	万公顷	10 800	10 800	10 800	10 800				
林业用地面积	万公顷	28 493	28 493	28 493	30 378	30 590	30 590	30 590	31 259
草地面积	万公顷	40 000	40 000	40 000	40 000				
活立木总蓄积量	亿米^3	136.2	136.2	136.2	136.2	149.1	149.1	149.1	164.3
森林面积	亿公顷	1.75	1.75	1.75	1.95	1.95	1.95	1.95	2.08
森林覆盖率	%	18.21	18.21	18.21	20.36	20.36	20.36	20.36	21.63
水资源总量	亿米^3	27 430	25 567	24 696	23 763	29 658	24 022	24 022	27 958
水力资源蕴藏量	亿千瓦	6.76	6.76	6.76	6.76	6.76	6.76	6.76	
其中：可开发量	亿千瓦	3.79	3.79	3.79	3.79	3.79	3.79	3.79	
内陆水域总面积	万公顷	1 747	1 747	1 747	1 747	1 747	1 747	1 747	
其中：可养殖面积	万公顷	675	675	675	675	675	675	675	
海水可养殖面积	万公顷	260	260	260	260	260	260	260	
其中：已养殖面积	万公顷	109	109	109	109	109	109	109	

注：① 森林资源为 1994—1998 年调查数；草地资源为 1991 年调查数；水面资源为 1988 年数；水力资源为 1999 年公报数；耕地面积为 1996 年农业普查数。

② 本表除国土面积、海域面积和降水量外，其余指标均未包括香港特别行政区、澳门特别行政区和台湾省数据。

《中国环境统计年报·2013》各章编写作者

第1章	统计调查企业基本情况	董广霞
第2章	废水	周冏、封雪
第3章	废气	王军霞、吕卓
第4章	工业固体废物	赵银慧
第5章	环境污染治理投资	王鑫、封雪
第6章	环境管理	董广霞
数据表汇总及整理		周冏、封雪、吕卓、董广霞、刘通浩、王军霞
汇稿		周冏、封雪、董文福、景立新

15

主要统计指标解释

ANNUAL STATISTIC REPORT ON ENVIRONMENT IN CHINA
2013

1. 工业企业污染排放及处理利用情况

【工业用水量】指报告期内企业厂区内用于工业生产活动的水量，它等于取水量与重复用水量之和。

【取水量】指调查年度企业厂区内用于工业生产活动的水量中从外部取水的量。根据《工业企业产品取水定额编制通则》(GB/T 18820—2002)，工业生产的取水量，包括取自地表水（以净水厂供水计量）、地下水、城镇供水工程，以及企业从市场购得的其他水（如其他企业回用水量）或水的产品（如蒸气、热水、地热水等），不包括企业自取的海水和苦咸水等以及企业为外供给市场的水的产品（如蒸气、热水、地热水等）而取用的水量，也不包括对天然水、污水、海水以及雨水、微咸水等类似水进行收集、处理后作为产品供应和利用而取用的水量。

工业生产活动用水包括主要工业生产用水、辅助生产（包括机修、运输、空压站等）用水和附属生产（包括厂内绿化、职工食堂、非营业的浴室及保健站、厕所等）用水；不包括 ① 非工业生产单位的用水，如厂内居民家庭用水和企业附属幼儿园、学校、对外营业的浴室、游泳池等的用水量；② 生活用水单独计量且生活污水不与工业废水混排的水量。

【重复用水量】指报告期内企业生产用水中重复再利用的水量，包括循环使用、一水多用和串级使用的水量（含经处理后回用量）。

指企业内部对工业生产活动排放的废水直接利用或经过处理后回收再利用的水量，不包括从城市污水处理厂回用的水量。每重复利用一次，则计算一次重复用水量。但锅炉、循环冷却系统等封闭式系统内的循环水不能计算重复用水量。重复用水量的计算原则：

（1）开放原则。即水的循环在开放系统进行，循环一次计算一次，但锅炉、循环冷却系统等封闭式系统内的循环水不能计算重复用水量。

（2）"源头"计算原则。对循环水来说，使用后的水，又回流到系统的取水源头，流经源头一次，计算一次。循环系统中的中间环节用水不得计算重复用水量。

（3）异地原则。对于非循环系统，根据不同工艺对不同水质的要求，在一个地方（工艺）使用过的水，在另外一个地方（工艺）中又进行使用，使用一次，计算一次。在同一地方（容器）多次使用的水，不得计算重复用水量。

（4）经过净化处理后的水重复再用，在任何情况下都按照重复用水计算。

【煤炭消耗量】指报告期内企业所用煤炭的总消耗量。

【燃料煤消耗量】指报告期内企业厂区内用做燃料的煤炭消耗量（实物量），包括企业厂区内生产、生活用燃料煤，也包括砖瓦、石灰等产品生产用的内燃煤，不包括在生产工艺中用做原料并能转换成新的产品实体的煤炭消耗量。如转换为水泥、焦炭、煤气、碳素、活性炭、氮肥的煤炭。

【燃料油消耗量（不含车船用）】指报告期内企业用做燃料的原油、汽油、柴油、煤油等各种油料总消耗量，不包括车船交通用油量。

【焦炭消耗量】指报告期内企业消耗的焦炭总量。

【天然气消耗量】指报告期内企业用做燃料的天然气消耗量。

【其他燃料消耗量】指报告期内企业除了煤炭、燃油、天然气等以外，用做燃料的其他燃料消耗量。其他燃料应根据当地的折标系数折算为标准煤后统一填报。

各类能源的参考折标系数表

能源种类		折标系数	能源种类		折标系数
原　　煤		0.714 3	煤焦油		1.142 9
洗精煤		0.900 0	粗苯		1.428 6
其他洗煤	洗中煤	0.285 7	原　　油		1.428 6
	煤泥	0.285 7～0.428 6	汽　　油		1.471 4
型　　煤		0.5～0.7	煤　　油		1.471 4
焦　　炭		0.971 4	柴　　油		1.457 1
焦炉煤气		0.571 4～0.614 3 千克标准煤/米³	燃料油		1.428 6
高炉煤气		0.128 6 千克标准煤/米³	热　　力		0.034 12 千克标准煤/百万千焦 0.142 86 千克标准煤/1 000 千卡
天 然 气		1.330 0 千克标准煤/米³	电　　力		0.122 9 千克标准煤/千瓦时
液化天然气		1.757 2	生物质能	大豆秆、棉花秆	0.543
液化石油气		1.714 3		稻秆	0.429
炼厂干气		1.571 4		麦秆	0.500
其他煤气	发生炉煤气	0.178 6 千克标准煤/米³		玉米秆	0.529
	重油催化裂解煤气	0.657 1 千克标准煤/米³		杂草	0.471
	重油热裂解煤气	1.214 3 千克标准煤/米³		树叶	0.500
	焦炭制气	0.557 1 千克标准煤/米³		薪柴	0.571
	压力气化煤气	0.514 3 千克标准煤/米³		沼气	0.714 千克标准煤/米³
	水煤气	0.357 1 千克标准煤/米³		—	—

注：除表中标注单位的能源外，其余能源折标系数单位均为：千克标准煤/千克。各地的能源折标系数由当地环保部门协调统计部门提供。调查对象也可根据燃料品质分析报告，自行折标填报。

【用电量】指报告期内企业的用电量，包括动力用电和照明用电。

【工业锅炉数】指报告期内企业厂区内用于生产和生活的大于 1 蒸吨（含 1 蒸吨）的蒸汽锅炉、热水锅炉总台数和总蒸吨数，包括燃煤、燃油、燃气和燃电的锅炉，不包括茶炉。

【其中 35 蒸吨及以上】指报告期内企业厂区内用于生产和生活的大于 35 蒸吨（含 35 蒸吨）的蒸汽锅炉、热水锅炉总台数和总蒸吨数。

【其中 20（含）～35 蒸吨】指报告期内企业厂区内用于生产和生活的大于 20 蒸吨（含 20 蒸吨）小于 35 蒸吨的蒸汽锅炉、热水锅炉总台数和总蒸吨数。

【其中 10（含）～20 蒸吨】指报告期内企业厂区内用于生产和生活的大于 10 蒸吨（含 10 蒸吨）小于 20 蒸吨的蒸汽锅炉、热水锅炉总台数和总蒸吨数。

【其中 10 蒸吨的以下】指报告期内企业厂区内用于生产和生活的小于 10 蒸吨的蒸汽锅炉、热水锅炉总台数和总蒸吨数。

【工业炉窑数】指报告期内企业生产用的炉窑总数，如炼铁高炉、炼钢炉、冲天炉、烘干炉窑、锻造加热炉、水泥窑、石灰窑等。

【废水治理设施数】指报告期内企业用于防治水污染和经处理后综合利用水资源的实有设施（包括构筑物）数，以一个废水治理系统为单位统计。附属于设施内的水治理设备和配套设备不单独计算。备用的、报告期内未运行的、已经报废的设施不统计在内。

只填报企业内部的废水治理设施，工业废水排入的城镇污水处理厂、集中工业废水处理厂不能算做企业的废水治理设施。企业内的废水治理设施包括一级、二级和三级处理的设施，如企业有 2 个排污口，1 个排污口为一级处理（隔油池、化粪池、沉淀池等），另一个排污口为二级处理（如生化处理），则该企业有 2 套废水治理设施；若该企业只有 1 个排污口，经由该排污口的废水先经过一级处理，再经二级（甚至三级）处理后外排，则该企业视为 1 套废水治理设施。即针对同一股废水的所有水治理设备均视为 1 套治理设施，针对不同废水的水治理设备可视为多套治理设施。

【废水治理设施处理能力】指报告期内企业内部的所有废水治理设施实际具有的废水处理能力。

【废水治理设施运行费用】指报告期内企业维持废水治理设施运行所发生的费用。包括能源消耗、设备维修、人员工资、管理费、药剂费及与设施运行有关的其他费用等。

【工业废水处理量】指经各种水治理设施（含城镇污水处理厂、工业废水处理厂）实际处理的工业废水量，包括处理后外排的和处理后回用的工业废水量。虽经处理但未达到国家或地方排放标准的废水量也应计算在内。计算时，如遇有车间和厂排放口均有治理设施，并对同一废水分级处理时，不应重复计算工业废水处理量。

【工业废水排放量】指报告期内经过企业厂区所有排放口排到企业外部的工业废水量。包括生产废水、外排的直接冷却水、废气治理设施废水、超标排放的矿井地下水和与工业废水混排的厂区生活污水，不包括独立外排的间接冷却水（清浊不分流的间接冷却水应计算在内）。

直接冷却水：在生产过程中，为满足工艺过程需要，使产品或半成品冷却所用与之直接接触的冷却水（包括调温、调湿使用的直流喷雾水）。

间接冷却水：在工业生产过程中，为保证生产设备能在正常温度下工作，用来吸收或转移生产设备的多余热量，所使用的冷却水（此冷却用水与被冷却介质之间由热交换器壁或设备隔开）。

【直接排入环境的】指废水经过工厂的排污口或经过下水道直接排入环境中，包括排入海、河流、湖泊、水库、蒸发地、渗坑以及农田等。对应的排水去向代码为 A、B、C、D、F、G、K。

【排入污水处理厂的】指企业产生的废水直接或间接经市政管网排入污水处理厂的废水量，包括排入城镇污水处理厂、集中工业废水处理厂以及其他单位的污水处理设施的废水量。对应的排水去向代码为 E、L、H。

【工业废水中污染物产生量】指报告期调查对象生产过程中产生的未经过处理的废水中所含的化学需氧量、氨氮、石油类、挥发酚、氰化物等污染物和砷、铅、汞、镉、六价铬、总铬等重金属本身的纯质量。它可采用产排污系数根据生产的产品产量或原辅料用量计算求得，也可以通过工业废水产生量和其中污染物的浓度相乘求得，计算公式是：

污染物产生量（纯质量）＝工业废水产生量×废水处理设施入口污染物的平均浓度（无处理设施可使用排口浓度）

计算砷、铅、汞、镉、六价铬、总铬等重金属污染物时，上述计算公式中"工业废水产生量"为产生重金属废水的车间年实际产生的废水量，"废水处理设施入口污染物的平均浓度"为该车间废水处理设施入口的年实际加权平均浓度，如没有设施则为车间排口的年实际加权平均浓度。

【工业废水中污染物排放量】指报告期内企业排放的工业废水中所含化学需氧量、氨氮、石油类、挥发酚、氰化物等污染物和砷、铅、汞、镉、六价铬等重金属本身的纯质量。它可采用产排污系数根据生产的产品产量或原辅料用量计算求得，也可以通过工业废水排放量和其中污染物的浓度相乘求得，计算公式是：

污染物排放量（纯质量）＝工业废水排放量×排放口污染物的平均浓度

（1）如企业排出的工业废水经城镇污水处理厂或工业废水处理厂集中处理的，计算化学需氧量、氨氮、石油类、挥发酚、氰化物等污染物时，上述计算公式中"排放口污染物的平均浓度"即为污水处理厂排放口的年实际加权平均浓度。如果厂界排放浓度低于污水处理厂的排放浓度，以污水处理厂的排放浓度为准。

（2）计算砷、铅、汞、镉、六价铬等重金属污染物时，上述计算公式中"工业废水排放量"为车间排放口的年实际废水量，"排放口污染物的平均浓度"为车间排放口的年实际加权平均浓度。

【工业废气排放量】指调查年度企业厂区内排入空气中含有污染物的气体的总量，以标准状态（273 开，101 325 帕）计。

工业废气排放总量＝燃料燃烧过程中废气排放量＋生产工艺过程中废气排放量

【废气治理设施数】指报告期末企业用于减少在燃料燃烧过程与生产工艺过程中排向大气的污染物或对污染物加以回收利用的废气治理设施总数，以一个废气治理系统为单位统计。包括除尘、脱硫、脱硝及其他的污染物的烟气治理设施。备用的、报告期内未运行的、已报废的设施不统计在内。锅炉中的除尘装置属于"三同时"设备，应统计在内。

【除尘设施数】指专门设计、建设的去除废气烟（粉）尘的设施。

【脱硫设施数】脱硫设施指专门设计、建设的去除废气二氧化硫的设施，具有兼性脱硫效果的设施，如湿法除尘等治理设施等其他可能具有脱硫效果的废气治理设施不计入脱硫设施。具有脱硫效果的生产装置，如制酸、水泥生产等不作为脱硫设施。

【脱硝设施数】指在治理设施中采用选择性催化还原技术（SCR）、选择性非催化还原技术（SNCR）及其联合技术或采用活性炭吸附进行烟气脱硝的设施。具有脱硝效果的生产装置，如水泥生产等不作为脱硝设施。

【废气治理设施处理能力】指报告期末企业实有的废气治理设施的实际废气处理能力。

【废气脱硫设施处理能力】指报告期末企业实有的废气脱硫设施的实际废气处理能力。

【废气脱硝设施处理能力】指报告期末企业实有的废气脱硝设施的实际废气处理能力。

【废气除尘设施处理能力】指报告期末企业实有的废气除尘设施的实际废气处理能力。

【废气治理设施运行费用】指报告期内维持废气治理设施运行所发生的费用。包括能源消耗、设备折旧、设备维修、人员工资、管理费、药剂费及与设施运行有关的其他费用等。

【二氧化硫产生量】指当年全年调查对象生产过程中产生的未经过处理的废气中所含的二氧化硫总质量。

【二氧化硫排放量】指报告期内企业在燃料燃烧和生产工艺过程中排入大气的二氧化硫总质量。工业中二氧化硫主要来源于化石燃料（煤、石油等）的燃烧，还包括含硫矿石的冶炼或含硫酸、磷肥等生产的工业废气排放。

【氮氧化物产生量】指当年全年调查对象生产过程中产生的未经过处理的废气中所含的氮氧化物总质量。

【氮氧化物排放量】指报告期内企业在燃料燃烧和生产工艺过程中排入大气的氮氧化物总质量。

【烟（粉）尘产生量】烟尘是指通过燃烧煤、石煤、柴油、木柴、天然气等产生的烟气中的尘粒。通过有组织排放的，俗称烟道尘。工业粉尘指在生产工艺过程中排放的能在空气中悬浮一定时间的固体颗粒。如钢铁企业耐火材料粉尘、焦化企业的筛焦系统粉尘、烧结机的粉尘、石灰窑的粉尘、建材企业的水泥粉尘等。烟（粉）尘产生量指当年全年调查对象生产过程中产生的未经过处理的废气中所含的烟尘及工业粉尘的总质量之和。

【烟（粉）尘排放量】指报告期内企业在燃料燃烧和生产工艺过程中排入大气的烟尘及工业粉尘的总质量之和。烟尘或工业粉尘排放量可以通过除尘系统的排风量和除尘设备出口烟尘浓度相乘求得。

【重金属产生量】指报告期调查对象生产过程中产生的未经过处理的废气中分别所含的砷、铅、汞、镉、铬及其化合物的总质量（以元素计）。

【重金属排放量】指报告期内企业在燃料燃烧和生产工艺过程中分别排入大气的砷、铅、汞、镉、铬及其化合物的总质量（以元素计）。

【一般工业固体废物产生量】是指未被列入《国家危险废物名录》或者根据国家规定的危险废物鉴别标准（GB 5085）、固体废物浸出毒性浸出方法（GB 5086）及固体废物浸出毒性测定方法（GB/T 15555）鉴别方法判定不具有危险特性的工业固体废物。根据其性质分为两种：

（1）第 I 类一般工业固体废物：按照 GB 5086 规定方法进行浸出试验而获得的浸出液中，任何一种污染物的浓度均未超过 GB 8978 最高允放排放浓度，且 pH 在 6～9 的一般工业固体废物；

（2）第 II 类一般工业固体废物：按照 GB 5086 规定方法进行浸出试验而获得的浸出液中，有一种或一种以上的污染物浓度超过 GB 8978 最高允许排放浓度，或者是 pH 在 6～9 的一般工业固体废物。

主要包括：

代码	名称	代码	名称
SW01	冶炼废渣	SW07	污泥
SW02	粉煤灰	SW08	放射性废物
SW03	炉渣	SW09	赤泥
SW04	煤矸石	SW10	磷石膏
SW05	尾矿	SW99	其他废物
SW06	脱硫石膏		

不包括矿山开采的剥离废石和掘进废石（煤矸石和呈酸性或碱性的废石除外）。酸性或碱性废石是指采掘的废石其流经水、雨淋水的 pH 小于 4 或 pH 大于 10.5 者。

冶炼废渣：指在冶炼生产中产生的高炉渣、钢渣、铁合金渣等，不包括列入《国家危险废物名录》中的金属冶炼废物。

粉煤灰：指从燃煤过程产生烟气中收捕下来的细微固体颗粒物，不包括从燃煤设施炉膛排出的灰渣。主要来自电力、热力的生产和供应行业和其他使用燃煤设施的行业，又称飞灰或烟道灰。主要从烟道气体收集而得，应与其烟尘去除量基本相等。

炉渣：指企业燃烧设备从炉膛排出的灰渣，不包括燃料燃烧过程中产生的烟尘。

煤矸石：指与煤层伴生的一种含碳量低、比煤坚硬的黑灰色岩石，包括巷道掘进过程中的掘进矸石、采掘过程中从顶板、底板及夹层里采出的矸石以及洗煤过程中挑出的洗矸石。主要来自煤炭开采和洗选行业。

尾矿：指矿山选矿过程中产生的有用成分含量低、在当前的技术经济条件下不宜进一步分选的固体废物，包括各种金属和非金属矿石的选矿。主要来自采矿业。

脱硫石膏：指废气脱硫的湿式石灰石/石膏法工艺中，吸收剂与烟气中 SO_2 等反应后生成的副产物。

污泥：指污水处理厂污水处理中排出的、以干泥量计的固体沉淀物。

放射性废物：指含有天然放射性核素，并其比活度大于 2×10^4 Bq/kg 的尾矿砂、废矿石及其他放射性固体废物（指放射性浓度或活度或污染水平超过规定下限的固体废物）。

赤泥：指从铝土矿中提炼氧化铝后排出的污染性废渣，一般还氧化铁量大，外观与赤色泥土相似。磷石膏：指在磷酸生产中用硫酸分解磷矿时产生的二水硫酸钙、酸不溶物，未分解磷矿及其他杂质的混合物。主要来自磷肥制造业。

其他废物：指除上述 10 类一般工业固体废物以外的未列入《国家危险废物名录》中的固体废物，如机械工业切削碎屑、研磨碎屑、废砂型等；食品工业的活性炭渣；硅酸盐工业和建材工业的砖、瓦、碎砾、混凝土碎块等。

一般工业固体废物产生量＝（一般工业固体废物综合利用量－其中的综合利用往年贮存量）＋一般工业固体废物贮存量＋（一般工业固体废物处置量－其中的处置往年贮存量）＋一般工业固体废物倾倒丢弃量

【一般工业固体废物综合利用量】指报告期内企业通过回收、加工、循环、交换等方式，从固体废物中提取或者使其转化为可以利用的资源、能源和其他原材料的固体废物量（包括当年利用的往年工业固体废物累计贮存量）。如用做农业肥料、生产建筑材料、筑路等。综合利用量由原产生固体废物的单位统计。

工业固体废物综合利用的主要方式有：

序号	综合利用方式	序号	综合利用方式
1	铺路	9	再循环/再利用不是用做溶剂的有机物
2	建筑材料	10	再循环/再利用金属和金属化合物
3	农肥或土壤改良剂	11	再循环/再利用其他无机物
4	矿渣棉	12	再生酸或碱
5	铸石	13	回收污染减除剂的组分
6	其他	14	回收催化剂组分
7	作为燃料（直接燃烧除外）或以其他方式产生能量	15	废油再提炼或其他废油的再利用
8	溶剂回收/再生（如蒸馏、萃取等）	16	其他有效成分回收

【一般工业固体废物贮存量】指报告期内企业以综合利用或处置为目的，将固体废物暂时贮存或堆存在专设的贮存设施或专设的集中堆存场所内的量。专设的固体废物贮存场所或贮存设施必须有防扩散、防流失、防渗漏、防止污染大气、水体的措施。

粉煤灰、钢渣、煤矸石、尾矿等的贮存量是指排入灰场、渣场、矸石场、尾矿库等贮存的量。

专设的固体废物贮存场所或贮存设施指符合环保要求的贮存场，即选址、设计、建设符合《一般工业固体废物贮存、填埋场污染控制标准》（GB 18599—2001）等相关环保法律法规要求，具有防扩散、防流失、防渗漏、防止污染大气和水体措施的场所和设施。

工业固体废物的贮存方式主要有：

序号	贮存方式
1	灰场堆放
2	渣场堆放
3	尾矿库堆放
4	其他贮存（不包括永久性贮存）

【一般工业固体废物处置量】指报告期内企业将工业固体废物焚烧和用其他改变工业固体废物的物理、化学、生物特性的方法，达到减少或者消除其危险成分的活动，或者将工业固体废物最终置于符合环境保护规定要求的填埋场的活动中，所消纳固体废物的量。

处置方式如填埋、焚烧、专业贮存场（库）封场处理、深层灌注、回填矿井及海洋处置（经海洋管理部门同意投海处置）等。

处置量包括本单位处置或委托给外单位处置的量。还包括当年处置的往年工业固体废物贮存量。

工业固体废物的主要处置方式有：

处置方式
围隔堆存（属永久性处置）
填埋
置放于地下或地上（如填埋、填坑、填浜）
特别设计填埋
海洋处置
经海洋管理部门同意的投海处置
埋入海床
焚化
陆上焚化
海上焚化
水泥窑共处置（指在水泥生产工艺中使用工业固体废物或液态废物作为替代燃料或原料，消纳处理工业固体或液态废物的方式）
固化
其他处置（属于未在上面5种指明的处置作业方式外的处置）
废矿井永久性堆存（包括将容器置于矿井）
土地处理（属于生物降解，适合于液态固废或污泥固废）
地表存放（将液态固废或污泥固废放入坑、氧化塘、池中）
生物处理
物理化学处理
经环保管理部门同意的排入海洋之外的水体（或水域）
其他处理方法

【一般工业固体废物倾倒丢弃量】 指报告期内企业将所产生的固体废物倾倒或者丢弃到固体废物污染防治设施、场所以外的量。倾倒丢弃方式如：

（1）向水体排放的固体废物；

（2）在江河、湖泊、运河、渠道、海洋的滩场和岸坡倾倒、堆放和存贮废物；

（3）利用渗井、渗坑、渗裂隙和溶洞倾倒废物；

（4）向路边、荒地、荒滩倾倒废物；

（5）未经环保部门同意作填坑、填河和土地填埋固体废物；

（6）混入生活垃圾进行堆置的废物；

（7）未经海洋管理部门批准同意，向海洋倾倒废物；

（8）其他去向不明的废物；

（9）深层灌注。

一般工业固体废物倾倒丢弃量计算公式是：

一般工业固体废物倾倒丢弃量＝一般工业固体废物产生量－一般工业固体废物贮存量－（一般工业固体废物综合利用量－其中的综合利用往年贮存量）－（一般工业固体废物处置量－其中的处置往年贮存量）

【危险废物产生量】 指当年全年调查对象实际产生的危险废物的量。危险废物指列入国家危险废物名录或者根据国家规定的危险废物鉴别标准和鉴别方法认定的，具有爆炸性、易燃性、易氧化性、毒性、腐蚀性、易传染性疾病等危险特性之一的废物。按《国家危险废物名录》（环境保护部、国家发展和改革委员会2008年部令第1号）填报。

【危险废物综合利用量】 指当年全年调查对象从危险废物中提取物质作为原材料或者燃料的活动中消纳危险废物的量。包括本单位利用或委托、提供给外单位利用的量。

【危险废物贮存量】指当年调查对象将危险废物以一定包装方式暂时存放在专设的贮存设施内的量。

专设的贮存设施应符合《危险废物贮存污染控制标准》（GB 18597—2001）等相关环保法律法规要求，具有防扩散、防流失、防渗漏、防止污染大气和水体措施的设施。

【危险废物处置量】指报告期内企业将危险废物焚烧和用其他改变工业固体废物的物理、化学、生物特性的方法，达到减少或者消除其危险成分的活动，或者将危险废物最终置于符合环境保护规定要求的填埋场的活动中，所消纳危险废物的量。处置量包括处置本单位或委托给外单位处置的量。

【危险废物倾倒丢弃量】指报告期内企业将所产生的危险废物未按规定要求处理处置的量。

2．工业企业污染防治投资情况

【污染治理项目名称】指以治理老污染源的污染、"三废"综合利用为主要目的的工程项目名称，或本年完成建设项目"三同时"环境保护竣工验收的项目名称。

【项目类型】按照不同的项目性质，污染治理项目分为3类，并给予不同的代码。

①老工业污染源治理在建项目；②老工业污染源治理本年竣工项目；③建设项目"三同时"环境保护竣工验收本年完成项目。

【治理类型】按照不同的企业污染治理对象，污染治理项目分为14类：

①工业废水治理；②工业废气脱硫治理；③工业废气脱硝治理；④其他废气治理；⑤一般工业固体废物治理；⑥危险废物治理（企业自建设施）；⑦噪声治理（含振动）；⑧电磁辐射治理；⑨放射性治理；⑩工业企业土壤污染治理；⑪矿山土壤污染治理；⑫污染物自动在线监测仪器购置；⑬污染治理搬迁；⑭其他治理（含综合防治）。

【本年完成投资及资金来源】指在报告期内，企业实际用于环境治理工程的投资额。投资额中的资金来源，是指投资单位在本年内收到的用于污染治理项目投资的各种货币资金，包括排污费补助、政府其他补助、企业自筹。各种来源的资金均为报告期投入的资金，不包括以往历年的投资。

本年污染治理资金合计＝排污费补助＋政府其他补助＋企业自筹

【竣工项目设计或新增处理能力】设计能力是指设计中规定的主体工程（或主体设备）及相应的配套的辅助工程（或配套设备）在正常情况下能够达到的处理能力。报告期内竣工的污染治理项目，属新建项目的填写设计文件规定的处理、利用"三废"能力；属改扩建、技术改造项目的填写经改造后新增加的处理利用能力，不包括改扩建之前原有的处理能力；只更新设备或重建构筑物，处理利用"三废"能力没有改变的则不填。

工业废水设计处理能力的计量单位为吨/天；工业废气设计处理能力的计量单位为（标态）米³/小时；工业固体废物设计处理能力的计量单位为吨/天；噪声治理（含振动）设计处理能力以降低分贝数表示；电磁辐射治理设计处理能力以降低电磁辐射强度表示（电磁辐射计量单位有电场强度单位：伏/米、磁场强度单位：安/米、功率密度单位：瓦/米²）。放射性治理设计处理能力以降低放射性浓度表示，废水计量单位为贝可/升，固体废物计量单位为贝可/千克。

3．农业源污染排放及处理利用情况

【治污设施累计完成投资】养殖场/小区当前可用于养殖污染治理设施的累计投资总额。

【新增固定资产】指报告期内交付使用的固定资产价值。对于新建治污设施，本年新增固定资产投资等于总投资；对于改、扩建治污设施，本年新增固定资产投资仅指报告期内交付使用的改、扩建部分的固定资产投资，属于累计完成投资的一部分。

4．城镇生活污染排放及处理情况

调查范围：

城镇生活包括"住宿业与餐饮业、居民服务和其他服务业、医院和独立燃烧设施以及城镇生活污染源"。

根据《关于统计上划分城乡的暂行规定》（国统字[2006]60号），城镇是指在我国市镇建制和行政区划的基础上，经以下规定划定的区域。城镇包括城区和镇区。

（1）城区是指在市辖区和不设区的市中，包括：①街道办事处所辖的居民委员会地域；②城市公共设施、居住设施等连接到的其他居民委员会地域和村民委员会地域。

（2）镇区是指在城区以外的镇和其他区域中，包括：①镇所辖的居民委员会地域；②镇的公共设施、居住设施等连接到的村民委员会地域；③常住人口在3 000人以上独立的工矿区、开发区、科研单位、大专院校、农场、林场等特殊区域。

城镇生活源的基本调查单位为地（市、州、盟），其所属的区、县城以及镇区数据包含在所在地（市、州、盟）数据中。

【城镇人口】是指居住在城镇范围内的全部常住人口。相关数据以统计部门公布数据为准。

【生活煤炭消费量】指报告期内调查区域用作城镇生活的煤炭总量，包括批发和零售贸易业、餐饮业、居民生活以及生活供热三个部分，以统计部门的能源平衡表数据为准。其中，批发和零售贸易业、餐饮业和居民生活煤炭消费量从能源平衡表直接获取；生活供热煤炭消费量需由能源平衡表中的供热总煤耗扣减工业供热煤耗得到。生活煤炭消费量计算公式为：

生活煤炭消费量＝批发和零售贸易业、餐饮业煤炭消费量＋居民生活煤炭消费量＋生活供热煤炭消费量

生活供热煤炭消费量＝供热总煤耗－工业供热煤耗（环境统计工业调查中4 430行业煤炭消费总量）

【生活用水总量】指报告期内调查区域住宿业与餐饮业、居民服务和其他服务业、医院以及城镇生活等的用水总量（含自来水和自备水），包括居民家庭用水和公共服务用水的总量两个部分（包括自来水和自备水），以城市供水管理部门的统计数据为准。

【居民家庭用水总量】指城镇范围内所有居民家庭的日常生活用水。包括城镇居民、农民家庭、公共供水站用水，以城市供水管理部门的统计数据为准。

【公共服务用水总量】指为城镇社会公共服务的用水。包括行政事业单位、部队营区和公共设施服务、社会服务业、批发零售贸易业、旅馆饮食业等单位的用水。

【城镇生活污水污染物产生量】指报告期内各类生活源从贮存场所排入市政管道、排污沟渠和周边环境的量。

【城镇生活污水污染物排放量】指报告期内最终排入外环境生活污水污染物的量，即生活污水污染物产生量扣减经集中污水处理设施去除的生活污水污染物量。

【城镇生活污水排放量】用人均系数法测算。

如果辖区内的城镇污水处理厂未安装再生水回用系统，无再生水利用量，则

城镇生活污水排放量＝城镇生活污水排放系数×城镇人口数×365

反之，辖区内的城镇污水处理厂配备再生水回用系统，其再生水利用量已经污染减排核查确认，则

城镇生活污水排放量＝城镇生活污水排放系数×城镇人口数×365－城镇污水处理厂再生水利用量

5. 机动车

调查范围：

以直辖市、地区（市、州、盟）为单位调查机动车保有量，以及其中分车型、分年度登记数量；机动车调查对象包括汽车和摩托车。

机动车保有量分类登记数量信息由直辖市、地区（市、州、盟）环保部门协调同级公安交通管理部门负责提供，调查表由各级环保部门填写。

大型载客汽车：车长大于等于6米或者乘坐人数大于等于20人。乘坐人数可变的，以上限确定。乘坐人数包括驾驶员；

中型载客汽车：车长小于6米，乘坐人数大于9人且小于20人；

小型载客汽车：车长小于6米，乘坐人数小于等于9人；

微型载客汽车：车长小于等于3.5米，发动机汽缸总排量小于等于1升；

重型载货汽车：车长大于等于6米，总质量大于等于12 000千克；

中型载货汽车：车长大于等于6米，总质量大于等于4 500千克且小于12 000千克；

轻型载货汽车：车长小于6米，总质量小于4 500千克；

微型载货汽车：车长小于等于3.5米，载质量小于等于1 800千克；

三轮汽车（原三轮农用运输车）：以柴油机为动力，最高设计车速小于等于50千米/小时，最大设计总质量不大于2 000千克，长小于等于4.6米，宽小于等于1.6米，高小于等于2米，具有三个车轮的货车；采用方向盘转向，由操纵杆传递动力、有驾驶室且驾驶员车厢后有物品存放的长小于等于5.2米，宽小于等于2.8米，高小于等于2.2米；

低速载货汽车（原四轮农用运输车）：以柴油机为动力，最高设计车速小于70千米/小时，最大设计总质量小于等于4 500千克，长小于等于6米，宽小于等于2米，高小于等于2.5米，具有四个车轮的货车；

普通摩托车：最大设计时速大于50千米/小时或者发动机汽缸总排量大于50毫升；

轻便摩托车：最大设计时速小于等于50千米/小时，发动机汽缸总排量小于等于50毫升。

6. 污水处理厂

填报范围：

城镇污水处理厂及其他社会化运营的集中污水处理单位，包括专业化的工业废水集中处理设施或单位填报本套表。

污水处理厂包括城镇污水处理厂、工业废（污）水集中处理设施和其他污水处理设施。

城镇污水处理厂：指城镇污水（生活污水、工业废水）通过排水管道集中于一个或几个处所，并利用由各种处理单元组成的污水处理系统进行净化处理，最终使处理后的污水和污泥达到规定要求后排放或再利用的设施。

工业废（污）水集中处理设施：指提供社会化有偿服务、专门从事为工业园区、联片工业企业或周边企业处理工业废水（包括一并处理周边地区生活污水）的集中设施或独立运营的单位。不包括企业内部的污水处理设施。

其他污水处理设施：指对不能纳入城市污水收集系统的居民区、风景旅游区、度假村、疗养院、机场、铁路车站以及其他人群聚集地排放的污水进行就地集中处理的设施。

污水处理厂调查中，不包括氧化塘、渗水井、化粪池、改良化粪池、无动力地埋式污水处理装置和土地处理系统处理工艺。

【污水处理厂累计完成投资】指至当年末，调查对象建设实际完成的累计投资额，不包括运行费用。

【新增固定资产】指报告期内交付使用的固定资产价值。对于新建污水处理厂，本年新增固定资产投资等于总投资；对于改、扩建污水处理厂，本年新增固定资产投资仅指报告期内交付使用的改、扩建部分的固定资产投资，属于累计完成投资的一部分。

【运行费用】指报告期内维持污水处理厂（或处理设施）正常运行所发生的费用。包括能源消耗、设备维修、人员工资、管理费、药剂费及与污水处理厂（或处理设施）运行有关的其他费用等，不包括设备折旧费。

【污水设计处理能力】指截至当年末调查对象设计建设的设施正常运行时每天能处理的污水量。

【污水实际处理量】指调查对象报告期内实际处理的污水总量。

【再生水利用量】指调查对象报告期内处理后的污水中再回收利用的水量。该水量仅指总量减排核定 COD 或氨氮削减量的用于工业冷却、洗涤、冲渣和景观用水、生活杂用的水量。

【工业用水量】指调查对象报告期内污水再生水利用量中用于工业冷却用水等工业方面的水量。

【市政用水（杂用水）】指调查对象报告期内污水再生水利用量中用于消防、城市绿化等市政方面的水量。

【景观用水量】指调查对象报告期内污水再生水利用量中用于营造城市景观水体和各种水景构筑物的水量。

【污泥产生量】指调查对象报告期内在整个污水处理过程中最终产生污泥的质量。折合为 80% 含水率的湿泥量填报。污泥指污水处理厂（或处理设施）在进行污水处理过程中分离出来的固体。

【污泥处置量】指报告期内采用土地利用、填埋、建筑材料利用和焚烧等方法对污泥最终消纳处置的质量。

【土地利用量】指报告期内将处理后的污泥作为肥料或土壤改良材料，用于园林、绿化或农业等场合的处置方式处置的污泥质量。

【填埋处置量】指报告期内采取工程措施将处理后的污泥集中堆、填、埋于场地内的安全处置方式处置的污泥质量。

【建筑材料利用量】指报告期内将处理后的污泥作为制作建筑材料的部分原料的处置方式处置的污泥质量。

【焚烧处置量】指报告期内利用焚烧炉使污泥完全矿化为少量灰烬的处置方式处置的污泥质量。

【污泥倾倒丢弃量】指报告期内不作处理、处置而将污泥任意倾倒弃置到划定的污泥堆放场所以外的任何区域的质量。

7. 生活垃圾处理厂（场）

填报范围：

所有垃圾处理厂（场）填报本表。

垃圾处理厂（场）包括垃圾填埋厂（场）、堆肥厂（场）、焚烧厂（场）和其他方式处理垃圾的处理厂（场）。其中，垃圾焚烧厂（场）不包括垃圾焚烧发电厂，垃圾焚烧发电厂纳入工业源调查。

【生活垃圾处理厂（场）累计完成投资】指至当年末调查对象建设实际完成的累计投资额，不包括运行费用。

【本年新增固定资产】指报告期内交付使用的固定资产价值。对于新建垃圾处理厂（场），本年新增固定资产投资等于总投资；对于改、扩建垃圾处理厂（场），本年新增固定资产投资仅指报告期内交付使用的改、扩建部分的固定资产投资，属于累计完成投资的一部分。

【本年运行费用】指报告期内维持垃圾处理厂（场）正常运行所发生的费用。包括能源消耗、设备维修、人员工资、管理费及与垃圾处理厂（场）运行有关的其他费用等，不包括设备折旧费。

【本年实际处理量】指报告期内对垃圾采取焚烧、填埋、堆肥或其他方式处理的垃圾总质量。

【渗滤液污染物排放量】指报告期内排放的渗滤液中所含的化学需氧量、氨氮、石油类、总磷、挥发酚、氰化物、砷和汞、镉、铅、铬等重金属污染物本身的纯质量。按年排放量填报。

表中所列出的污染物，调查对象要根据监测结果或产排污系数如实填报。

【焚烧废气污染物排放量】指报告期内垃圾焚烧过程中排放到大气中的废气（包括处理过的、未经过处理）中所含的二氧化硫、氮氧化物、烟尘和汞、镉、铅等重金属及其化合物（以重金属元素计）的固态、气态污染物的纯质量。按年排放量填报。

表中所列出的污染物，调查对象要根据监测结果或产排污系数如实填报。

8. 危险废物（医疗废物）集中处理（置）厂

填报范围：

所有危险废物（医疗废物）集中处理（置）厂填报本表。集中式危险废物处理（置）厂指统筹规划建设并服务于一定区域专营或兼营危险废物处理（置）且有危险废物经营许可证的危险废物处置厂。

【危险废物（医疗废物）集中处理（置）厂累计完成投资】指至当年末，调查对象建设实际完成的累计投资额，不包括运行费用。

【本年新增固定资产】指报告期内交付使用的固定资产价值。对于新建危险废物（医疗废物）处置厂，本年新增固定资产投资等于总投资；对于改、扩建危险废物（医疗废物）处置厂，本年新增固定资产投资仅指报告期内交付使用的改、扩建部分的固定资产投资，属于累计完成投资的一部分。

【本年运行费用】指报告期内维持危险废物处置厂正常运行所发生的费用。包括能源消耗、设备维修、人员工资、管理费及与危险废物处置厂运行有关的其他费用等，不包括设备折旧费。

【危险废物设计处置能力】指调查对象设计建设的每天可能处置危险废物的量。

【本年实际处置危险废物量】指报告期内调查对象将危险废物焚烧和用其他改变危险废物的物理、化学、生物特性的方法，达到减少已产生的危险废物数量、缩小危险废物体积、减少或者消除其危险成分的活动，或者将危险废物最终置于符合环境保护规定要求的填埋场的活动中，所消纳危险废物的量。

【处置工业危险废物量】指调查对象报告期内采用各种方式处置的工业危险废物的总量。医疗废物集中处置厂不得填写该项指标。

【处置医疗废物量】指调查对象报告期内采用各种方式处置的医疗废物的总量。

【处置其他危险废物量】指调查对象报告期内采用各种方式处置的除工业危险废物和医疗废物以外其他危险废物的总质量，如教学科研单位实验室、机械电器维修、胶卷冲洗、居民生活等产生的危险废物。医疗废物集中处置厂不得填写该项指标。

【本年危险废物综合利用量】指报告期内调查对象从危险废物中提取物质作为原材料或者燃料的活动中消纳危险废物的量。

【渗滤液污染物排放量】指报告期内排放的渗滤液中所含的汞、镉、铅、总铬等重金属和砷、氰化物、挥发酚、石油类、化学需氧量、氨氮、总磷等污染物本身的纯质量。按年排放量填报。

表中所列出的污染物，调查对象要根据监测结果或产排污系数核算结果如实填报。

【焚烧废气污染物排放量】指报告期内危险废物焚烧过程中排放到大气中的废气（包括处理过的、未经过处理）中所含的二氧化硫、氮氧化物、烟尘和汞、镉、铅等重金属及其化合物（以重金属元素计）的固态、气态污染物的纯质量。按年排放量填报。

9．环境管理

【环保系统机构数/人数】指环保系统行政主管部门及其所属事业单位、社会团体设置情况，包括机构总数和在编人员总数。

【环保行政主管部门机构数/人数】指各级人民政府环境保护行政主管部门设置情况，不包括各类开发区等非行政区环保主管部门。

【环境监测站机构/人数】统计独立设置的核与辐射环境监测机构以外的环境监测机构设置情况。在机构数中应注明加挂了核与辐射环境监测站牌子的机构数，以及未加挂牌子但承担了核与辐射环境监测职能的机构数；在总人数中注明承担核与辐射环境监测职能的人数。

【环境监察机构数/人数】指独立设置的环境执法监察机构设置情况。

【核与辐射环境监测站机构数/人数】统计独立设置的核与辐射环境监测站机构设置情况。

【科研机构数/人数】指独立设置的环境科研机构设置情况。

【宣教机构数/人数】指独立设置的环境宣传教育机构设置情况。

【信息机构数/人数】指独立设置的环境信息机构设置情况。

【环境应急（救援）机构数/人数】指环保部门建立或与其他部门共同建立的专兼职环境应急救援机构数、人数。

【当年受理行政复议案件数】指调查年度内环保部门受理的所有行政复议案件数（含非环保案件），包括已受理但未办结的案件；但不包括非本统计年受理而在本统计年内办理或办结的案件。

【本级行政处罚案件数】

【电话/网络投诉数；电话/网络投诉办结数】开通12369环保举报热线电话和网络的地区，按受理情况填报；未开通地区，可根据当地环保部门的公开举报电话和网络信箱的投诉情况填报。只填报本级受理的投诉数，不填报上级转到下级处理的投诉数。

【来访总数】指各级环保部门接待上访人员的数量。同时统计批次、人次。

【来信、来访已办结数量】

【承办的人大建议】指国家、省、市、县环保部门承办的本年度本级人大代表建议数总和。

【承办的政协提案数】指国家、省、市、县环保部门承办的本年度本级政协提案数总和。

【本级环保能力建设资金使用总额】是指本级使用的、当年已完成的用于提高环境保护监督管理能力和环境科技发展能力的基本建设和购置固定资产的投资。具体包括：各级环境保护行政主管部门、各类环境保护事业单位、各类环境监测站点等的基本建设投资、购置固定资产的投资及监测、监察等环境监管运行保障经费（含办公业务用房和科研用实验室的建设费用；不含行政运行费用、生活福利设施的建设费用）。

环境保护事业单位包括：环境监测、环境监察、环境应急、固体废物管理、环境信息、环境宣传、核与辐射、环境科研以及环保部门驻外和派出机构等。

其中：

（1）监测能力建设资金使用总额是指各级环境监测机构、各类环境监测站点等当年已完成的基本建设投资和购置固定资产的投资总额。

（2）监察能力建设资金使用总额是指各级环境监察机构、污染源监控中心、污染源自动监控设施等当年已完成的基本建设投资和购置固定资产的投资总额。

（3）核与辐射安全监管能力建设资金使用总额是指各级核与辐射安全监管机构、核与辐射自动监测站点等当年已完成的基本建设投资和购置固定资产的投资总额。

（4）固体废物管理能力建设资金使用总额是指各级固体废物管理机构当年已完成的基本建设投资

和购置固定资产的投资总额。

（5）环境应急能力建设资金使用总额是指各级环境应急管理机构当年已完成的基本建设投资和购置固定资产的投资总额。

（6）环境信息能力建设资金使用总额是指各级环境信息机构、环境信息网络当年已完成的基本建设投资和购置固定资产的投资总额。

（7）环境宣教能力建设资金使用总额是指各级环境宣教机构、培训基地等当年已完成的基本建设投资和购置固定资产的投资总额。

（8）环境监管运行保障资金使用总额是指各级监测、监察、核与辐射安全、宣教等机构开展污染源与总量减排监管、环境监测与评估、环境信息等业务发生的环境监管运行保障经费。

【清洁生产审核当年完成企业数】是指本地区在调查年度完成清洁生产审核并通过评估或验收的企业数量。各地区在清洁生产审核的管理一般有两种情况：评估和验收分开实施，则统计该地区当年完成评估的企业数量；评估和验收同时进行，则统计该地区当年完成验收的企业数量。

【强制性审核当年完成数】是指本地区在调查年度完成强制性清洁生产审核并通过评估或验收的企业数量。各地区在强制性清洁生产审核的管理一般有两种情况：评估和验收分开实施，则统计该地区当年完成评估的企业数量；评估和验收同时进行，则统计该地区当年完成验收的企业数量。

【应开展监测的重金属污染防控重点企业数】指按照《重金属污染综合防治"十二五"规划》要求开展监测的重金属污染防控重点企业家数。重金属污染防控重点企业指《重金属污染综合防治"十二五"规划》中的 4 452 家重金属排放企业，主要考核重点企业废气、废水中重点重金属污染物达标排放情况，若企业关闭则不纳入统计基数。《重金属污染综合防治"十二五"规划》要求各地对 4 452 家重点企业每两个月开展一次监督性监测，重点企业应实现稳定达标排放。

【重金属排放达标的重点企业数】指重点重金属污染物达标排放的重点企业数。由各地环保部门按照《重金属污染综合方式"十二五"规划》要求对重点企业实施每两个月一次的监督性监测，涉及企业实际排放的铅、汞、镉、铬、砷等 5 种重点重金属中的一种或多种按排放标准评价其达标排放情况。若一次监测不达标则视为该企业不达标；未开展重金属监督性监测视为不达标。

【已发放危险废物经营许可证数】截至报告期末，由各级环境保护行政主管部门依法审批发放并处于有效期内的危险废物经营许可证数量，包括综合经营许可证和收集经营许可证。

【具有医疗废物经营范围的许可证数】指在"已发放危险废物经营许可证数"指标的统计范围内，核准经营危险废物类别包括 HW01 医疗废物的危险废物经营许可证数。

【监测用房面积】指开展环境监测工作所需的实验室用房、监测业务用房、大气、水质自动监测用房等的面积。

【监测业务经费】指完成常规监测、专项监测、应急监测、质量保证、报告编写、信息统计等工作所需经费，不含自动监测、信息系统运行费和仪器设备购置费等。

【监测仪器设备台套数及原值总值】指基本仪器设备、应急环境监测仪器设备和专项监测仪器设备等的数量。监测仪器设备的原值总值，是指通过政府采购、公开招投标或其他采购方式购买的各类监测仪器设备的购置金额。

【空气监测点位数】指位于本辖区、为监测所代表地区的空气质量而设置的常年运行的例行监测点位的数量。

【其中：国控监测点位数】指位于本辖区、由国家批准纳入国家城市环境空气质量监测网络的空气监测点位数。

【酸雨监测点位数】指位于本辖区、为了掌握当地、区域或全国酸雨状况，了解酸雨变化趋势而设立的例行采集降水样品的点位数。

【沙尘天气影响空气质量监测点位数】指位于本辖区、为反映监测沙尘天气对我国城市环境空气质量的影响，由环境保护部专门设立的环境空气质量监测点位数。

【地表水水质监测断面数】指位于本辖区、为反映地表水水质状况而设置的监测点位数。一般包括背景断面、对照断面、控制断面、削减断面等。

【其中：国控断面数】指位于本辖区、由国家组织实施监测的，为反映水体水质状况而设置的监测点位数。

【地表水集中式饮用水水源地监测点位数】指位于本辖区、为反映地表水集中式饮用水水源地水质状况而设置的监测点位数。

【近岸海域环境功能区点位数】指位于本辖区、为反映近岸海域功能区环境质量而布设的监测点位数。

【近岸海域环境质量点位数】指位于本辖区、为反映近岸海域环境质量而布设的环境监测点位数。

【开展污染源监督性监测的重点企业数】指按照相关要求开展污染源监督性监测的重点企业家数，以省、市、县本级实施监督性监测的企业数为准，委托监测的企业由受托方统计。

重点企业：由各级政府部门监控的占辖区内主要污染物排放负荷达到一定比例以上的，以及其他根据环境保护规划或者专项环境污染防治需要而列入监控范围的排污单位，包括国控、省控、市控企业和其他企业数。

污染源监督性监测：环境保护行政主管部门所属环境监测机构对辖区内的污染源实施的不定期监测，包括对污染源污染物排放的抽查监测和对自动监测设备的比对监测。

【已实施自动监控国家重点监控企业数】根据污染源自动监控工作进展情况，至本报告期末在环保部门污染源监控中心已经实现自动监控的国家重点监控企业数。

【已实施自动监控国家重点监控企业中水排放口数】已实施自动监控的国家重点监控企业中，环保部门污染源监控中心能够实施自动监控的水排放口数。

【已实施自动监控国家重点监控企业中气排放口数】已实施自动监控的国家重点监控企业中，环保部门污染源监控中心能够实施自动监控的气排放口数。

【COD 监控设备与环保部门稳定联网数】已实施自动监控的国家重点监控企业中，其 COD 自动监控设备正常运行、自动监控数据（浓度和排放量）能通过数据采集与传输设备与环保部门污染源监控中心稳定联网报送的企业数。

【NH$_3$-N 监控设备与环保部门稳定联网数】已实施自动监控的国家重点监控企业中，其 NH$_3$-N 自动监控设备正常运行、自动监控数据（浓度和排放量）能通过数据采集与传输设备与环保部门污染源监控中心稳定联网报送的企业数。

【SO$_2$ 监控设备与环保部门稳定联网数】已实施自动监控的国家重点监控企业中，其 SO$_2$ 自动监控设备正常运行、自动监控数据（浓度和排放量）能通过数据采集与传输设备与环保部门污染源监控中心稳定联网报送的企业数。

【NO$_x$ 监控设备与环保部门稳定联网数】已实施自动监控的国家重点监控企业中，其 NO$_x$ 自动监控设备正常运行、自动监控数据（浓度和排放量）能通过数据采集与传输设备与环保部门污染源监控中心稳定联网报送的企业数。

【排污费解缴入库户数】指调查年度经对账、实际解缴国库的排污费所对应的户数，同一排污者分期分批计征或解缴排污费的不重复计算户数。

【排污费解缴入库户金额】指调查年度经对账、实际解缴国库的排污费所累计金额。

【自然保护区个数】指全国（不含香港特别行政区、澳门特别行政区和台湾地区）建立的各种类型、不同级别的自然保护区的总数相加之和，应包括国家级、省级和市（县）级自然保护区。

【自然保护区个数国家级】至报告期末，经国务院批准建立的自然保护区的个数。

【自然保护区个数省级】至报告期末，经省（自治区、直辖市）人民政府批准建立的自然保护区的个数。

【自然保护区面积】指全国（不含香港特别行政区、澳门特别行政区和台湾地区）建立的各种类型、不同级别的自然保护区的总面积相加之和，应包括国家级、省级和市（县）级自然保护区。

【自然保护区面积国家级】至报告期末，国家级自然保护区的面积总和。

【自然保护区面积省级】至报告期末，省（自治区、直辖市）级自然保护区的面积总和。

【生态市、县建设个数】

【国家级生态市个数】至报告期末，环境保护部对外正式公告的国家级生态市个数。

【国家级生态县个数】至报告期末，环境保护部对外正式公告的国家级生态县个数。

【省级生态市个数】至报告期末，获得省政府或省环境保护主管部门命名的省级生态市个数。

【省级生态县个数】至报告期末，获得省政府或省环境保护主管部门命名的省级生态市个数。

【农村生态示范建设个数】

【国家级生态乡镇个数】指环境保护部公告命名的国家级生态乡镇数量。

【国家级生态村个数】指环境保护部公告命名的国家级生态村数量。

【国家有机食品生产基地数量】指符合《国家有机食品生产基地考核管理规定》相关要求的有机食品生产基地的数量。

【当年开工建设的建设项目数量】指一切基本建设项目和技术改造项目，包括饮食服务等三产项目（区域开发已列入规划环评管理，不再计入建设项目）。由省级环保部门根据统计部门数据填报。

【执行环境影响评价制度的建设项目数量】指调查年度开工并依法履行环境影响评价制度的建设项目数。由省级环保部门填报。

【当年审批的建设项目环境影响评价文件数量】指调查年度审批的建设项目数（包括新建项目、改扩建项目）。按审批权限统计，分别由国家、省级、地市级、县级环保部门对审批管理的项目进行统计。委托下级审批的项目，由受委托部门统计。

【当年审批的建设项目投资总额】指调查年度审批的建设项目工程总投资的汇总数额。

【当年审批的建设项目环保投资总额】指调查年度审批的建设项目环保投资的汇总数额。

【当年审查的规划环境影响评价文件数量】指调查年度审查的规划环境影响评价文件数量。按审查权限统计，分别由国家、省级、地市级环保部门对负责审查的规划环境影响评价文件进行统计。

【当年完成环保验收项目数】指调查年度完成环保验收的建设项目数。按审批权限统计，分别由国家、省级、地市级、县级环保部门对负责验收管理的项目进行统计。委托下级验收的项目，由受委托部门统计。

【环保验收一次合格项目数】指执行了环保"三同时"制度，经环保部门验收一次合格的建设项目数。

【经限期改正验收合格项目数】指未执行环保"三同时"制度，经限期改正后通过环保验收的建设项目数。

【工业企业当年完成环保验收项目总投资】指调查年度工业企业完成环保验收项目工程总投资的汇总数额。

【工业企业当年完成环保验收项目环保投资】指调查年度工业企业完成环保验收项目环保投资的汇总数额。

【突发环境事件次数】指调查年度发生突发性环境事件的次数。若突发环境事件涉及两个以上省（自治区、直辖市），由事发地省级环保部门负责上报。特别重大、重大、较大、一般环境事件的级别划分参考《突发环境事件信息报告办法》（环保部 17 号令）。

【当年开展的社会环境宣传教育活动数/人数】指以环保部、省级和地市级环保部门或环境宣教机构本级名义主办的面向社会的环境宣传教育活动，主要包括重要纪念日环保宣传纪念活动、公众参与环保宣传活动、环保主题大型文艺演出、环保影视作品公映等形式。人数为参加该活动的人数。

【环境教育基地数】环保部、省级和地市级环保部门或环境宣教机构命名的环境教育基地。